International Studies in the History of Mathematics and its Teaching

Series Editors

Alexander Karp
Teachers College, Columbia University, New York, NY, USA

Gert Schubring
Universität Bielefeld, Bielefeld, Germany
Universidade Federal do Rio de Janeiro, Rio de Janeiro, Brazil

The International Studies in the History of Mathematics and its Teaching Series creates a platform for international collaboration in the exploration of the social history of mathematics education and its connections with the development of mathematics. The series offers broad perspectives on mathematics research and education, including contributions relating to the history of mathematics and mathematics education at all levels of study, school education, college education, mathematics teacher education, the development of research mathematics, the role of mathematicians in mathematics education, mathematics teachers' associations and periodicals.

The series seeks to inform mathematics educators, mathematicians, and historians about the political, social, and cultural constraints and achievements that influenced the development of mathematics and mathematics education. In so doing, it aims to overcome disconnected national cultural and social histories and establish common cross-cultural themes within the development of mathematics and mathematics instruction. However, at the core of these various perspectives, the question of how to best improve mathematics teaching and learning always remains the focal issue informing the series.

More information about this series at http://www.springer.com/series/15781

Alexander Karp
Editor

Eastern European Mathematics Education in the Decades of Change

Springer

Editor
Alexander Karp
Teachers College
Columbia University
New York, NY, USA

ISSN 2524-8022   ISSN 2524-8030 (electronic)
International Studies in the History of Mathematics and its Teaching
ISBN 978-3-030-38746-4   ISBN 978-3-030-38744-0 (eBook)
https://doi.org/10.1007/978-3-030-38744-0

© Springer Nature Switzerland AG 2020
This work is subject to copyright. All rights are reserved by the Publisher, whether the whole or part of the material is concerned, specifically the rights of translation, reprinting, reuse of illustrations, recitation, broadcasting, reproduction on microfilms or in any other physical way, and transmission or information storage and retrieval, electronic adaptation, computer software, or by similar or dissimilar methodology now known or hereafter developed.
The use of general descriptive names, registered names, trademarks, service marks, etc. in this publication does not imply, even in the absence of a specific statement, that such names are exempt from the relevant protective laws and regulations and therefore free for general use.
The publisher, the authors, and the editors are safe to assume that the advice and information in this book are believed to be true and accurate at the date of publication. Neither the publisher nor the authors or the editors give a warranty, expressed or implied, with respect to the material contained herein or for any errors or omissions that may have been made. The publisher remains neutral with regard to jurisdictional claims in published maps and institutional affiliations.

Cover Caption: Cover image is based on the cuneiform tablet, known as Plimpton 322 (collections of Columbia University, New York)

This Springer imprint is published by the registered company Springer Nature Switzerland AG
The registered company address is: Gewerbestrasse 11, 6330 Cham, Switzerland

Introduction

This book is about recent history. The late 1980s saw dramatic changes in the countries of Eastern Europe, which had previously belonged to the so-called Warsaw Pact, which provided a formal basis for their subjugation to and effective occupation by the Soviet Union. During those years, Soviet troops stayed in their barracks, and some time later the Soviet Union itself fell apart. A new age began.

Strange as it may seem, although an enormous number of publications have been devoted to the political, economic, and even cultural changes that have ensued, the manner in which mathematics education has changed has been little studied and documented, even though there has been no shortage of emotional outcries. Meanwhile, prior to the collapse of the Soviet bloc, Eastern European mathematics education had held in high esteem around the world (see, for example, Swetz 1978). Izaak Virzsup has stressed that, although he was "very anti-Communist and anti-dictatorial," he always spoke of "the superb work in mathematics education in the Soviet Union" (Roberts 2010, pp. 56–57). The good reputation of Eastern European mathematics education was confirmed at least to a certain extent when Soviet and, more broadly, Eastern European mathematicians began pouring across the newly opened borders, and in the words of the prominent French mathematician Pierre Cartier accomplished what Stalin could not do with all his army: they conquered the world (Senechal 1998).

Worldwide respect and recognition, however, did not lead to any large-scale attempt in the West to preserve the existing achievements in mathematics education in any way (and much less, to adopt them), even where, as in Germany, this would not have required learning a foreign language. The author of this introduction remembers talking to young West German Slavists in the summer of 1990, who noted with disappointment and helplessness that even that little which had been quite praiseworthy in East Germany was being destroyed before their eyes—the wave of mass dissatisfaction with everything in general was already rolling without discrimination. In certain countries and to a certain extent, mathematics education also found itself in the path of such a wave. Nor should it be forgotten that many Eastern European countries went through a period of severe economic crisis, which necessarily had an impact on education in general and mathematics education in particular.

It would be altogether mistaken to conclude, however, that recent decades have always brought about a decline in all things. The general liberalization that occurred in the different countries, even if it unfolded in different ways, and in some places was quite limited and short-lived, inevitably exerted a positive influence on the development of mathematics education, if only because it ushered in new possibilities and a greater openness. Generally speaking, it is impossible to measure the state of education in a country only by looking at the purchasing power of a teacher's salary, as is occasionally done in one way or another; although, of course, to forget about this purchasing power is also impossible. Very many other factors must be borne in mind: a country's legislation (written and unwritten); its methodological development, including established traditions and openness to new ideas; the inclinations, demands, and interests of its population; the presence or absence of a qualified teaching workforce; the possibility of expressing and spreading various opinions about methodology and the organization of education; and much else besides.

This book was written in order to help readers begin to make sense of what actually happened over the years that have passed, without confining the account to either lamentations or shouts of enthusiasm that need no basis in facts. For a serious analysis, reliably established facts are indispensable, and—let us say at once what will be repeated many times in various ways throughout this book—the reliably established is in very short supply. A great deal of information was not collected in time, while that which perhaps *was* established is not always accessible, and that which is presented as established in fact merely expresses one or another political position. It goes without saying that not everything is accounted for in this book.

The history of mathematics education, including its recent history, seems to us indissolubly linked with social-political and economic history, although, of course, it cannot be reduced to the latter. At times, the social-political exerts an influence in complex ways and often with a great delay. Nonetheless, the dramatic recent history of Eastern European countries—a history of hopes and disappointments, despair and happiness—will not be complete without an examination of what has taken place in education, including mathematics education. One would like to think that the study of what has taken place will also be useful to mathematics educators, if only because it will provide another opportunity to recognize just how complex and contradictory are the processes taking place in the sphere of mathematics education.

This book contains seven chapters, six of which are devoted to specific countries, while the last of which represents an attempt to bring everything that has been said together (to repeat, an attempt whose success thus far can only be limited—such studies must be continued). The countries are dealt with in alphabetical order chapter by chapter: the Czech Republic by Alena Hošpesová and Jarmila Novotná; Germany by Regina Bruder; Hungary by János Gordon Győri, Katalin Fried, Gabriella Köves, Vera Oláh, and Józsefné Pálfalvi; Poland by Marcin Karpiński, Ewa Swoboda, and Małgorzata Zambrowska; Russia by Alexander Karp; and Ukraine by Vasyl O. Shvets, Valentyna G. Bevz, Oleksandr V. Shkolnyi, and Olha I. Matiash.

Alexander Karp

References

Roberts, David. 2010. Interview with Izaak Wirszup. *International Journal for the History of Mathematics Education* 5(1): 53–74.

Senechal, M. 1998. The continuing silence of Bourbaki. An interview with Pierre Cartier, June 18, 1997. *The Mathematical Intelligencer* 1: 22–28.

Swetz, Frank. 1978. *Socialist Mathematics Education.* Southampton: Burgundy Press.

Contents

1 Development of Mathematics Education in the Czech Republic (1989–2018): From a Search for Structure to Mathematical Literacy.................................... 1
Alena Hošpesová and Jarmila Novotná

2 Traditions and Changes in the Teaching and Learning of Mathematics in Germany 45
Regina Bruder

3 The Traditions and Contemporary Characteristics of Mathematics Education in Hungary in the Post-Socialist Era..................................... 75
János Gordon Győri, Katalin Fried, Gabriella Köves, Vera Oláh, and Józsefné Pálfalvi

4 Changes in Polish School Mathematics Education in the Years 1989–2019.. 131
Marcin Karpiński, Ewa Swoboda, and Małgorzata Zambrowska

5 Russian Mathematics Education After 1991 173
Alexander Karp

6 Ukraine: School Mathematics Education in the Last 30 Years...... 229
Vasyl O. Shvets, Valentyna G. Bevz, Oleksandr V. Shkolnyi, and Olha I. Matiash

7 In Lieu of a Conclusion.. 275
Alexander Karp

Author Index 291

Subject Index... 297

Contributors

Valentyna G. Bevz Dragomanov National Pedagogical University, Kyiv, Ukraine

Regina Bruder Technical University of Darmstadt, Darmstadt, Germany

Katalin Fried Eötvös Loránd University, Budapest, Hungary

János Gordon Győri Eötvös Loránd University, Budapest, Hungary

Alena Hošpesová University of South Bohemia České Budějovice, České Budějovice, Czechia

Alexander Karp Teachers College, Columbia University, New York, NY, USA

Marcin Karpiński The School of Education Polish-American Freedom Foundation and University of Warsaw, Warsaw, Poland

Gabriella Köves Károli Gáspár University of the Reformed Church in Hungary, Budapest, Hungary

Olha I. Matiash Vinnytsia State Pedagogical University, Vinnytsia, Ukraine

Jarmila Novotná Faculty of Education, Charles University, Praha 1, Czechia

Vera Oláh János Bolyai Mathematical Society, Budapest, Hungary

Józsefné Pálfalvi Eötvös Loránd University, Budapest, Hungary

Oleksandr V. Shkolnyi Dragomanov National Pedagogical University, Kyiv, Ukraine

Vasyl O. Shvets Dragomanov National Pedagogical University, Kyiv, Ukraine

Ewa Swoboda The State Higher School of Technology and Economics in Jarosław, Jarosław, Poland

Małgorzata Zambrowska The Maria Grzegorzewska University, Warsaw, Poland

About the Authors

Valentyna G. Bevz is a doctor of pedagogical sciehces (doctor habil.) and a full professor in the Mathematics and Methodology of Mathematics Teaching Department at the National Pedagogical Dragomanov University. Her research is related to innovations in education, teaching of mathematics, and history of mathematics in higher and secondary education. She has more than 400 publications. Also, she is the editor-in-chief of the scientific and methodological journal *Mathematics in the Native School* and founder and organizer of the all-Ukrainian student scientific and practical conference "History of Science for Future Teachers of Mathematics."

Regina Bruder studied mathematics and physics teacher education in Potsdam and has more than 10 years of school experience in East and West Germany. She earned her doctorate in mathematics in 1975 and habilitated in 1988. Since 2001 she has been a professor in the subject of didactics of mathematics at the Technical University of Darmstadt. Her work focuses on developments in task-based teaching concepts and their implementation in multiannual teaching projects, in particular for problem-solving learning, dealing with heterogeneity in the classroom, and computer-aided teaching and learning of mathematics. Her empirical educational research focuses on competence development, measuring competency in mathematics in school, and on the transition from school to university based on activity theory. She is interested in the role that digital learning environments and the inclusion of games play in education as well as in teacher training. She has received two research awards and three teaching awards.

Katalin Fried has been engaged with teacher training at Eötvös Loránd University in Budapest since 1983. She has a PhD in mathematics. As an associate professor she is teaching several subjects related to mathematics and mathematical didactics to preservice mathematics teachers.

János Gordon Győri is an associate professor of education at Eötvös Loránd University in the faculty of education and psychology and holds a position at the Institute of Intercultural Psychology and Education in Budapest, Hungary. He is an acclaimed expert on a number of topics in education like mathematics education, culture and education, gifted education, teachers' learning and professional development, shadow education, artificial intelligence in education, and many others. In the past 20 years he published a dozen books as well as more than a hundred articles and chapters, and he has presented at many international and national conferences.

Alena Hošpesová is an associate professor in the faculty of education of the University of South Bohemia in České Budějovice. Her research interests focus mainly on primary school teacher education (pre- and in-service) and ways to improve teachers' competencies (via self- and joint reflection, problem posing, and methods of inquiry-based education). She is an active member of the Czech and the international research communities: she was a member of the International Committee of The International Group for the Psychology of Mathematics Education (PME); she was one of the organizers of Thematic Working Groups on teacher education at several recent conferences of ERME (European Society for Research in Mathematics Education); and she is a member of the informal community called the Learners' Perspective Study. She is a member of several councils of PhD studies in didactics of mathematics in the Czech Republic, and she is supervisor of many PhD students.

Alexander Karp is a professor of mathematics education at Teachers College, Columbia University. He received his PhD in mathematics education from Herzen Pedagogical University in St. Petersburg, Russia, and also holds a degree from the same university in history and education. Currently, his scholarly interests span several areas, including the history of mathematics education, gifted education, mathematics teacher education, and mathematical problem solving. He served as the managing editor of the *International Journal for the History of Mathematics Education* and is the author or editor of over one hundred publications, including over thirty books.

Marcin Karpiński is Lecturer at the School of Education at the Polish-American Freedom Foundation and the University of Warsaw. His main areas of interest are curriculum studies and international comparison of mathematics teaching methods; using large-scale assessment of students' mathematical competences for improving teaching methods in mathematics; and teacher training methods. From 2010 to 2016 he has been a leader of the Mathematics Education Section at the Educational Research Institute. He has been coauthor of the National Curriculum for Mathematics and author of mathematics textbooks for primary, lower secondary, and secondary schools. He is a member of the Polish research team for PISA, TIMSS, and PIAAC.

Gabriella Köves has been engaged with teacher training at Károli Gáspár University of the Reformed Church in Budapest, Hungary, as a professor. She teaches mathematics and mathematical didactics. She earned her PhD in mathematics didactics, conducting a historical study of Hungarian elementary schools' mathematics textbooks. She has also worked on a number of mathematics textbooks.

Olha I. Matiash is a full professor in the faculty of mathematics, physics, and technology of Vinnytsia Mykhailo Kotsiubynskyi State Pedagogical University, as well as head of the Department of Algebra and Methodology of Mathematics Teaching. Her research interests are focused on the theory and methodology of mathematics teaching at school and on the ways of improving the efficiency of training future mathematics teachers. She also has considerable experience in teaching mathematics at school. She also organized the International Scientific and Practical conference titled "Problems and Prospects of Mathematics Teacher Training" (2009, 2012, 2015, 2018) and organized the publication of thematic collections "Methodological Search of Mathematics Teachers." She is the chairman of the Educational and Methodological Board of the Ministry of Education as well as its Science of Training Future Ukrainian Teachers Board (2016–2018).

Jarmila Novotná is a professor in the faculty of education of Charles University in Prague. She has an HDR (habilitation *à diriger des recherches*) at l´Université Bordeaux 2 Segalen, France. She is *chercheur titulaire* at CeDS—Université Bordeaux Segalen. Her main fields of interest are didactical conditions for transforming students' models of activities that lead to grasping knowledge and skills, pre- and in-service training of mathematics teachers, and transfer of research results into practice. She is an active member of the Czech and international research communities: she was a member of the International Committee of The International Group for the Psychology of Mathematics Education (PME) and the European Society for Research in Mathematics Education (ERME); she has been a member or chair of IPCs of several important international scientific events; she was a member of the IPC (International Program Committees) of International Commission on Mathematical Instruction, Study 15 "The Professional Education and Development of Teachers of Mathematics" and ICMI Study 23 "Primary Mathematics Study on Whole Numbers"; she was a member of International Congress on Mathematical Education Survey Team 10 "The Professional Development of Mathematics Teachers"; she has been a member of editorial boards of several international journals; and she is a member of an informal community called the Learners' Perspective Study and Lexicon. She is a member of several councils of PhD studies in didactics of mathematics in the Czech Republic and abroad, and she has been the supervisor of PhD students.

Vera Oláh graduated from Eötvös Loránd University in 1977 as a mathematician and then worked as a computer scientist in information technology. From 1992 until 2002 she was the editor-in-chief of *KöMaL*, and since then she has been involved in Bolyai Society activities as a member of the Education Committee. She has been the vice president of the Society since last year. Between 2005 and 2018 she worked in the Educational Authority and in its background institutions, participating in EU projects and working on developments in mathematical education and its methodology. From 2016 she has worked in the Alfred Rényi Institute of Mathematics, a division of the Hungarian Academy of Sciences, as editor-in-chief of an online mathematical journal. She is now a retired expert of mathematics education.

Józsefné Pálfalvi is a retired associate professor of mathematics education in the faculty of science at Eötvös Loránd University, in Budapest, Hungary. In the 1970s and 1980s she undertook the extension of Tamás Varga's Complex Mathematics Education Experiment program for secondary education. For 16 years she was the head of the Department of Mathematics at the School of Education at Eötvös Loránd University. For 25 years, and with the help of her colleagues, she organized the Varga Tamás Methodological Conferences, which played a pivotal role in saving the teachings and spirit of Tamás Varga. For many decades Józsefné Pálfalvi was also active in in-service teacher education in mathematics, and she also served as a senior reviewer of a number of mathematics curricula and textbooks. Today her research concerns the history of Hungarian mathematics education.

Oleksandr V. Shkolnyi is a full professor in the Mathematics and Methodology of Mathematics Teaching Department at National Pedagogical Dragomanov University (Kyiv, Ukraine). At the beginning of his scientific career, he studied fractal distribution of probabilities on the complex plane and defended a PhD thesis devoted to this topic. Since 2005, he has focused on pedagogical and methodological investigations that relate to the methodology of organizing and carrying out nationwide standardized assessments of pupils' achievements in mathematics in Ukraine. On this topic he published more than 20 textbooks, about 100 articles, a monograph, and defended his doctoral dissertation (Dr. Habil.) in 2015. He is also employee, expert, and certified developer of tests for the Ukrainian Center of Educational Quality Assessment. He works with gifted students and with teachers of mathematics as part of their continuing education.

Vasyl O. Shvets is a full professor in the Mathematics and Methodology of Mathematics Teaching Department at National Pedagogical Dragomanov University (Kyiv, Ukraine). He has been engaged in research and teaching activities for more than 45 years. His research is concentrated on the applied direction of the school's mathematics courses, the control and assessment of pupils' achievements, and the study of approximate computing in secondary and high school. He is the author and coauthor of Ukrainian State Programs in Mathematics for secondary and higher education, the first State Standard of Mathematical Education, and several textbooks on mathematics as well as methodological guides for teachers using those books for secondary and higher education. He was also one of the main organizers of the International Scientific and Practical Conferences titled "Actual Problems of the Theory and Methodology of Teaching Mathematics," which took place in 2004, 2007, 2011, and 2017.

Ewa Swoboda is a retired professor of the faculty of mathematics and natural sciences of the University of Rzeszów. Currently, she continues work as a lecturer, training future teachers. She is author of textbooks for primary school mathematics; editor-in-chief of *Didactica Mathematicae* (from 2012), which is published annually and was established in 1981 by Anna Zafia Krygowska; and member of the

Board of the Polish Mathematical Society, since 2013. The main topics of her research are related to children's development of geometrical thinking, communication in the mathematics classroom, and teachers' mathematical education.

Małgorzata Zambrowska is an assistant professor at Maria Grzegorzewska University in its faculty of mathematics teaching. Her main areas of interest are the history of methods of teaching geometry to young children; influence of national examination systems on the level of student competence in mathematics; and teacher training methods. From 2010 to 2016 she has been a member of the Mathematics Education Section at the Educational Research Institute. She has coauthored the research design, items, and data analysis for two projects: the Nationwide Test of Skills of Third, Fifth, and Ninth Graders as well as "Mathematical Competences of Primary and Lower Secondary School Students." She is also a member of the Polish research team for TIMSS and PIAAC.

Chapter 1
Development of Mathematics Education in the Czech Republic (1989–2018): From a Search for Structure to Mathematical Literacy

Alena Hošpesová and Jarmila Novotná

Abstract Education has always played an important role in the history of the region of the current Czech Republic. Five phases of development after 1989 can be differentiated: (a) deconstruction (1990–1991)—characterized by opposition against the current state; (b) partial stabilization (1991–2000)—decision-making based on what needed to be addressed and analysis of the state of education; (c) system reconstruction (2001–2004)—the beginning of the National Program for the Development of Education (2001) and subsequent Educational Framework Programs; (d) implementation (2005–2015)—elaboration of school education programs, and (e) the current period. This chapter describes the characteristics of these phases, legislative documents, and the development of approaches to the concept of mathematics education, mathematics teacher education, textbooks, and forms of assessment. This chapter concludes with a reflection on the problems with mathematical education in the Czech Republic and the outlook for the future.

Keywords Development of mathematics education in the Czech Republic · History · Mathematics education · Mathematics teacher education · International comparison · Curricular reform after 2001

A. Hošpesová
University of South Bohemia České Budějovice, České Budějovice, Czechia
e-mail: hospes@pf.jcu.cz

J. Novotná (✉)
Faculty of Education, Charles University, Praha 1, Czechia
e-mail: jarmila.novotna@pedf.cuni.cz

J. A. Colmenius: "... Our first wish is for the full power of development to reach all of humanity, not of one particular person or a few or even many, but of every single individual, young and old, rich and poor, noble and ignoble, men and women – in a word, of every human being born of Earth, with the ultimate aim of providing education to the entire human race regardless of age, class, sex and nationality.

Secondly, our wish is that every human being should be rightly developed and perfectly educated, not in any limited sense, but in every respect that makes for the perfection of human nature ..." (Pampaedia or Universal Education)

1 Introduction

The main goal of this chapter is to describe the development that mathematics education has undergone in the territory of the Czech Republic since the so-called Velvet Revolution. The Velvet Revolution is a term used to describe the changes the country underwent from November 17 until December 29, 1989, that had the result of overthrowing the communist regime and installing a pluralistic democratic system. This chapter also summarizes the current state of (mathematics) education in the Czech Republic.

The former Czechoslovakia split into two independent countries in 1993, the Czech and Slovak Republics. The focus of this chapter, when describing the years after 1993, is on the Czech Republic. A lot of attention is paid to the positive changes in that period. However, the authors also pinpoint some of the prevailing problems in the area of mathematics education.

This chapter is divided into several parts. It presents an educational background on the topic of Czech mathematics education, including a brief account of the historical development of the territory of the current Czech Republic. The reader will become familiar with the milestones of Czech history and see the roots of contemporary mathematical education in the Czech Republic. The following section focuses on legislation in the area of education and changes it underwent after 1989 and 1993. It also presents information on various concepts in mathematics education, the content of school mathematics, mathematics textbooks, the use of ICT in mathematics education, and assessment. Without doubt, the role of teachers in, and their impact on, mathematics education is immense, which is why much attention is paid to mathematics teacher training. This chapter also presents information on the relationship of the general public to education and on the opportunities it has to influence what is happening, for example, in the form of public discussions in various forums. The conclusion of this chapter presents possible perspectives on the field of mathematics education as implied by the present state as they are perceived by the authors of this chapter.

2 Background

2.1 History

The history of the region known as the Czech lands is very rich. This territory has changed hands many times and has been known by a variety of different names. With respect to this chapter, let us state here only that the Lands of the Bohemian Crown were incorporated into the Austrian (Habsburg) Empire in 1526 and later became part of the Austro-Hungarian Monarchy. In 1918, after World War I, the Czechoslovak Republic was established. It was a multinational state of the Czechs (the majority population), Germans, Slovaks, and minorities (Hungarians, Russians, Poles).

The territory of Czechoslovakia was reduced considerably in 1938 when the country lost Sudetenland to Nazi Germany, the eastern part of Cieszyn Silesia and a part of Northern Slovakia to Poland, and a part of Carpathian Ruthenia and Eastern and Southern Slovakia to Hungary. Slovakia proclaimed independence on March 14, 1939. The rest of Czechoslovakia was occupied by Nazi troops on March 15, 1939, and on March 16 the creation of the Protectorate of Bohemia and Moravia as a part of the German Reich was proclaimed.

World War II ended with the defeat of Germany, and thus Czechoslovakia was re-established, although Carpathian Ruthenia became part of the Soviet Union. Three million German inhabitants of Czechoslovakia were expelled from the country and sent to Germany and Austria on the basis of presidential decrees (the Beneš decrees) with the consent of the allies. The postwar Czechoslovak Republic was dependent on the Soviet Union. Reforms were implemented that led to a gradual elimination of private ownership and ensured a monopoly of power. The country was taken over by the communists in a coup d'état in February 1948, and the country became part of the Soviet bloc.

The 1960s witnessed a time of liberalization that culminated in 1968 in the so-called Prague Spring. However, the reform movement was suppressed by the invasion of the Warsaw Pact armies on August 21, 1968. Russian occupying troops stayed in the country until 1991. The Prague Spring and invasion in 1968 were followed by 20 years of the so-called normalization. The communist regime became less and less popular in the 1980s. The Velvet Revolution overthrew the communist regime. The first free elections in 1990 brought a victory for democratic parties. The country started the process of liberalization and oriented itself toward western democratic countries. Since then its economy has been privatized and new political parties have been established.

One of the quickly developing problems of the new democratic country was its multinational structure. The problem resulted in a splitting of the country. Czechoslovakia ceased to exist at the end of 1992 without a plebiscite after negotiations failed between the top political representatives of both parts of the country. The Czech and Slovak Republics came into existence on January 1, 1993. The countries were quickly internationally recognized as successor states of the former

Czechoslovakia. The Czech Republic entered NATO on March 12, 1999. The 2003 plebiscite confirmed that the country would enter the EU, which happened on May 1, 2004. In 2007, the Czech Republic became a part of the Schengen Area. The Czech Republic is a member of the so-called Visegrád group that came to existence in the time of the former Czechoslovakia (February 15, 1991), which now defends the common interests of the Czech Republic, Slovakia, Poland, and Hungary in the European Union.

2.2 Educational Tradition in the Region of Today's Czech Republic

Education has always played an important role in the history of the region of the Czech Republic. Gergelová Šteigrová (2011) highlighted as historical milestones the foundation of Charles University in Prague and the life and work of Jan Amos Comenius.

The University in Prague, the affiliation of one of the authors, was founded in 1348 by Charles IV, the King of Bohemia and Roman Emperor. At that time, it was the first *Studium generale* north of the Alps and east of Paris. It was modeled on universities in Bologna and Paris, and within a very short time it achieved international recognition. Mathematics education was, according to testimony, on a higher level than mathematics education at the University of Paris (Hankel 1874 in Mikulčák 2010). At the bachelor's level, it included the processes of practical addition, subtraction, multiplication, and division. On the master's level, students focused on the fundamentals of Euclid. The quality of the university was important not only for the development of mathematical thinking, but it also affected the level of education in the whole region. The rector of the university published school rules with curricula and appointed some teachers. It is not surprising that population groups in the region that would not have received any education in other countries were often able to read and write (e.g., Hussite women in the fifteenth century) (Mikulčák 2010).

This tradition was part of the background of the personality of Jan Amos Comenius (1592–1670), one of the greatest Czech thinkers. He gained a reputation not only as a theorist in pedagogy but also wrote several textbooks for different school levels and formulated a number of recommendations for mathematics education from pre-school to secondary school levels. He emphasized that pupils should learn arithmetic and geometry comprehensively (Mikulčák 2010). For example, in his treatise *Schola Ludus* ("Playful School" 1654) he showed how pupils should get to know and interpret basic mathematical and geometrical concepts through a form of dramatic improvisation. General pedagogical principles were developed in *Didactica magna* ("The Great Didactic" 1657) and *Orbis Pictus* ("The World in Pictures" 1658).

2.3 Mathematics Education in the Newly Established Czechoslovakia

If we want to understand the situation of the system of education after 1989, several facts from earlier in the twentieth century must be mentioned. The newly established Czechoslovak Republic (1918) inherited the educational system of Austrian-Hungarian Empire. Although the Minor Education Act (1922) was passed, the system of education did not transform much. Lower secondary schools remained a part of the educational system and served the function of a social filter: a consequence of school fees and demands on pupils' cultural level was that only about 40% of pupils were able to finish their studies (Walterová 2011). Any attempts to innovate the teaching of mathematics were in the hands of outstanding primary school teachers. In secondary schools, the process was supported by the Union of Czechoslovak Mathematicians and Physicists. The Union is one of the oldest (and still existing) learned societies in our territory. It was founded in 1862. Its original mission was to improve the teaching of physics and mathematics at all levels and types of schools and to support and develop these disciplines. The members of the union were mostly teachers from upper secondary schools and institutions of higher education. The minority were university teachers and scientists.

However, pedagogical reform inspired by American sources was gradually gaining importance in the interwar Czechoslovakia, which resulted in the establishment of several reformed schools in the 1930s. The development of reformist ideas was stopped during the period of World War II, because this was the period of German pressure on Czech schools.

After World War II, a renaissance of reformist ideas had no time to develop. The communist coup d'état in 1948 brought a gradual increase of influence from the Soviet Union on the system of education. However, this influence was not only negative. The positive effects were publications of translations of methodological (mathematical) literature from Russian and the availability of a large number of academic works in Russian. Also, Russian translations of French, English, and German works were very valuable as they were not available in Czech or Slovak.

However, Soviet influence also brought the new conception of undifferentiated (uniform) schools, which in practical terms meant the replacement of several streams of education (secondary grammar schools, *stadt schule*, *real schule*) by one type of school with the same demands on pupils of the same age. The intervention was a political move and was interpreted by the communist leaders as providing equity in education for all pupils. Compulsory school attendance lasted 9 years. The first 5 years were called the national school and the following 4 years were called the middle school. This was followed by a third level, which was not compulsory. This level included 3-year-long vocational schools, 4–5-year-long professional schools, and 4-year-long gymnasiums (grammar schools). Students at professional schools and gymnasiums finished by passing the Maturita (school leaving examination). The view of the status of the Maturita examination was gradually changing. Although it was still the necessary condition for being accepted to university, it was not the only condition for being accepted. Its content and form were affected by the contemporary social and political changes. The communist government tried to build a regime based on the Soviet model. The school was to be connected to the everyday life of the working class. The Maturita exam was no longer to be an elite exam. The goal was to change it to make it accessible and passable to a greater number of working-class students and politically reliable supporters of the regime so that they could start their studies at university. In consequence, the prestige and level of the exam dropped significantly. It was also possible to pass the exam by studying only part time while working and simultaneously studying at a so-called evening school or in the 1-year "working class prep" course (Morkes 2003). Mathematics was an elective subject of the oral part of the Maturita exam.

However, it became obvious that the uniform requirements did not reflect the needs of everyone—weaker pupils could not cope with them and talented pupils were not developed sufficiently. The political decision made in reaction to the economic crisis of the 1950s was to have students graduate quickly and then have them participate in "building communism" as soon as possible. This resulted in the establishment of the 11-year school system, which followed the Soviet model. Compulsory school attendance was shortened to 8 years. The following 3 years were elective, offering upper secondary general education. Before this concept was passed into law, it was subject to a wider discussion. Teachers who disagreed with its concepts were forced to leave their jobs. Shortening the length of education by 2 years required a change in curriculum, from a cyclical model to one organized in a linear fashion. The systematic teaching of disciplines started in the 7th grade and finished in the 11th grade. There was no review before the school leaving exam. This meant that pupils who left school after 8 years of compulsory education did not have a complex, comprehensive education and lacked a lot of knowledge that was taught from the 9th to 11th grades (e.g., analytical geometry, congruence, solid geometry).

The 11-year school system focused on providing education in the sciences (mathematics, physics, and chemistry). Language education (with the exception of Russian) remained compulsory but was not strong. Also, the status of biology declined. The 11 years of school were concluded by the Maturita exam; the written part of it was done in Czech and Russian and the oral part included compulsory mathematics. The Maturita exam in mathematics had a time limit of 15 min.

The system changed in the 1960s. Compulsory school attendance was restored to 9 years. The system diverted from the Russian model. Compulsory education was provided at what was known as 9-year basic schools, which had 5-year-long primary and 4-year-long lower secondary levels. Upper secondary studies lasted from 2 to 5 years. The 3-year-long upper secondary comprehensive school was primarily a preparation for university studies. However, its curriculum also included work practice in an industry. This work practice was later omitted from the curriculum, and the schools began to specialize in arts or sciences. Studies concluded with the Maturita exam. The compulsory part was Czech (or Slovak), Russian, or elected subjects including mathematics.

Gymnasiums (grammar schools) were re-established after 1968. Their curricula came out of a historical tradition but were also inspired by international trends. In the beginning this resulted in the improved quality of upper secondary education. However, *normalization* (the period of political oppression after the unsuccessful reform movement in 1968) resulted in a decline in the prestige of gymnasiums when the so-called dorm schools for the working class were established in 1974. Their purpose was to prepare young, talented, politically conscious and loyal members of the working class for university studies. The acceptance criteria were political-engagement based. The schools were 1-year long and concluded by the Maturita exam. Their graduates were accepted to universities without entrance exams. These schools ceased to exist as late as 1984.

The 1978 Act of Education introduced a 10-year compulsory school attendance. Practically, it meant that the lower secondary school level was shortened by 1 year, which was followed by two compulsory years of upper secondary education. All students were obliged to study at some kind of upper secondary school—gymnasiums, professional, or vocational schools. Vocational school students could sit for the Maturita exam and then be accepted to universities. Members of the working class could also sit for the Maturita exam in one of the shorter evening schools or for distant studies. These measures were in line with the vision of the Communist party that the majority of the young should graduate from an upper secondary school. The content of the Maturita exam changed. Czech language study was still compulsory. At certain types of schools (technical schools), Maturita in mathematics became compulsory at certain periods.

The tradition of high-quality education did not vanish completely even during the hard times. There were a number of excellent teachers at schools, thanks to whom the level of education remained good (Novotná et al. 2019). This holds true especially for mathematics education, in which a great interest in quality was retained. It was crucially important that school mathematics education was in the spotlight of several scientists in the field of mathematics. Let us name here at least B. Bydžovský and E. Čech. Čech was the author of a number of demanding textbooks for secondary schools, which emphasized the structure of mathematics. He also formulated the demand that schools should develop pupils' activity, develop their interest in mathematics, their language, and their ability to think independently. The textbooks presented mathematics as a system (Boček and Kuřina 2013).

An important turning point came in the 1960s. Innovation in mathematics education was gradually breaking through thanks to the influence of similar efforts in Europe and the United States as well as by the Kolmogorov reform in the Soviet Union. Czechoslovakia, like many other countries, had its period of New Math—another terminology was used of course. (For more details, see Jelínek and Šedivý 1982a, b).

A specialized department was established in the Institute of Mathematics, Czechoslovak Academy of Sciences, *Section for Didactics of Mathematics*. Its work was characterized by the effort to connect research in teaching practices with curricular research, both on theoretical and on practical levels. Emphasis was put on connections between school mathematics and everyday reality, geometrization, and learner-centered approaches where the teacher's role was to support discovery (i.e., experimenting, problem-based teaching, genetic approach). The aim of the Section for Didactics of Mathematics was to reduce fact-based teaching, to develop a psychological-genetic approach, to put emphasis on working methods in mathematics, on laboratory and problem-based approaches, and to reinforce algorithmic components (Vondrová et al. 2015a, b). The collaboration with university teachers and researchers in the field of mathematics by the staff team of the Section for Didactics of Mathematics was supported as a fundamental direction of research. The team cooperated with experimental schools too.

The partial results and experience from this experiment were used in applied research conducted by the Research Institute of Education. The role of the institute was to modify the didactical system of school mathematics from the primary to upper secondary school levels with respect to the current state of mathematical disciplines. Its role was also to introduce the basic elements of logic and to get rid of isolation in various parts of mathematics by linking curricular content on the basis of knowledge of sets and by preparing pupils to understand mathematical structures (Kabele 1968). The system was tested in select primary and lower secondary schools. Then in 1976 a set of textbooks was designed and "set" mathematics started to be taught in all schools (Kabele 1968), first at the secondary level (described in more detail in e.g., Mikulčák 1967), then in grades 1–9. Given the scope and radical nature of the changes, mathematics became the flagship of the innovation project with the publication of Ministerstvo školství (1976) which gave it its informal name of "sets."

This was also the time when more attention turned to mathematically gifted pupils. There were more than 100 classes for extended mathematics education at basic schools. Special learning materials were produced for them. Mathematically gifted pupils were also taken care of within the framework of the Mathematical Olympiads. The stimulus for this was very weak results from Czechoslovak participants at the International Mathematical Olympiad in the 1960s and the beginning of the 1970s. The cause of these failures was not worse knowledge or abilities of the Czechoslovak participants in comparison to their competitors from countries with traditionally good results, such as the Soviet Union or Hungary; the cause was inadequate preparation and lack of experience at competitions. That is why a correspondence seminar was established. It resulted in an improvement especially in the years 1979–1983 (Mikulčák 2007).

Although it is often said that pedagogy was strongly subordinate to ideology, the modernization movement was inspired not only by the Soviet Union but also by different sources. Tichá (2013) discussed how, for example, Bruner, Papy, Freudenthal, and the results of the International Congress on Mathematical Education (ICME) in 1988 inspired the work of the Section for Didactics of Mathematics.

3 Legislation Before and After 1989

3.1 Selected Legal Anchoring Before 1989

Compulsory 6-year school attendance was introduced for boys and girls in Bohemia and Moravia in 1774. It was extended to compulsory 8-year school attendance in 1869 in the Austro-Hungarian Empire, which also included today's territory of the Czech Republic. The national school consisted of institutionally separated levels: a 5-year general school (*Volkschule* in German) and a 3-year municipal school (*Stadtschule* in German); boys, not girls, were allowed to study at different types of secondary schools (grammar schools, gymnasium, real gymnasium) instead of municipal schools. The only public secondary school girls could attend were pedagogical schools. Other secondary schools for girls were private. Up until 1918, education for girls was not supported by the state. Curricula for *Volksschule* were based upon basic subjects (reading, writing, counting, measuring, physical education, religious education), to which biology, geography, and history were added in *Stadtschule*.

When Czechoslovakia was established (1918), it adopted the laws of the Austro-Hungarian state. Only partial changes were made. Predominant in the countryside were 1–2-year general schools, which allowed the establishment of schools in even the smallest villages. The presence of rural schools helped the development of literacy. The school system in larger towns was composed of 5-year general schools and then followed up by 3-year municipal schools.

The end of World War II was followed by a very short period of partial changes and was terminated by the Act of Education in April 1948. The 5-year national school was followed by a 4-year middle school. A higher level of general education was provided by 4-year secondary grammar schools. Technical education was provided by various types of 4-year technical schools (concluded by the *Maturita* [graduation] examination). Skilled workers were educated in vocational schools. Mathematical education officially proclaimed to follow Soviet models. For example, a conference of the Union of Czechoslovak Mathematicians and Physicists on a new conception of teaching mathematics was held in August 1949. There E. Čech introduced the participants to the principles of scientific quality in teaching mathematics. His vision was for school mathematics to be built on the structure of mathematics as science.

3.2 Development of Education and Legislation After 1989

After 1989, ideological dependence, unilateralism, dogmatism, pressure, emphasis on discipline, and formalism in education were strongly criticized. Kotásek (2004) differentiated four phases of the initial period:

1. Deconstruction (1990–1991)—characterized by resisting against the current state
2. Partial stabilization (1991–2000)—decision-making on what needs to be addressed and analyzing the state of education

3. System Reconstruction (2001–2004)—White paper (National Program for the Development of Education) (Kotásek et al. 2001) and subsequent Educational Framework Programs
4. Implementation (2005) —adoption of the Education Act.

3.2.1 Deconstruction of the Socialist System of Education (1990–1991)

At the beginning of this period, tight ideological control ended and the state monopoly of education was broken down. The management of the education system was consecutively decentralized mainly by delegating a number of decision-making powers to municipalities, schools, and their headmasters. Schools had more freedom in the allocation of lessons to each subject. Several private and denominational schools were established. Most mathematics teachers rejected uniform textbooks based on set mathematics, looked for other ways of interpreting the topics, and asked for new textbooks. The challenge was taken up by several teams of authors. New series of textbooks were developed quickly and quickly became used in practice at schools.

The recognition of the rights of pupils (or their parents) to choose their educational path according to their abilities and interests as an integral part of the overall liberalization process started to be taken into account.

Many aspects of teachers' work changed. On the one hand, teachers gained significant space for their independent creative work in preparation of the content of individual subjects, in the choice of teaching methods, and the choice of textbooks. On the other hand, it meant greater burdens on teachers; many teachers had to deal with a temporary loss of their professional competence (pedagogical as well as technical), which necessitated partial, and sometimes even full, requalification. For example, Krajčová and Münich (2018) report the insufficient development of ICT literacy among older teachers at the time.

Regarding the organization of the system, a note should be made that compulsory 10-year school attendance (discussed earlier) was abolished after the year 1989, after the Velvet Revolution, when the country returned to the model of 9 years of compulsory school attendance: 5 years of primary and 4 years of lower secondary levels at basic schools.

Upper secondary school education builds on basic school education. There are gymnasiums (grammar schools), technical and professional secondary schools, and vocational schools. Generally, the length of study is 4 years (sometimes, but very rarely, longer), if the school ends with Maturita. There are a wide range of apprentice schools preparing students mostly for various crafts that do not end with Maturita.

The 1990s saw the restoration of 8- and 6-year grammar schools, which previously existed between World War I and II. These schools prepare their students for university studies. This means very early differentiation of pupils. Pupils who are interested in studying at these schools are admitted if they pass the entrance exams after the fifth and seventh grades of school attendance.

3.2.2 Partial Stabilization (1991–2000)

This period focused on the adjustment of the existing curriculum and the creation of space for the implementation of new curricular projects. The Ministry of Education, Youth and Sport[1] initiated a process in which the existing curriculum, called Basic School, was modified by a group of authors in 1996 (Jeřábek 1998). What was added in this document was an explicit definition of fundamental curricular content (referred to as "What the pupil should have mastered"). The minimum number of lessons for each educational area was specified. The rest of the lessons, the so-called disposable lessons, allowed schools to specialize. This change did not have much impact on the content of mathematics education.

Apart from the project Basic School, there were other projects ratified by the Ministry too. Each school could make a decision as to which of the projects to follow.

The Association of Teachers of Primary Education created a project called National School (MŠMT 1997), alternative school curricula for compulsory school education (i.e., first to ninth grade). The ideas that came out of it were that it was essential to differentiate education according to the decisions of a particular school to the maximum degree possible and to adapt education to the needs of individual pupils. In contrast to the above described Curricula for Basic School, less time was given to mathematics (about 20% less). As it was known, the common teaching content (compulsory minimum content for all schools at the given level) was specified.

Similar projects were the programs called Educational program Municipal School (MŠMT 1996) (grades one through five) and the follow-up Civic School (grades six through nine). These projects did not only define what a pupil should and can master but also present model problems.

In Table 1.1, we present how the introduction to the concept of numbers in the first grade is treated in the different curricular projects.

Besides those projects, other traditional alternative programs (Waldorf pedagogy, Montessori, Dalton) were introduced in the Czech Republic. Montessori mathematics especially became very attractive to a number of parents.

There is another fact that must be mentioned here. The mid-1990s witnessed the establishment of the first lower secondary grammar schools, which represented a more academic strand of education. The tradition of these schools goes back to the interwar years and already at those times they were criticized for their elitist nature (Mikulčák 2010).

Nationwide discussion on the concept of education in the Czech Republic, description of problems, and the definition of future development was based on

---

[1] In the following text, it is just called the Ministry.

Table 1.1 The concept of numbers in different curricula

National school	Basic school	Municipal school
Knowledge of numbers 0–20 Position of numbers on a number line Orientation on number line Writing numbers to 20 Comparing numbers and signs of inequality	**Content** Numerical series Relations greater than, smaller than, and equals Numbers Symbols >, <, =, +, - ***** Counting objects in a given set. Choosing various specific sets with a given number of elements Reading and recording numbers Comparing numbers Solving and posing word problems involving the comparison of numbers Solving and posing word problems and addition and subtraction **What the pupils should have mastered** Count the elements of a given set to 20 (including 20). Create a specific set (beads, marbles, etc.) with a given number of elements up to 20 (including 20). Compare numbers and sets of elements up to 20. Read and record numbers from 0 to 20 Solve word problems requiring the comparison of numbers from 0 to 20 **Examples of extending contents** Measuring lengths using segments (e.g., centimeters)	**Pupils should master (level Z)** Figures and numbers. Reading and recording numbers. Numbers as amounts and positions comparing numbers Guessing the amount The principle of the decimal system Mathematical notation and expression. The needed mathematical symbols. **Activities, tools, inspiration** Marking and comparing numbers on a number line and in the context of measuring. Arrow diagrams. Estimating the number of objects and numerical figures. In the number 3724, the number 37 also expresses the number of hundreds. **Pupils can master (level R)** Negative numbers in real life (thermometer, lift, bank account). Roman numerals. **Increasing accent** Flexibility of the use of the decimal system (decomposition and models). Notion of magnitude of numbers. Use of natural language and its gradual precision. **Decreasing accent** Decimal positional notation system Formal isolation of domains of numbers

analyses conducted in the Czech Republic (MŠMT 1998) and abroad (OECD 1997; Čerych 1999b). It came out of the study "České vzdělání a Evropa: strategie rozvoje lidských zdrojů při vstupu ČR do EU" (Czech Education and Europe: Strategy for the Development of Human Resources on Entering the EU, Čerych 1999a). The process ended with the creation of the White Paper (Kotásek et al. 2001).

3.2.3 The Period of System Reconstruction (2001–2004)

The period started with the issuing of the *National program for the development of education in the Czech Republic: White Paper* (Kotásek et al. 2001).

White Paper (2001) and Act of Education (2004)

The White Paper represented a proposal for a system of education that defined the ideological background, general goals, and actionable programs by which the development of the education system in the intermediate-term should be directed. It was a binding foundation on which all specific action plans of the Ministry were to be planned. The direction of Czech educational policy was expressed by the main strategic lines that are presented schematically in Fig. 1.1.

The White Paper did not focus on individual subjects; but it also defined the basic principles of education. From the viewpoint of mathematics education, most important was the focus on the needs of life in a knowledge-based society. This allowed the financing of a number of projects focusing on the development of mathematical literacy and inquiry-based mathematics (and science) education. The White Paper was planned to be the fundamental document for the medium-term horizon (until about 2010).

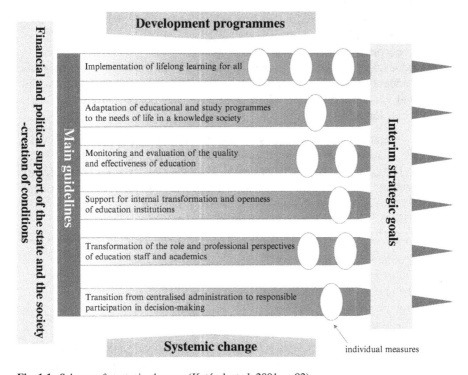

Fig. 1.1 Schema of strategic changes (Kotásek et al. 2001, p. 92)

The White Paper was followed by the new Act of Education (Parliament 2004) which came into force in 2005. It legitimized the scheme of the system of education presented (in Fig. 1.2). The Act of Education stated that the Ministry would issue *Framework Education Programs* (FEP) for each field of education in preschool, primary, secondary, art, and language education. The Act also stated that each school (its headmaster and teachers) was responsible for the creation of their own School Education Programs (SEP). Thus, a two-level curriculum was enacted (the schema is presented in Fig. 1.3).

Framework for Education Programs (FEP) and School Education Programs (SEPs)

FEPs define the compulsory content, range, and conditions of education. They are necessary for the development of School Education Programs (SEPs), the assessment of educational outcomes, and for the development and certification of textbooks and other teaching texts. SEPs are developed based on school levels while reflecting their particular regional needs. They allow schools to make maximum use of local materials and personal conditions.

Janík et al. (2010) summarize that FEPs create the following innovations:

- At the level of organization and management of an education system, the development and implementation of curricula becomes decentralized.
- At the level of conception and goals of school education, emphasis is put on the development of key competencies.
- The concept of inclusive education is promoted.
- At the level of school educational content, they aim to surpass encyclopedic approaches of education through the enrichment of curricula by adding new educational content, structuring educational areas and fields, and creating cross-curricular links.

If we compare FEP BE with previous curricular projects, we see that the main difference is in the formulation of educational outcomes as competencies. This approach is used in all FEPs and puts emphasis on the process of life-long learning. The schematic in Fig. 1.4 shows the structure of FEP for basic education (pupils aged 6–15) (FEP BE English translation from 2007).

The FEP BE demands the development of:

- learning competencies
- problem-solving competencies
- communication competencies
- social and personal competencies
- civil competencies
- working competencies

Development of competencies is the general goal, which is not linked explicitly to any of the subjects. Disciplinal development is expected to be supported in the solution of various types of problems across subjects.

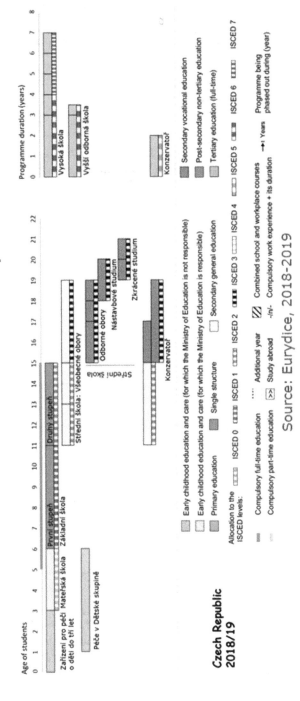

Fig. 1.2 Structure of the Czech educational system, (Source: Euridice, 2018–2019. https://eacea.ec.europa.eu/national-policies/eurydice/content/czech-republic_en)

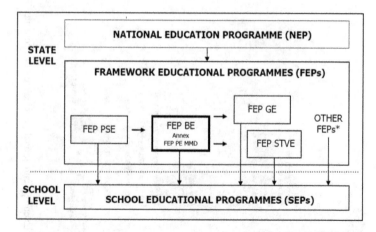

Fig. 1.3 Structure of Czech curricular materials (Source: FEP BE 2007, p. 6)

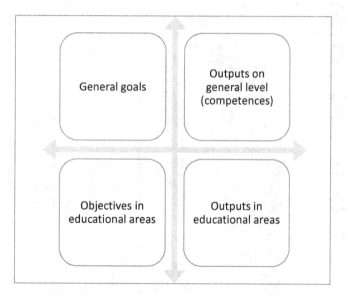

Fig. 1.4 Structure of FEP BE

How are the demands of these general outcomes formulated? Let us illustrate this using the problem-solving competency that is defined in the FEP for preschool education, the FEP BE, and the FEP for upper secondary education (grammar schools). The components of the problem-solving competence copy to a certain degree the stages of solving a problem (Polya 1945). The various content of areas of school education means that different subjects offer different opportunities for the development of problem-solving competency. In general, topics in the area of mathematics and science provide ample opportunities for solutions to rational-logic problems and deterministic problems.

One of the problems is that the competences are formulated very vaguely (Straková et al. 2009). Also, its gradation (for 6-, 15-, and 19-year-old pupils) is in our opinion not convincing. Table 1.2 presents a comparison of requirements for problem-solving competency.

How to form, shape, and develop pupils' key competencies? Setting key competencies as target categories of basic education is problematic. For one thing, their formulation is rather vague (see the examples in Table 1.2). Also, student achievement is difficult to assess (Janík et al. 2010, p. 24). Mathematics teachers are used to focusing on educational outcomes with achievements that are verifiable and quantifiable, i.e., based on knowledge of curricular content and the ability to work with this content. This stems from the very essence of mathematics, which has a solid knowledge structure (Müller and Steinbring 2004).

Another question is how a teacher should work with these competencies when planning, conducting, and evaluating a lesson. It is recommended that teachers should first work with the educational content part of the FEP BE related to educational areas (EA). Here they find characteristics of the EA, objectives of the EA, and educational content for each EA.

The questions still open are how well teachers have managed and continue to manage to fulfill the goals defined by the FEP and what the impact of the introduction of FEP is on the performance of Czech pupils. We think an answer to these questions can be sought in the results of Czech pupils in international comparative studies. These are presented in more detail in Sect. 7.

3.2.4 Implementation (2005–2015)

This period was characterized by the elaboration of school education programs (SEPs). SEPs were obligatory documents. Preparation of the SEP was presented as:

> ... manifestation of the school's pedagogical autonomy as well as its responsibility for teaching methods and outcomes. For this reason, the individual parts of the SEP are prepared with the participation of all of the teachers of the school in question, who are also co-responsible for implementing the SEP in the conditions of their school." (FEP BE 2007, p. 212).

Groups of teachers handled the newly gained autonomy on different levels (Janík and Knecht 2007). Some schools invited support from the outside and worked on the SEP in teams. In other schools, the responsibility was delegated to one teacher for each subject. The SEP for each subject was developed in isolation from other subjects, often very formally. Pedagogical research focusing on this period agrees that the reform was not well understood by teachers who had very contradictory attitudes toward it (Janík and Knecht 2007).

The first stage of this period, when FEPs were prepared, transitioned to the stage when FEPs were implemented by schools (2007–2015). Even this period is marked by contradictory attitudes. Some teachers speak of the reform as a very unpleasant matter that bore a lot of extra, unpaid work and unnecessary formalities and had no

Table 1.2 Comparison of components of problem-solving competence for FEP Preschool Education, FEP BE, and FEP in secondary grammar schools

FEP Preschool Education (6 years old) 2018, p. 11 and 12	FEP BE (15 years old) 2007, p. 11	FEP Upper Secondary (19 years old) 2007, p. 9
A child/student finishing the pre-school/basic/upper secondary education		
… takes notice of events and problems in their environment. …has a positive echo to their active interest that is a good motivation for solving further problems and situations	… notices the most various problem situations at school and outside of school; recognizes and understands a problem; reflects on discrepancies and their causes; considers and plans ways to address problems while employing his/her judgment and experience	… recognizes a problem, elucidates its nature, divides it into parts
… decides problems up to their ability; tries to decide known and repeated situations independently (through imitation or repetition) and the more exigent ones with support and aid of an adult	… is able to find information useful for solving problems; identifies identical, similar, and different features; applies the knowledge acquired to discover various solutions; is not discouraged by a failure should there be one and persistently seeks conclusive solutions to a problem	…forms hypotheses, proposes gradual steps, considers the application of various methods when solving problems or verifying a hypothesis. … applies appropriate methods and prior knowledge and skills when solving problems; apart from analytical and critical thinking, the pupil uses also creative thinking while employing imagination and intuition. … is open to using various methods when solving problems, considers a problem from various sides
… decides the problems according to their immediate experience, proceeds via trial and error, proves, experiments, spontaneously brings out new solutions to problems, searches for different possibilities and variants, has their own original ideas and applies previous experiences	… solves problems independently; selects suitable ways to solve problems; uses logical, mathematical and empirical methods when solving problems	
… can decide between virtual and real problems, uses logic, mathematic and empiric procedures, grasps simple algorithms for solving various tasks and situations, and uses them for further ones. … states precise ideas about numbers, expresses sizes, uses numerical and mathematic concepts and perceives elementary mathematic relations	… tests in practice the correctness of problem-solving methods and applies sound practices when addressing similar or new problem situations; monitors progress when overcoming problems	… interprets critically the acquired knowledge and findings and verifies them; finds arguments and evidence for claims, formulates, and defends well-founded conclusions

(continued)

Table 1.2 (continued)

FEP Preschool Education (6 years old) 2018, p. 11 and 12	FEP BE (15 years old) 2007, p. 11	FEP Upper Secondary (19 years old) 2007, p. 9
… differentiates between function solutions leading to an aim and non-functional solutions and is able to make a choice between them. …understands to avoid a problem solution that does not lead to a target and that timely and prudent solutions are an advantage; realizes that they are able to affect situations due to their actions. … is not afraid to fail when not only their success is appreciated but also their endeavors	… thinks critically; makes prudent decisions and is able to defend them; realizes the responsibility for decisions; is able to evaluate the results of their decisions	… considers the possible advantages and disadvantages of the individual solution variants, including the assessment of their risks and consequences

positive impact on education. Teachers did not oppose the idea of educational reform, but their interpretation of the objectives of the reform was different from the expected aims (Straková 2010). Questionnaire surveys (Janík et al. 2011) confirm that teachers failed to understand that the aim of the reform was to reinforce autonomy of schools, to support decentralization, to allow schools to take part in the creation of curricula, and innovation at the level of educational outcomes and content. On the other hand, teachers appreciated the opportunity to choose methods of work more freely.

In this period, the Analysis of Fulfilment of the Objectives of the National Program for the Development of Education (Straková et al. 2009) was expanded. A group of experts invited by the Ministry stated in this document that the objectives formulated in the White Paper were not achieved. Some problems were caused by a too general formulation of goals. Some goals were not ordered hierarchically from the point of view of importance, and causal links predominated. Only in some cases was operationalization (i.e., expression by measurable signs) of the goals of individual measures done. Indicators of achievement for the goals were largely absent. Inequity in education was described as one of the persistent problems of the Czech system of education because the performance and results of pupils in the system were closely related to the economic and social status of their families.

Early diversification of pupils into a more academic stream of education (the so-called lower secondary grammar schools) also received negative public assessment as hindering equity in the Czech education system. The OECD found that:

> A majority of Czech parents from a low socio-economic background send their children to local schools, while only a small number of parents from higher socio-economic background choose to do so. School choice can further increase inequities if mechanisms are not in place to lessen the negative effects. Czech pupils who struggle in school are often streamed into special schools with reduced curricula. Streaming pupils based on academic

ability and early tracking can negatively impact their educational outcomes if education pathways do not provide high curricular standards and opportunities to transfer between tracks. Early tracking occurs at age 11 (compared to the OECD average of 14), and differentiation of educational pathways has increased. Research shows a strong relation between tracks chosen and socio-economic background. At age 11, 13% of the pupil cohort, mostly from high socio-economic backgrounds, enters lower secondary grammar schools. (OECD 2013, p. 6)

The above-presented conclusions were confirmed by the results of Czech pupils in the international comparative PISA (Programme for International Student Assessment) study of 2015 (Blažek and Příhodová 2016). Analysis of results achieved in tests of mathematical literacy shows that Czech schools replicate social inequality. Secondary analysis of PISA 2015 results (ČŠI 2017, p. 20) shows that if a pupil with a lower socioeconomic background (SEB) attends a school with pupils of above-average SEB, i.e., the average SEB of the school is high, the pupil achieves better results than if they attend a school with a below-average SEB. This relation is very strong among the population of 15-year olds. In other words, if a significant group of pupils leave elementary school and go to lower secondary grammar schools at a relatively early stage, the results of the whole population are negatively affected because SEB in the elementary school declines.

3.2.5 System Reconstruction (2015 and Further)

The phases formulated by Kotásek (2004) can be amended with the current phase, which is referred to as the period of *system reconstruction*. In 2014, the White Paper was replaced by new strategic material, Strategy of Educational Policy of the Czech Republic until 2020 (MŠMT 2014). In contrast to the many priorities declared in the White Paper, this material focuses only on a few strategic goals:

– reduce inequality in education.
– promote quality teaching and teachers as a key condition of education.
– institute responsible and effective management of the education system.

The integration of children with special needs started; however, a lack of funds, teaching assistants, and training of teachers in this area made the process very painstaking. The problems of children from socially and culturally disadvantaged backgrounds persist because support for their education fails and social inequality is replicated. Also, the number of pupils with a limited command of Czech is increasing. This problem has not yet been solved. Simultaneously to all these changes, the FEP BE is currently being modified.

More information about the situation in Czech schools during this period can be found in (Dvořák et al. 2010, 2015).

4 Conception of Mathematics Education

The current conception of mathematics education is influenced by the ideas of constructivism, which have been adapted to the conditions of teaching mathematics in Czech schools by Hejný and Kuřina (2015). They developed the conception of didactic constructivism (Hejný and Kuřina 2015) which comes out of the idea of mathematics as a human activity that, through solving tasks and problems, leads to looking for connections, creating concepts, generalizing statements, and proving them. Part of this activity is the development of mathematical models of reality. This approach was further developed by other authors (Stehlíková 2004; Stehlíková and Cachová 2006; Molnár et al. 2008; FEP BE 2007). The teacher's role is to encourage pupils' activity. In places where pupils can work with their existing knowledge and experience, they are encouraged to use them. The teacher steps in at places where recapitulation and the summary of knowledge, sample solutions, and deduced algorithms are necessary. It is the teacher who knows the pupils best and can make decisions about methods and procedures for the given lesson (Hejný et al. 2004). The aim is to guide the teachers to use such teaching methods and forms of work that, according to their long-term experience, raise pupils' interest in new knowledge, encourage dialog between the teacher and their pupils, enable the individualization of teaching and introduction of a variety of modern technologies, and stimulate pupils' teamwork. Success is not measured by the amount of mastered encyclopedic knowledge, but by the ability to solve problems and to respond adequately to unexpected situations. The FEP states that "Owing to its activity-driven and practical nature using appropriate methods, education motivates pupils to continue learning and leads them to a learning activity and to finding that it is possible to seek, discover, create and find suitable ways of solving problems" (FEP BE 2007, p. 9).

The diagnostic criterion is not the speed and reliability with which pupils are able to imitate the teacher's mathematical activity but the level of understanding they have achieved on a scale of cognitive mechanism. Individualization is a two-way process. One aspect is taking into account different cognitive types of pupils (e.g., girls' approaches are often different from boys' approaches), and the other one is connected to the speed of progress in their appropriation and grasping of concepts and procedures. Individualization with respect to different cognitive styles is enabled by a variety of model spectra and the accepting of individual algorithms.

4.1 Characteristics of the Educational Area of Mathematics and Its Application in FEP BE

Curricular documents express explicitly the demands on pupils' activity in mathematics lessons:

> In basic education, the educational area of Mathematics and its application is based primarily on activities typical for working with mathematical concepts and for using mathematics

in real-life situations. It provides the knowledge and skills necessary for practical life and facilitates the acquisition of mathematical literacy. (FEP BE 2007, p. 27)

This clearly shows that the understanding of the sense of pursuing mathematics was gradually transforming and was modified to indicate the development of a mathematical literacy. Because of this indispensable role, mathematics permeates all basic education and creates the preconditions for further successful learning.

Education for mathematical literacy places an emphasis on a thorough understanding of basic ways of thinking in regard to mathematical concepts and their interaction. Pupils gradually learn various mathematical concepts, algorithms, terminology, and symbols, as well as methods for their application.

The educational content of this area is divided into four thematic areas in (FEP BE 2007):

- Numbers and numerical operations in stage one, which is followed up and expanded upon in stage two with the thematic area of numbers and variables
- Dependencies, relations, and working with data
- Two- and three-dimensional geometry
- Non-standard application problems and tasks.

Pupils learn to use technological devices (calculators, computer software, and various types of educational software) and to use some other aids, which enable pupils with difficulties in performing numerical calculations or in geometrical techniques to succeed. Students also improve in critical work through their use of informational sources (FEP BE 2007).

Instruction in this educational area focuses on the formation and development of key competencies by guiding pupils toward the application of mathematical knowledge and skills in practical activities, the development of memory, and the cultivating of combinatory and logical thinking. It also guides them to apply mathematical knowledge toward abstract and precise thinking, acquire a repository of mathematical tools, effectively use acquired mathematical skills, gain experience in the use of mathematical modeling, use precisely and succinctly the language of mathematics, include mathematical symbols, perform analyses, co-operate, and trust in their own problem-solving skills and abilities.

4.2 Problems with Teaching Mathematics

Representatives of the Union of Czech Mathematicians and Physicists stated that the overall state of mathematics teaching at our schools is not good. Mavrou and Meletiou-Mavrotheris (2014, p. 507) claim that:

> The methods of teaching mathematics in schools have been identified as contributing to the falling interest in mathematics. Empirical classroom research over several decades shows that, with some notable exceptions, mathematics instruction has been characterized by traditional, abstract formulation ... Ideas are presented in an overly theoretical and abstract manner without sufficient opportunities for students to engage in problem-solving and experimentation.

In recent years, mathematics has begun to be viewed as a meaning-making activity. Some educational leaders and professional organizations in mathematics have been advocating for the adoption of more active learning environments that motivate learners and encourage them through authentic inquiry to establish the relevance and meaning of mathematical and scientific concepts (Samková et al. 2016). They have been stressing the fact that the core of school mathematics and science should no longer be the teaching of techniques and calculations that computers can do much faster and more reliably. Instead, the core should be the development of problem-solving skills that students will need to effectively live and function in a highly complex society. This shift is being reflected in educational policies and official curricula which advocate pedagogical approaches that support inquiry-based learning of mathematics.

Despite the extensive calls for learner-centered, inquiry-based pedagogical models, changing teaching practices is proving to be quite difficult. There is strong evidence that the implementation of inquiry-based teaching and learning of mathematics is not fully accepted. The authors' observations from Czech classrooms confirm that pupils and teachers often prefer problems in which the use of an adequate solving algorithm is evident. They prefer problems where they can determine the algorithm without any doubt. Thus, the burden of the tricky and difficult search for understanding of the problem is elevated, and the teacher's role is simplified to the mere discovery of the places where pupils make mistakes and to evaluation the "correctness" of their solutions (Novotná 2010).

In recent years, a lot of attention in research has been paid to critical areas of mathematics education from the point of view of teachers (Rendl et al. 2013) and of learners (Vondrová et al. 2015b). In the case of teachers, the authors tried to pinpoint what areas teachers consider critical, which pupil problems they consider as critical, and how they try to prevent them. What turned out to be difficult was to distinguish clearly whether teachers see pupils' problems as stemming from their conceptual understanding or rather their mastery of procedures. Teachers did not speak about more cognitively demanding skills, such as finding a solving strategy, argumentation, and reasoning. Also, they did not mention higher levels of conceptual understanding, such as explicit connections of the concept with the procedures in problem-solving, using different means of representations, and application in a variety of contexts.

In the case of the pupils, among what was discovered were deficiencies in the mental representation of a continuum of rational numbers, conceptual understanding of an algebraic expression as an object to be manipulated, breach of relationship between theoretical and spatial graphic spaces when interpreting and using a picture in geometry, and the conceptual understanding of measure in geometry. A strong tendency to use formulas in geometry and give preference to calculations over reasoning was identified (Rendl et al. 2013). Word problems manifested as one of the most critical sites of difficulty. Pupils also struggled with grasping a text, mathematization of situations when passing from specific to abstract representations, and making records or sketches. Pupils often perceived a graphical representation as a tool for explanation or clarification, but not for solving a problem.

5 Content of Mathematics Education and Textbooks

The textbook is often considered an important source of a pupil's (but also a teacher's) knowledge because:

- It is one of the program projects of education.
- It represents an important stage in didactical transformation of cultural contents into school education (Moraová 2018a, b).
- It is an inseparable part of a teacher's and pupil's everyday activity in lessons and outside of lessons too, as it is usually easily available to pupils (in comparison to other items of the multimedia system of didactic aids and tools).

The traditional conception of a textbook as a tool for handing over curricular content to pupils is gradually being overcome. Textbooks are now conceived of in a more complex way, as the mediator or facilitator of pupils' learning based on their own activities (Moraová 2018a, b).

At present, Czech teachers, or more often headmasters (Moraová 2014a, b), can choose between more than ten sets of textbooks for basic education, all of which have the official certificate of the Ministry. In contrast, the period before 1989 during the communist regime was a time when schools were allowed only one prescribed set of textbooks (no other had the official certification of the Ministry). Now there is a wide range of textbooks both with and without certification of the Ministry. The list of certified textbooks is available at the Ministry website. The selection of textbooks used by a school is up to the decision of the school (its headmaster). The Act of Education allows schools to use textbooks without official certification on the condition that they meet the demands of Framework Education Program and School Education Program. The headmaster must be able to explain why they have chosen an uncertified textbook, including why and how it better suits the school.

We will limit ourselves to mentioning only one set of textbooks—the set of mathematics textbooks of Hejný and his team on elementary education. This set of textbooks has sparked discussion both in professional and in parental communities. Hejný's concept of teaching mathematics is based on scheme-building (its theoretical background is presented in Hejný 2014). The basis for building schemes of mathematical knowledge is the creation of collections of isolated and generic models.[2] The process of solving a problem is connected to the activation of some of these schemes, their intertwining, and restructuration. Mathematics education is based on a set of tasks and problems that together form learning environments (LE). Basic properties of learning environments were formulated by Wittmann (2001) who stressed that they enable us to formulate a series of problems which help a pupil to understand deep ideas of mathematics. Hejný added three more requirements: "connection to a pupil's life experience, long-term nature (usable for pupils

---

[2] An isolated model is used in the case of future knowledge. A generic model is created from the process of generalization from a collection of isolated models (Hejný 2014).

of different ages, at best from grade 1 to grade 12), and differentiated nature (catering to needs of individual pupils)" (Hejný 2012, p. 46). Hejný and his coauthors created about 20 learning environments, for example:

> "– area starts with generic models of the area of square, rectangle, triangle, ... within the environments 'tessellation,' 'paper folding,' 'grid paper,' 'geoboard,' and 'stick shapes';
>
> – small natural numbers start with generic models of address, status, operator of change and operator of comparison within the environments 'stepping,' 'money,' 'pebbles,' 'rhymes,' 'ladder,' ...;
>
> – fractions start with generic models of 'one half,' 'one quarter,' 'divide into halves,' 'equal sharing,' ... within the environments 'pizza', 'stick,' and 'chocolate'." (Hejný 2012, p. 47)

What the authors of this series of textbooks stress is an increase in pupils' interest in mathematics. A comparison of the results of pupils educated using the Hejný method and other methods show no significant differences (ČŠI 2017).

Mathematics textbooks deserve the attention they are due. As research studies show, planning and conducting a mathematics lesson often depends largely on the mathematics textbook. This means pupils spend a great deal of time working with a textbook and inhabiting its world. Most research focusing on mathematics textbooks analyzes their mathematical content and didactical treatment, including the proportion of textual and non-textual elements, language difficulty, and correspondence to Framework Education Programs. However, there are many more aspects that can be analyzed and studied. As Moraová (2018a, b) states, a textbook is not just a pedagogical document; it is a cultural artifact created in and for a specific society with specific cultural norms. This is crucially important with respect to whether mathematics is perceived by pupils as important not only at school but also in everyday life. It also affects whether the particular textbook will motivate the learners to gain new mathematical knowledge and skills.

6 ICT in Mathematics Education

A relatively new issue in mathematics education is the use of information technology in mathematics classrooms and its impact on teaching, methods, techniques, and approaches to mathematical content. Recent studies in mathematics education show that despite many national and international incentives whose goal is the integration of ICT into mathematics classrooms, such integration in schools is underdeveloped (Cox and Marshall 2007; García-Campos and Rojano 2008; Černochová and Novotná submitted for publication). Furthermore, the speed of the introduction of new technologies into mathematics classrooms is significantly slower than the speed of their development.

There are many reasons why the introduction of ICT into classrooms is a slow and painstaking process (Jančařík and Novotná 2011a). First, ICT offers a wide range of resources, which in consequence makes teachers unsure which of them to use as well as when and how to use them. Also, there is a lack of information on the potential advantages and dangers of the use of ICT activities in lessons. Another reason is that teachers do not have any prior experience with the use of ICT in mathematics lessons from their own school years and pre-service training, which causes them to feel incompetent in using the devices. Unlike the new generation of pupils, who are digital natives, most teachers do not feel at home and comfortable in the area and have no hands-on experience with mathematical software (Černochová, Novotná submitted for publication).

ICT has a huge potential impact on teaching mathematics (Artigue 2002; Ruthven 2007; Vaníček 2010), but it is not always used for the benefit of the learners. Examples from practice show that in many cases teachers use technology "for show" (Jančařík and Novotná 2011a, b). These theatrical examples contribute very little to the development of mathematical knowledge and may be even counterproductive. Use of information technology in education should always be governed by the principle of efficiency. It is appropriate to use information technologies only in situations in which they really bring benefit, open new perspectives, or significantly decrease the amount of time needed for technical calculations.

As digital technology enters various areas of our lives and a wide range of activities, the extent of what we must be able to do with it increases. In schools, work with information technology is very often limited to one subject (named differently in different schools—information and communication technology, computing, informatics). However, it is difficult to develop pupils' skills in this area separately from other schoolwork. The concept of ICT in the Framework Education Program for Elementary Education is now being transformed according to the Strategy of Digital Education 2020 to make it reflect the needs of society and the labor market more. One of the priorities is to "improve pupils' competence in the area of work with information and digital technologies." The new conception of digital literacy penetrates all educational areas. New educational resources are presently being developed for children and pupils and piloted in selected schools. The new approach to the development of digital technology plans the integration of work with digital technologies in all subjects across the curricula. The expected outcome for the competence of pupils is to allow them to work with technology confidently, safely, critically, and creatively. This competency should be developed by all teachers in schools (MŠMT 2018).

7 Assessment

When discussing assessment, we have to distinguish between two areas. One of them is assessment in the classroom and the other is the high-stakes testing that provides the basis for evaluation of efficiency in the system of education.

7.1 Assessment in a Classroom

The predominant form of assessment in the Czech educational environment is summative (assessment of learning) (Žlábková and Rokos 2013). Formative assessment (assessment for learning) is less common. It still remains more or less on a theoretical level. For example, there are publications focusing on formative assessment whose authors present various examples of how to use methods of formative assessment, but the majority of them focus on the arts and humanities (Starý and Laufková et al. 2016). Mathematics teachers do not have sufficient support. The available examples of good practice are random and do not provide a systematic basis for the development of teachers' assessment skills and competence.

7.2 High-Stakes Testing

The testing of the whole population in selected cohorts has little tradition in the Czech Republic. Czech School Inspection started performing this kind of testing several years ago, and there have been several rounds of selective testing of different cohorts and different educational outcomes and competences (ČŠI 2017).

Apart from these, there are two exams in which large group of pupils participate: entrance exams to upper secondary schools and the state school exit (Maturita) exam.

7.2.1 Centralized Entrance Exams to Upper Secondary Schools with the Maturita Exam

On entering an upper secondary school whose graduates pass the state Maturita exam, the applicants must pass an entrance exam. The exam consists of a written test in mathematics and Czech language. Since 2017 the entrance exams have been centralized, organized by the Centrum pro zjišťování výsledků vzdělávání (The Center for Assessment of Educational Outcomes).

Although the tests are centralized, their results may not be used for assessing the quality of education or for any other form of comparison between schools. Before these entrance exams, many pupils take part in massive preparatory courses organized by private companies. Thus, pupils do not have equal opportunities under the exam.

7.2.2 School Exit Examination (Maturita)

The school exit examination after secondary school—the so called Maturita (from the word mature—the exam on the threshold of adulthood) has had a long tradition in our country. The basis for the Maturita exam was the standardized state exams opening the door to universities to those candidates who passed successfully. This type of exam came to the Czech territory in 1849. The Maturita has always taken the form of an oral (and in some subjects written) exam sat in several subjects. Czech language and literature became compulsory exam subjects when Czechoslovakia was established (1918). Mathematics has never been compulsory for all types of upper secondary schools.

The changes in 1989 initiated a discussion on introducing a state Maturita exam (centralized didactical tests). One of the incentives was a recommendation from the OECD. It was expected that the results of a state Maturita exam could replace university entrance exams. There were several failed attempts to start a centralized exam. This ended when the state Maturita exam was put into practice in the spring of 2011. The Maturita has two levels of difficulty: basic and advanced. The Act of Education defines the Maturita exam as having two parts—the profile part organized by each school and the centralized part identical for all schools in the country. The centralized part consists of two compulsory exams: one in Czech language and literature and another either in a foreign language (taught at the school) or mathematics.

The issue of mathematics as a part of the Maturita exam and pupils' performance on it has attracted much attention. The Maturita in mathematics was planned to become compulsory at secondary grammar schools in 2020/2021 and the following year in all upper secondary schools. The only exception were art schools and schools for nurses and social workers. Up till now, a majority of school leavers choose a foreign language rather than mathematics. In 2018, mathematics was selected by 23.4% of graduates of which 22.3% failed the didactical test. The policymakers and public are currently discussing whether the compulsory Maturita in mathematics should or should not be postponed or not introduced.

7.2.3 International Comparative Studies

The Czech Republic had already joined the TIMSS (Trends in International Mathematics and Science Study) in 1995 and later got involved in PISA and other surveys and studies. In the beginning, the performance of Czech pupils was well above the average of the OECD countries in both tested populations (11- and 14-year-old pupils). The Czech Republic was by far the best performing post-communist country. However, the following testing brought disappointment. The test results of 8th graders on the TIMSS grew worse after 1999 (Tomášek et al. 2008). In the research study PISA, the performance drop of the Czech Republic between 2003 and 2009 was the greatest among all the countries that took part in both of the cycles (Palečková et al. 2010). In mathematics, the fourth graders dropped to the below-average zone in the 2007 TIMSS and their decline in comparison to 1995 was the greatest compared to all European and OECD countries involved in both surveys (Fig. 1.5). Czech's fourth graders struggled the most with problems on fractions and decimals, which was a consequence of including this curricular content in higher grades than in other countries (Potužníková et al. 2014). This does not mean that Czech pupils were significantly better in mastering other topics. Even

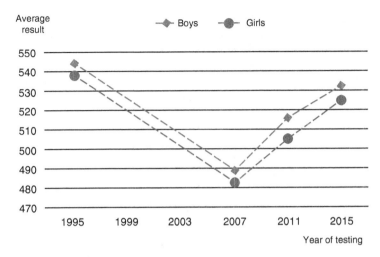

Fig. 1.5 Comparison of results of Czech 11-year-old boys and girls in mathematics TIMSS over the past 20 years. (Taken from Tomášek et al. 2016, p. 11.) NB: Only 14-year-old pupils were tested in 1999. The Czech Republic did not take part in the testing in 2003

if all problems involving fractions and decimals were to be excluded from analysis, the performance of Czech pupils would remain only average (Dvořák 2010).

It is difficult to determine the reasons for this drop in Czech pupils' performance. It is very likely a combination of several factors. For example, one of them may be pupils' lack of interest in mathematics or unwillingness to engage with it (Basl 2009), a result of insufficient development of mathematical reasoning that is an obstacle to mastering more difficult topics in higher grades (Hejný and Jirotková et al. 2013), or changes in society and in the place of mathematics within curricula (Dvořák 2010). The performance of pupils in mathematics improved in 2011 and improved again in 2015. However, the achieved results were still worse than in 1995.

One of the possible reasons for the decline may also be the addition of one extra year to the primary school level (from 4 to 5 years); this was the case for fraction problems. Another reason may be the introduction of FEP; the result of which was that much more attention at schools had to be paid to administrative demands than to the quality of teaching. Also, the FEP defines the required minimum of knowledge and skills.

These comparative studies provided some information that help to characterize the system of education. PISA 2000 implies that the Czech Republic is a country where pupils' performance is very strongly dependent on the socioeconomic status of their parents (Straková 2009).

A survey of fifth and ninth grade pupils' results in Czech, mathematics, foreign languages, and areas from other subjects was conducted in 2016–2017. The least successful subject was mathematics in the ninth grade, where every third pupil was able to solve only two-fifths of the test problems correctly. The mathematics results of the fifth graders corresponded to the expectations. The results differed regionally. The most successful group were pupils from Prague. As expected, pupils from lower secondary grammar schools performed on average better than elementary school pupils. The results of fifth graders in mathematics were not affected by the extent to which teachers used the techniques of the so-called Hejný method (it is not used by 57% of the surveyed teachers, and it is more widespread in non-state schools). The results are more closely connected to what the teacher regards as the main goal of teaching mathematics (helping pupils find positive attitudes to mathematics or teaching the entirety of the FEP) (ČŠI 2017).

8 Mathematics Teacher Education

8.1 Pre-service Mathematics Teacher Education

The first educational institutions for teachers were established in 1774 (Mikulčák 2010). The Austro-Hungarian Education Act of 1869 introduced teacher-training institutes. Students of teacher-training institutes graduated after taking the Maturita, but the graduates were not fully qualified for the teaching profession. Having passed the examination, they next had to undergo a prescribed practical training. Then they passed an examination in pedagogical competence and acquired professional qualification. Since the nineteenth century, teachers in primary and lower secondary schools have tried to elevate their training to the university level. Charles University, and later other universities, initiated teacher training courses (Novotná 2019).

After the establishment of Czechoslovakia in 1918, the Ministry as well as the public started to oppose the concept of university education for primary school teachers. In 1946, the government set down provisions for the establishment of faculties of education, mostly within universities. In 1953, faculties of education were changed to higher schools of education. Three of these (located in Prague, Olomouc, and Bratislava) were later renamed pedagogical colleges. In 1959, 18 institutes of education were set up, bearing the title regional universities. At the time, pre-service primary teacher education was 3 years long and 4 years for lower secondary school. Upper secondary school teachers had to get their training at universities (e.g., at faculties of natural sciences, arts, or the Faculty of Mathematics and Physics of Charles University). Pedagogical institutes existed until 1964 when faculties of education were brought into existence again (Mareš 2007).

Until 1989, teacher training was the same at all faculties in the country which trained future teachers. All the institutions had the same study plans, curricula, textbooks, number of lessons, and student duties. Primary school teachers were generalists, and secondary school teachers studied to teach two subjects. The possible combinations of the two subjects were fixed; students could not choose a combination of their personal preferences or interests. Each education faculty had a list of allowed subject combinations. The most common combinations of mathematics were with physics, chemistry, biology, geography, and physical education. For example, a combination of mathematics and Czech language was not offered; it was regarded as too time demanding. The number of lessons and duties was divided into three more or less equal parts: one-third was the common core subjects (especially pedagogy and pedagogical psychology), and the other two-thirds were the two studied subjects. Teaching practice was part of the common core subjects.

The basic change in university teacher education, which took place after 1989, opened the way for faculties to define the content and methodology of teacher training independently. Much discussion was generated by questions on the issues of teaching practice and the relationship between teaching a specific subject and the common core (pedagogical and psychological components) of teacher training. A lot of attention was paid to the teacher's work with pupils (diagnostics, communication, evaluation, etc.) (Beneš and Rambousek 2007).

The traditional way of training prospective teachers of mathematics emphasized scientific knowledge. Nowadays, facing the necessity of preparing teachers for a new flexible school system, the focus is much more on the didactic aspects of teacher training. The starting point is to state which parts of mathematics are necessary for future mathematics teachers.

Not all faculties in the Czech Republic have moved toward this new system of teacher training. Some of them stress only on pure mathematical content and tend to ignore the didactic aspects. In the study model for lower and upper secondary school teacher training, there is a great variety in the organization of the program. For example, noticeable differences can be found in the content and extent of knowledge that pre-service teachers have to master in subjects at the master's level, concerning how much attention is paid to the school teaching practice as well as how much emphasis is put on the common core studies (e.g., pedagogy, psychology, Czech language, foreign language, human biology, philosophy of education, educational technology, and introduction to logic). The attention paid to common core subjects has, in general, been reduced, and the focus has been shifted to studied disciplines, both the subject content knowledge and the pedagogical content knowledge (in the sense of Shulman 1987)

The basic legislative framework for the training of teachers and other education staff is the Higher Education Act of 1998. Based on that act, the responsibility to define the content and organization of teacher training programs was handed over to each relevant higher education institution. Internal regulations of individual teacher training colleges thus set out curriculum content and its organization for both standard and alternative ways of obtaining the teacher qualification. However, each study program, even the programs for life-long teacher education, had to be accredited. Starting in 1990 the accreditation was done by the Ministry, whose advising body was the Accreditation Commission. The Accreditation Commission [Akreditační komise][3] was established in 1990 by the law and was closed in 2016. It took care of the quality of higher education and carried out comprehensive evaluations of teaching, scientific research, development and innovative work, and artistic

---

[3] Statute for the Accreditation commission available from http://www.msmt.cz/areas-of-work/tertiary-education/statute-of-the-accreditation-commission?highlightWords=soubor

and other creative activities of higher education institutions. The chief means of achieving these objectives were by:

- evaluating the activities of higher education institutions and the quality of accredited activities and publishing the results of such evaluations
- reviewing other issues affecting the system of higher education, when asked to do so by the Minister, and expressing its standpoint on these issues.

As per the government's decision in September 2016, the Accreditation Commission was replaced by the National Accreditation Authority for Higher Education [Národní akreditační úřad pro vysoké školství][4]. This is an independent body with the power to make decisions without any approval from the Ministry.

The qualification needed for the teaching profession is given by Act No. 561/2004 (2004). The original demand was that all teachers would have to be graduates of pre-service teacher education programs. However, a lack of qualified teachers in some regions resulted in the modification of this original condition in its amendments. There are now more ways of gaining the needed qualification. However, the condition that all teachers must be graduates of a master's degree program remains. Several faculties offer programs oriented at the obtaining, extension, and deepening of teacher qualification (not offering an academic degree but guaranteeing the pedagogical qualification). Teacher education which is not provided in higher education institutions, but at lower levels of the education system, is regulated by the Education Act and related decrees (Eurydice 2016).

8.2 University Programs Preparing Teachers for Their Profession

In 2017, the Ministry accepted the Framework of Requirements on Study Programs providing qualifications for regulated teaching professions (MŠMT 2017). It is a methodological document that helps to evaluate university programs which prepare teachers for their profession. The material is relevant for study programs preparing pre-primary, primary, lower secondary, and upper secondary schools as well as programs training special pedagogues, psychologists, and pedagogical staff for after-school clubs and other childcare institutions. The material formulates general grounds on whose basis the Ministry assesses study programs.

Fulfilling the framework requirements guarantees that graduates of the study program will be adequately equipped for regulated teaching professions. The main goal for the division of the individual components of the study program is to define a well-balanced proportion between each of them. The bottom limit in each of the

---

[4] Government regulation No. 274/2016 coll. on standards for accreditation in higher education available from http://www.msmt.cz/vzdelavani/vysoke-skolstvi/preklad-zakona-o-vysokych-skolach?highlightWords=accreditation

Table 1.3 Proportion of components of professional teacher training (MŠMT 2017)

Component	Pre-primary	Primary	Lower secondary	Upper secondary
	in % of study plan			
Teaching propaedeutic	23–30	26–32	20–25	20–25
Subject component with the subject didactics (for lower secondary education divided into first and second subject and both subject didactics; for upper secondary divided into first and second subject)	45–50	50–55	25–30 25–30 10–15	25–33 25–33
Teaching practice	10–15	10–15	8–10	8–10
Work on bachelor/diploma thesis	5–10	5–10	5–10	5–10

components determines the minimum and is binding. The top limit can be exceeded; it is only a recommendation. The proportion of the individual components of the studies is given in percent, credits, and hours according to the European Transfer and Accumulation System. The teaching propaedeutic component involves pedagogical-psychological training, special pedagogical training, general didactic, developmental psychology, and other disciplines (e.g., ICT in education, foreign language). Teaching practice is perceived as supervised and reflected, both in the form of lesson observations, and ongoing, continuous teaching practice.

Table 1.3 presents the distribution of individual components for pre-primary, primary, lower secondary and upper secondary teacher training.

Examples of the differences between the designs of mathematics teacher training programs are presented in Novotná (2019).

8.3 Teacher Professional Development

The in-service training of education staff includes study programs for unqualified teachers which lead to formal teacher qualification and programs leading to other qualifications (such as the school advisor qualification) or to professional development courses (for continuing education). This in-service training is delivered within the lifelong learning system at higher education institutions, in establishments for the in-service training of education staff, or other institutions accredited by the Ministry. Additionally, this type of education can be gained through self-study.

In-service teacher education is not compulsory in the Czech Republic. In the mandatory[5] pedagogical documentation of schools, there is a 1-year school plan for in-service teacher education. At the end of the school year, this information must be published in an annual school report. Schools have, in their budgets, finances for

---

[5] Although in-service teacher education is not compulsory in the Czech Republic, the headmaster is expected to encourage and keep a record of participation of the teachers who attend in-service training seminars and courses.

covering the costs of the training. At this point, there is nothing that would force or motivate in-service teachers to participate in in-service teacher education if they are already fully qualified teachers. The in-service teacher education plan is the headmasters' responsibility, and it is up to the headmasters to motivate their teachers to participate. However, if the teachers are not motivated, their attendance at the training will be purely formal. There is a lack of quality control regarding in-service teacher training courses. Most of them are run by private bodies despite the need to have the seminars accredited by the Ministry.

9 Discussion and Public Opinion

Discussions on educational issues take place at many different levels. They are not just direct "face-to-face" discussions. They are in the media, in the press, in special journals, magazines, websites, and also in the non-expert media. The discussions focus on all aspects of education, such as educational content, forms, methods, and financing of schools. The Czech population has traditionally shown a lot of interest in education. Voices of the people are, however, far from unified. On the contrary, many contradictory opinions are often expressed.

The Ministry, which prepares strategic documents in cooperation with other departments and educational experts, permits a public debate on the drafts of proposed documents. Discussions are organized on drafts that are presented to the professional public. All inhabitants of the Czech Republic have the right to express their opinion on the issue and its solution. However, the Ministry or other government institutions may not consider these opinions as relevant and may not take them into account. The web pages of the Ministry provide contacts where inquiries from the press and media may be addressed. The website also offers Ministry press releases, both for the media and the general public. Various professional and interest-based organizations that focus on education use this as an opportunity to express their views on educational issues. The website also provides an archive of Ministry press releases since 2009.

Illustration 1: The Ministry organizes discussions on possible changes to the FEP, on state Maturita exams, on entrance exams for upper secondary schools, on the possibility of the introduction of the cut-off score (i.e., the minimum score needed to be accepted to a study program concluded by the state Maturita exam), or on support of novice teachers, etc.

Discussions, round tables, and public hearings on educational issues with education experts and others are also organized by the Committee on Science, Education, Culture, Youth, and Sports of the Parliament of the Czech Republic. Members of professional organizations are also invited to participate in these discussions. In the case of mathematics education, the corresponding organization is the Society of Mathematics Teachers of the Union of Czech Mathematicians and Physicists. Conclusions and unresolved issues from these discussions are usually published.

Illustration 2: On March 14, 2018, the Committee on Science, Education, Culture, Youth, and Sports of the Parliament of the Czech Republic held a meeting. One of the discussed issues was teaching mathematics at the primary and secondary school levels. Representatives of the Union of Czech Mathematicians and Physicists were invited to the meeting. Most of the discussion focused on the use of the Hejny method (https://www.h-mat.cz/en) in teaching mathematics. The discussion continued (and in fact still continues) even after the meeting finished, both in the press and on other platforms. The society of teachers of mathematics of the Czech Union of Mathematicians and Physicists published its opinion on its webpage (the official opinion is available in Czech from https://suma.jcmf.cz/_files/200000120-64c5b65c62/stanisko%20SUMA%20k%20metod%C4%9B%20Hejn%C3%A9ho.pdf).

Discussions on current issues in Czech education also are held at another level—at the country's regions and regional governments. Regions are responsible for education on the regional level; they establish the majority of upper secondary schools. Also, meetings of regional departments of education with the press and the public are organized at this level. At these meetings, it is quite common to hear the opinions of companies and businesses because they are the potential employers in the region and they often take part in the training of future graduates. In addition, MPs or representatives of the Ministry are invited to these meetings. If this level is compared to the level of the national government, we can state that regional authorities have a clearer idea of the situation in the terrain of their region than national government officials. However, their role is not to prepare strategic documents. Regions also need to know the views not only of professional groups but also of the public. The public is therefore invited to contribute to these discussions, although not much attention is paid to their opinion. The situation always changes in a pre-election period, when all the parties try to get votes by promising to improve education in their regions. However, many of these promises remain only promises, and their implementation after the election never happens.

Illustration 3: Discussions at the regional level usually do not focus on individual subjects, i.e., they do not focus exclusively on mathematics. They address issues of high relevance for the particular region. For example, the Vysočina region initiated a discussion on upper secondary school education. The Liberec region initiated a discussion on projects in education, on school and kindergarten capacity, and on improving the quality of schools. If many regions regard the same issue as highly relevant, the regional representatives may propose to the Ministry a discussion of the issue on the national level.

The third important level is the local level. Local authorities are responsible for elementary and preschool education. These are the levels closest to the average citizen. The public is invited to join in the discussions. The stimulus for the discussion might be a local issue or issues coming from above. Although there is no relevant research focusing on this aspect, we can assume that there is less interest for making strategic decisions about the system than in making decisions on local issues.

Citizens have the right to attend the meetings of local councils. However, they are allowed to enter discussions only with the consent of the local council. The extent to which citizens are active in these meetings varies greatly, from intense involvement to a lack of interest.

A burning issue at all levels is who should be making decisions about education. The answer to this question is tied to the question of who bears the responsibility for education. Responsibilities of two types are to be considered: individual and social responsibilities. Individual responsibility is born by pupils/students themselves and their parents. However, neither pupils nor parents have much freedom to make decisions in the current system. The question is how to adjust the decision-making process to get pupils and parents more involved.

Social responsibility is connected especially (but not exclusively) with the role of the state in education. The state should, for example, set conditions under which parents and their children can make responsible, sensible decisions, and it should guarantee equity. However, it is not easy to answer how this should be achieved (Pol 2007; Štefflová and Švancar 2004).

The platform through which the public can influence what is happening at schools is the school board. The public can also join open discussion forums to express its opinions on what is happening in the education system. An example of such a forum is the methodological portal www.rvp.cz, where there are for example:

- a discussion module that contains news, such as what is new in forums and in the community, online meetings, and forums focusing on a selected subject.
- archives of discussions on the methodological portal.
- a guide through the updated FEP.
- links to other websites of interest for teachers and educators (see e.g., https://www.suma.jcmf.cz/).

Documents connected to education and the educational system, such as archives of press releases before 2010 and articles from various areas of education (topics include teachers, the state Maturita exam, Inclusive Education, the Comprehensive National Entrance Exam, Education Reform, and Amendment to the Act of Education), are available at the portal www.eduin.cz. The portal also publishes articles on current issues in education, opinions, and discussions.

There are also several professional organizations and societies. They may organize discussions on education from their own initiatives or may be invited to participate in discussions on educational issues. Some of these organizations, e.g., the Union of Czech Mathematicians and Physicists, are very active. The Union not only takes part in discussions but also frequently initiates them.

10 Conclusions

Let us conclude this chapter by drawing attention to several general issues that are not linked exclusively to mathematics education, although they have an impact on it. Since 1989, the Czech educational system has evolved significantly from a centralized, unified system to an open and pluralist system. The responsibility for education is shared by the Ministry and the regional and local authorities. Private persons and institutions are able to enter the domain of educational debate and policy.

One of the events that generated discussion on education and was at the beginning of the ongoing changes was the significant drop in Czech pupils' performance both in the TIMSS and the PISA from 1995 to 2007 (see Fig. 1.5). In order to stop this decline and bring about improvement in pupils' performance on an international scale, it was recommended that pupils actively participate in constructing their own knowledge. Also, the search for new teaching approaches was encouraged.

As stated in the text above, interest in educational issues is quite intensive at all levels of the Czech Republic. The website www.eduin.cz (see l. 1223) hosts a discussion that was begun by the initiative "Education first" that comes out of the idea that "education is a basic value of our society." This initiative not only names problems but also proposes possible solutions. The following are considered the fundamental problems of our system (and are not listed according to their priority):

- The world changes very fast and the Czech system of education lags behind.
- People see education as an urgent issue.
- Czech pupils have the greatest dislike of schooling out of all the OECD countries.
- The role of the teacher is changing, but neither teachers not the public are ready for this change.
- The Czech Republic invests nearly the smallest portion of its GNP into primary and secondary education compared to other OECD countries.
- Teachers are growing old but only about 40% of graduates from faculties of education go on to teach.
- Salaries for headmasters and teachers are below the national average.

The initiative recommends that the good parts of the Czech system of education should be preserved. This is the extensive network of public universities, high level of literacy among the population, the relatively good level of autonomy among schools and headmasters, a low early school leaving rate, and the activity of parents whose aim is to establish new and innovative schools. Teachers and headmasters should be supported by a change in the system of benefits and a better career system. Some of the issues that the initiative perceives as problematic are uniform state entrance exams (they were introduced into the system in 2016–2017), the current organization of the state Maturita exam (upper secondary school leaving exam), and the Educational Staff Law.

Let us stress here that no positive change can happen unless due attention is paid to teacher training. Until 1989, the system of life-long teacher education was very formal and in terms of the professional development of teachers not very functional. This system of in-service training was abolished in 1991 and has so far not been replaced by a well-structured system. In-service teacher education is not compulsory in the Czech Republic. The in-service teacher education plan is the headmasters' responsibility, and it is up to the headmasters to motivate their teachers to participate. If the teachers are not motivated, their attendance is largely a waste of time (Novotná 2019).

It also seems that there is a lack of control over the quality of offered in-service teacher training courses. Most of them are run by private bodies, and although the seminars must be accredited by the Ministry, many of them seem to be run for profit rather than for improving teaching at schools (Novotná 2019).

The last regulation that defined in-service teacher education is from September 2005. A new legislative framework is still in the preparation phase, and it is not clear when it will be approved. However, it promises a career system for teachers that would incorporate in-service teacher training as a prerequisite to career growth and promotion. The prepared career system is designed to help increase the prestige of the teaching profession in society, stimulate interest in the teaching profession among young people, and to give new and experienced teachers a perspective on professional growth throughout their professional lives. The career system will be based on the principle of interdependence in professional development, the career system, remuneration, the rights and obligations of teachers to develop their professional skills, and the right to choose possible career paths.

The authors of this chapter are aware of the fact that in this chapter they could present only a small part of what could be said about the development and trends of the Czech educational system. Interesting results could be received by a deeper analysis of doctoral studies in the theory of education, as well as by analysis of educational organizations, their requirements, students' motives for enrolling in courses, etc.

However, our study of the organization of the educational system and teacher education and its changes since 1989 points toward some important issues for further research: on the one hand, the influence of stereotypes, on the other hand, the effort to prepare teachers open to new educational approaches and teaching methods flexible when including new approaches into their own teaching, and shifting toward a pupil-centered type of teaching.

The new challenges do not come from the educational system only, but from the outside world. For example, the multicultural dimension of modern society constitutes one of the most significant changes to have influenced schools in many European countries, especially at the primary and middle school level. The teaching profession is all the more difficult because the teacher is usually not sufficiently prepared to deal with a new classroom context as well as the added challenge of reaching pupils of a migrant background, coming from countries with different cultures and languages. The teacher is seldom aware of the need to rethink and, if necessary, modify his/her methodological and pedagogical approach. Additionally, the teachers do not know how to use the multicultural dimension in a constructive way when teaching. This attitude is even more evident among mathematics teachers, who often consider their subject universal and culture-free.

An important discussion on the future of the Czech system of education took place on February 29, 2019. About 200 major actors in the system of education in the Czech Republic participated. This was the official, formal opening of the Preparation for the Strategy of Educational Policy of the Czech Republic until 2030. The aim of the conference was to open to the public and experts a debate on key challenges and visions of education in the Czech Republic as well as to look for ways of achieving it. Part of the conference focused on revisions to the current framework of education programs. The new strategy should answer all the key problems of today's system of education. It will define the vision, priorities, goals, and measures in the area of educational policy in the Czech Republic for the period extending to the horizon of 2030. The process of making the draft of Strategy 2030 should be open and transparent. The aim is to take into account the results of consultations with the greatest possible number of actors in the educational system of the Czech Republic.

References

Artigue, Michèle. 2002. Learning mathematics in a CAS environment: the genesis of a reflection about instrumentation and the dialectics between technical and conceptual work. *International Journal of Computers for Mathematics Learning* 7: 245–274.

Basl, Josef. 2009. *Širší souvislosti mezinárodního srovnání výsledků českých žáků v matematice a přírodních vědách* [Broader context of comparison of Czech students' results in mathematics and science]. Socioweb06. http://www.socioweb.cz/index.php?disp=teorie&shw=407&lst=108. Accesses 12 April, 2019.

Beneš, Pavel, and Vladimír Rambousek, eds. 2007. *60 let pedagogických fakult* [60 years of faculties of education]. Pilsen: Koniáš. (In Czech and English).

Blažek, Radek, and Silvie Příhodová. 2016. *Mezinárodní šetření PISA 2015. Národní zpráva. Přírodovědná gramotnost* [International survey PISA 2015. National report. Science literacy]. Prague: Česká školní inspekce.

Boček, Leo, and František Kuřina. 2013. *Ovlivnili vyučování matematice. Eduard Čech 1893–1960* [They influenced mathematics teaching. Eduard Čech 1893–1960]. Prague: Matfyzpress. Persistent URL: http://dml.cz/dmlcz/501209. Accessed 12 April 2019.

Černochová, Miroslava, and Jarmila Novotná. Submitted for publication. Report on ICT in Education of the Czech Republic. In *Report on ICT in Education of China and CEECs*.

Čerych, Ladislav. 1999a. *České vzdělání a Evropa: strategie rozvoje lidských zdrojů v České republice při vstupu do Evropské unie* [Czech education and Europe. Strategy for the development of human resources when entering the European Union]. Prague: Sdružení pro vzdělávací politiku.

———. 1999b. *Priority pro českou vzdělávací politiku: mimořádné zasedání Výboru pro vzdělávání OECD v Praze 26.-27. dubna 1999* [Priorities for Czech educational policy]. Prague: Ústav pro informace ve vzdělávání.

Cox, Margaret J., and Gail Marshall. 2007. Effects of ICT: Do we know what we should know? *Education and information technologies* 12 (2): 59–70.

ČŠI. 2017. *Výběrové zjišťování výsledků žáků na úrovni 5. a 9. ročníků základních škol ve školním roce 2016/2017. Závěrečná zpráva* [Selective survey of 5th and 9th grader results in 2016/2017. Final Report]. 2017. Prague: Česká školní inspekce. [Czech School Inspectorate]. https://www.csicr.cz/getattachment/17f8e265-b04f-4459-a1063aecbf735ca0/Vyberove-zjistovani-vysledkuzaku-na-urovni-5-a-9-rocniku-ZSzaverecna-zprava.pdf. Accessed 7 May 2018.

Dvořák, Dominik. 2010. *Ve kterých úlohách TIMSS naši žáci nejméně uspěli (a proč)* [On which TIMSS Questions were our Pupils the Least Successful? (and why)]. Odborný seminář k matematickému vzdělávání [Scientific Seminar about Mathematics Education]. https://slideplayer.cz/slide/11219374/. Accessed 12 April 2019.

Dvořák, Dominik, Karel Starý, Petr Urbánek, Martin Chvál, and Eliška Walterová. 2010. *Česká základní škola. Vícepřípadová studie* [Czech basic school. A multiple case study]. Prague: Karolinum.

Dvořák, Dominik, Karel Starý, and Petr Urbánek. 2015. *Škola v globální době. Proměny pěti českých základních škol* [The school in a global age: The Changes of five Czech basic schools]. Prague: Karolinum.
Eurydice. 2016. *Czech Republic: Upper secondary and post-secondary non-tertiary education.* https://webgate.ec.europa.eu/fpfis/mwikis/eurydice/index.php/Czech-Republic:Upper_Secondary_and_Post-Secondary_Non-Tertiary_Education. Accessed 12 April 2019.
FEB BE (The Framework Education Programme for Basic Education). 2007. http://www.msmt.cz/vzdelavani/zakladni-vzdelavani/framework-education-programme-for-basic-education. Accessed 12 April 2019.
FEP Preschool Education (The Framework Education Programme for Preschool Education). 2018. http://www.msmt.cz/file/45304/. Accessed 14 September 2019.
FEP Upper Secondary (The Framework Education Programme for Upper Secondary Education). 2007. http://www.nuv.cz/file/159. Accessed 14 September 2019.
García-Campos, Montserrrat, and Teresa Rojano. 2008. Appropriation processes of CAS: A multidimensional study with secondary school mathematics teachers. In *PME 32 and PME-NA XXX*, ed. Olimpia Figueras, José Luis Cortina, Silvia Alatorre, Teresa Rojano, and Armando Sepúlveda, vol. 1, 260. Mexico: Cinvestav-UMSNH.
Gergelová Šteigrová, Leona, ed. 2011. *The education system in the Czech Republic*. Prague: MŠMT.
Hankel, Hermann. 1874. *Zur Geschichte der Mathematik in Alterthum und Mittelalter*. Leipzig: B. G. Teubner.
Hejný, Milan. 2012. Exploring the cognitive dimension of teaching mathematics through scheme-oriented approach to education. *Orbis Scholae* 6: 41–55. https://doi.org/10.14712/23363177.2015.39.
———. 2014. *Vyučování matematice orientované na budování schémat: aritmetika 1. stupně* [The teaching of mathematics based on the building of schemas: Elementary arithmetic]. Prague: Univerzita Karlova.
Hejný, Milan, and František Kuřina. 2015. *Dítě, škola a matematika: konstruktivistické přístupy k vyučování* [Children, school, and mathematics: Constructivist approaches to teaching]. Prague: Portál.
Hejný, Milan, Jarmila Novotná, and Naďa Stehlíková. 2004. *Dvacet pět kapitol z didaktiky matematiky* [Twenty-five chapters on mathematics education]. Prague: Univerzita Karlova, Pedagogická fakulta.
Hejný, Milan, Darina Jirotková, et al. 2013. *Úlohy pro rozvoj matematické gramotnosti: Utváření kompetencí žáků na základě zjištění šetření PISA 2009* [Problems for the development of mathematical literacy: formation of pupils' competences based on PISA 2009 findings]. Prague: Česká školní inspekce.
Jančařík, Antonín, and Jarmila Novotná. 2011a. "For show" or efficient use of ICT in mathematics teaching? In *Proceedings of the 10th International Conference for Technology in Mathematics Education (ICTMT 10)*, ed. Marie Joubert, Alison Clark-Wilson, and Michael McCabe, 166–171. Portsmouth: University of Chichester.
———. 2011b. Potential of CAS for development of mathematical thinking. In *Aplimat 2011*, ed. Monika Kováčová, 1375–1384. Bratislava: STU in Bratislava.
Janík, Tomáš, and Petr Knecht. 2007. *Pedagogický výzkum a kurikulární reforma české školy* [Pedagogical Research and Curricular Reform of the Czech School]. http://www.ped.muni.cz/weduresearch/publikace/0004.pdf. Accessed 12 April 2019.
Janík, Tomáš, Josef Maňák, Petr Knecht, and Jiří Němec. 2010. Proměny kurikula současné české školy: vize a realita [Transformation of Contemporary School Curriculum: Vision and Reality]. *Orbis scholae* 4: 9–35. https://doi.org/10.14712/23363177.2018.109.
Janík, Tomáš, Petr Knecht, Petr Najvar, Michaela Píšová, and Jan Slavík. 2011. Kurikulární reforma na gymnáziích: výzkumná zjištění a doporučení [Curricular Reform at Grammar Schools: Research Findings and Recommendations]. *Pedagogická orientace* 21 (4): 375–415.
Jelínek, Miloš, and Jaroslav Šedivý. 1982a. 25 let modernizačního hnutí ve školské matematice [25 years of the modernisation movement in school mathematics]. *Pokroky matematiky, fyziky*

a astronomie 27 (5): 282–289. Persistent URL: http://dml.cz/dmlcz/137785 Accessed 12 April, 2019.

———. 1982b. 25 let modernizačního hnutí ve školské matematice. Dokončení [25 years of the modernisation movement in school mathematics. Conclusion]. *Pokroky matematiky, fyziky a astronomie* 27 (6): 335–344. Persistent URL: http://dml.cz/dmlcz/138156 Accessed 12 April 2019.

Jeřábek, Jaroslav. 1998. *Vzdělávací program Základní škola* [Educational program Basic school]. Prague: Fortuna.

Kabele, Jiří. 1968. Nové pojetí matematiky a rýsování na ZDŠ [New conception of mathematics and drawing at Basic Nine-year School]. *Matematika ve škole* 18 (1967/68): 580–600.

Kotásek, Jiří, ed. 2001. *National Programme for the Development of Education in the Czech Republic: White Paper*. Prague: Institute for Information on Education. http://www.msmt.cz/dokumenty/bila-kniha-narodni-program-rozvoje-vzdelavani-v-ceske-republice-formuje-vladni-strategii-v-oblasti-vzdelavani-strategie-odrazi-celospolecenske-zajmy-a-dava-konkretni-podnety-k-praci-skol. Accessed 12 April 2019.

———. 2004. Budoucnost školy a vzdělávání [Future of school and education]. In *Úloha školy v rozvoji vzdělanosti*, ed. Eliška Walterová, 442–492. Brno: Paido.

Krajčová, Jana, and Daniel Münich. 2018. *Intelektuální dovednosti českých učitelů v mezinárodním a generačním srovnání* [Czech teachers' intellectual skills in international and generational comparison]. Prague: Národohospodářský ústav AV ČR. https://idea.cerge-ei.cz/files/IDEA_Studie_10_2018_Intelektualni_dovednosti_ceskych_ucitelu/files/downloads/IDEA_Studie_10_2018_Intelektualni_dovednosti_ceskych_ucitelu.pdf. Accessed 20 April 2019.

Mareš, Jiří. 2007. Šedesátiletí pedagogických fakult: Hledání svébytnosti [Sixty years of faculties for education: looking for independence]. *Pedagogika* LVII: 312–325.

Mavrou, Katerina, Maria Meletiou-Mavrotheris. 2014. Flying a math class? Using web-based simulations in primary teacher training and education. In *Handbook of research on transnational higher education*, eds. Mukerji Siran, and, Tripathi Purnendu, 506–532. Hershey (USA): IGI Global.

Mikulčák, Jiří. 1967. K nové koncepci střední všeobecně vzdělávací školy [About the new conception of the general secondary school]. *Matematika ve škole* 18 (1967/68): 601–604.

———. 2007. Jak se vyvíjela pedagogika matematiky ve druhé polovině 20. století [How did pedagogy develop in the second half of the 20th century]. In *Matematika v proměnách věků, V*, ed. Martina Bečvářová and Jindřich Bečvář, 249–315. Prague: Matfyzpress. http://dml.cz/dmlcz/400897 Accessed 12 April 2019.

———. 2010. *Nástin dějin vyučování v matematice (a také školy) v českých zemích do roku 1918* [Outline of the history of mathematics education (and also of school) in Czech Lands until 2018]. Prague: Matfyzpress.

Ministerstvo školství (Czechoslovakia). 1976. *Další rozvoj československé výchovně vzdělávací soustavy* [Further development of the Czechoslovak system of education]. Prague: SPN.

Molnár, Josef, Slavomíra Schubertová, and Vladimír Vaněk. 2008. *Konstruktivismus ve vyučování matematice* [Constructivism in teaching mathematics]. Olomouc: Univerzita Palackého v Olomouci.

Moraová, Hana. 2014a. Pre-service mathematics teachers' preferences when selecting their coursebook. In *Proceedings of the international conference on mathematics textbook research and development (ICMT-2014)*, ed. Keith Jones, Christian Bokhove, Geoffrey Howson, and Lianghuo Fan, 351–356. Southampton: University of Southampton.

———. 2014b. *Strategie vzdělávací politiky České republiky do roku 2020* [Strategy of the educational policy of the Czech Republic until 2020]. Prague, MŠMT. http://www.msmt.cz/uploads/Strategie_2020_web.pdf. Accessed 12 April 2019.

———. 2018a. *Nematematický svět učebnic matematiky pro 6. ročník základních škol a v oblasti finanční matematiky* [Non-mathematical world of mathematics textbook for 6th grade of lower secondary schools and in the area of financial literacy]. Prague: Univerzita Karlova, Pedagogická fakulta.

———. 2018b. *Vyhlášení pokusného ověřování učebních materiálů na podporu rozvoje digitální gramotnosti dětí a žáků MŠ, ZŠ a SŠ* [Announcement of experimental piloting of teaching materials for supporting development of children's and students' digital literacy]. http://www.msmt.cz/vzdelavani/zakladni-vzdelavani/vyhlaseni-pokusneho-overovani-digitalni-gramotnost?highlightWords=informa%C4%8Dn%C3%AD+technologie. Accessed 12 April 2019.

Morkes, František. 2003. *Historický přehled postavení maturitní zkoušky a analýza jejích funkcí*. [Historical overview of the status of the Maturita exam and an analysis of its functions]. Prague: Ústav pro informace ve vzdělávání.

MŠMT. 1996. *Vzdělávací program Obecná škola, pojetí obecné školy, učební osnovy obecné školy* [Educational program Municipal school, Conception of Municipal school, curriculum of Municipal school]. Prague: Portál.

———. 1997. *Vzdělávací program Národní škola: vzdělávací program pro 1.-9. ročník základního vzdělávání*. [Educational program National school: Educational program for the 1st to 9th Grade of basic education]. Prague: Státní pedagogické nakladatelství.

———. 1998. *Školství na křižovatce: výroční zpráva o stavu a rozvoji výchovně vzdělávací soustavy v letech 1997–1998* [Education on the crossroads: Annual report on the state and development of education in the years 1997–1998]. Prague: MŠMT.

———. 2014. *Strategy for education policy of the Czech Republic until 2020*. http://www.vzdelavani2020.cz/images_obsah/dokumenty/strategy_web_en.pdf. Accessed 7 March 2020.

———. 2017. *Rámcové požadavky na studijní programy, jejichž absolvováním se získává odborná kvalifikace k výkonu regulovaných povolání pedagogických pracovníků* [Framework requirements for study programs, the completion of which gives professional qualifications to pursue the regulated professions of teachers]. http://www.msmt.cz/vzdelavani/vysoke-skolstvi/ramcove-pozadavky-na-studijni-programy-jejichz-absolvovanim-1?highlightWords=MSMT+21271%2F2017. Accessed 11 April 2019.

———. 2018. *Strategie digitálního vzdělávání do roku 2020* [Strategy of Digital Education 2020]. http://www.msmt.cz/uploads/DigiStrategie.pdf. Accessed 7 March 2020.

Müller, Gerhard N., Heinz Steinbring, and Erich Chr. Wittmann (Eds.). 2004. *Arithmetik als Prozess*. Seelze: Kallmeyersche Verlagsbuchhandlung, Germany: Kallmeyer.

Novotná, Jarmila. 2010. *Study of solving word problems in teaching of mathematics. From atomic analysis to the analysis of situations*. Saarbrücken, Germany: LAP LAMBERT Academic Publishing.

———. 2019. Learning to teach in the Czech Republic: Reviewing policy and research trends. In *Knowledge, policy and practice in teacher education. A cross-national study*, ed. Maria Teresa Tatto and Ian Menter, 39–59. London (GB): Bloomsbury Academic.

Novotná, Jarmila, Marie Tichá, and Naďa Vondrová. 2019. Czech and Slovak Research in didactics of mathematics: Tradition and a glance at present state. 2019. In *European traditions in didactics of mathematics*, ed. Werner Blum, Michèle Artigue, Maria Alessandra Mariotti, Rudolph Strässer, and Marja Van den Heuvel-Panhuizen, 187–212. Cham (Ch): Springer Open. https://doi.org/10.1007/978-3-030-05514-1_7.

OECD. 1997. *Thematic review of the transition from initial education to working life – Czech Republic. Background report*. http://www.oecd.org/education/skills-beyond-school/1908234.pdf . Accessed 17 October 2019.

———. 2013. *Education Policy Outlook: Czech Republic*. www.oecd.org/czech/EDUCATION%20POLICY%20OUTLOOK Accessed 12 April 2019.

Palečková, Jana, Vladislav Tomášek, and Josef Basl. 2010. *Hlavní zjištění výzkumu PISA 2009. Umíme ještě číst?* [Main findings from the PISA 2009 research. Do we still know how to read?] Prague: Ústav pro informace ve vzdělávání. https://www.csicr.cz/getattachment/cz/O-nas/Mezinarodni-setreni-archiv/PISA/PISA-2009/narodni-zprava.pdf. Accessed 12 April 2019.

Parliament 2004. *Act No. 561/2004. Collection of law, on pedagogical staff and on the amendment to some other acts*. http://www.msmt.cz/documents-1/act-no-561-2004-collection-of-law-on-pre-school-basic-secondary-tertiary-professional-and-other-education-the-education-act-as-amended?highlightWords=ACT+561%2F2004 Accessed 20 April 2019.

Pol, Milan. 2007. Škola vedená, řízená a spravovaná [School led, regulated and administered]. *Pedagogika roč. LVII* [online] [cit. 2018-05-20]. http://pages.pedf.cuni.cz/pedagogika/files/2014/01/P_2007_3_02_%C5%A0kola_213_226.pdf Accessed 20 April 2019.

Polya, George. 1945. *How to Solve it*. New Jersey: Princeton University Press.

Potužníková, Eva, Veronika Lokajíčková, and Tomáš Janík. 2014. Mezinárodní srovnávací výzkumy školního vzdělávání v České republice: zjištění a výzvy [International comparative research on school education in the Czech Republic: findings and challenges]. *Pedagogická orientace* 24 (2): 185–221. https://journals.muni.cz/pedor/article/view/618 Accessed 12 April 2019.

Rendl, Miroslav, Naďa Vondrová, Lenka Hřibková, Darina Jirotková, Jaroslava Kloboučková, Ladislav Kvasz, Anna Páchová, Isabella Pavelková, Irena Smetáčková, Eliška Tauchmanová, and Jana Žalská. 2013. *Kritická místa matematiky na základní škole očima učitelů* [Critical places of mathematics—investigating teachers' discourse]. Prague: Univerzita Karlova, Pedagogická fakulta.

Ruthven, Kenneth. 2007. Teachers, technologies and the structures of schooling. In *Proceedings of the 5th Congress of the European Society for research in mathematics education*, ed. Demetra Pitta-Pantazi and George Philippou, 52–67. Cyprus: University of Cyprus.

Samková, Libuše, Alena Hošpesová, and Marie Tichá. 2016. Role badatelsky orientované výuky matematiky v přípravě budoucích učitelů 1. stupně ZŠ. [The role of inquiry based mathematics education in the education of future primary school teachers.]. *Pedagogika* 66 (5): 549–569. http://pages.pedf.cuni.cz/pedagogika/ Accessed 12 April 2019.

Shulman, Lee. 1987. Knowledge and teaching: Foundations of the New Reform. *Harvard Educational Review* 57 (1): 1–23. https://doi.org/10.17763/haer.57.1.j463w79r56455411.

Starý, Karel, Veronika Laufková, Kateřina Novotná, Jana Stará, Vít Šťastný, and Zuzana Svobodová. 2016. *Formativní hodnocení ve výuce* [Formative assessment in teaching]. Prague: Portál.

Štefflová, Jaroslava, and Radmil Švancar. 2004. Školská rada nese pečeť povinnosti [School board bears a seal of duty]. *Učitelské noviny [online]* 41. http://www.ucitelskenoviny.cz/?archiv&clanek=3880. Accessed 20 April 2019.

Stehlíková, Naďa. 2004. Konstruktivistické přístupy k vyučování matematice. In *Dvacet pět kapitol z didaktiky matematiky* [Twenty-five chapters from mathematics education], ed. Milan Hejný, Jarmila Novotná, and Naďa Stehlíková, 11–21. Prague: UK-PedF.

Stehlíková, Naďa, and Jana Cachová. 2006. Konstruktivistické přístupy k vyučování a praxe [Constructivist approaches to teaching and practice]. [CD ROM.] In *Podíl učitele matematiky ZŠ na tvorbě ŠVP*. Prague: JČMF.

Straková, Jana. 2009. Vzdělávací politika a mezinárodní výzkumy výsledků vzdělávání v ČR [Educational policy and international research on educational outcomes in the Czech Republic]. *Orbis scholae* 3 (3): 103–118.

———. 2010. Postoje českých učitelů k hlavním prioritám vzdělávací politiky [Czech teachers' attitudes towards main priorities of educational policy]. In *Učitel v současné škole*, ed. Hana Krykorková, Růžena Váňová, et al., 303–313. Prague: Univerzita Karlova, Filosofická fakulta.

Straková, Jana, et al. 2009. *Analýza naplnění cílů Národního programu rozvoje vzdělávání v České republice (Bílé knihy) v oblasti předškolního, základního a středního vzdělávání* [Analysis of fulfilling the objectives of the National program of education development in the Czech Republic (White Paper) in the domain of pre-school, basic and secondary education]. Prague: MŠMT.

Tichá, Marie. 2013. Modernizace vyučování matematice v letech 1965–1985. Ohlédnutí za prací Kabinetu pro modernizaci vyučování matematice MÚ ČSAV [Modernisation of mathematics education in the years 1965–1985. Looking back at the work of the department for modernisation of teaching mathematics of the Mathematical Institute of the Czechoslovak Academy of Sciences]. *Orbis scholae* 7 (1): 119–130.

Tomášek, Vladislav, Josef Basl, Iveta Kramplová, Jana Palečková, Dagmar Pavlíková. 2008. *Výzkum TIMSS 2007. Obstojí čeští žáci v mezinárodní konkurenci?* [TIMSS 2007 Research. will

Czech pupils succeed in international competitions?] Prague: ÚIV. https://www.csicr.cz/getattachment/cz/O-nas/Mezinarodni-setreni-archiv/TIMSS/TIMSS-2007/Narodni-zprava-2007.pdf Accessed 12 April 2019.

Tomášek, Vladislav, Josef Basl, and Svatava Janoušková. 2016. *Mezinárodní šetření TIMSS 2015. Národní zpráva* [International research TIMSS 2015. National report]. Prague: Česká školní inspekce.

Vaníček, Jiří. 2010. Mathematics and project-based learning employing technologies. *South Bohemia Mathematical Letters* 17 (1): 77–92.

Vondrová, Naďa, Jarmila Novotná, and Marie Tichá. 2015a. Didaktika matematiky: historie, současnost a perspektivy s důrazem na empirické výzkumy [Mathematical education: History, present and perspectives with the focus on empirical research]. In *Oborové didaktiky: vývoj—stav—perspektivy*, ed. Iva Stuchlíková, Tomáš Janík, et al., 93–122. Brno: Masarykova univerzita.

Vondrová, Naďa, Miroslav Rendl, Radka Havlíčková, Lenka Hříbková, Anna Páchová, and Jana Žalská. 2015b. *Kritická místa matematiky základní školy v řešeních žáků* [Critical places of basic school mathematics in students' solutions]. Prague: Karolinum.

Walterová, Eliška. 2011. Vývoj primární a nižší sekundární školy v českém kontextu [Development of primary and lower secondary schools in the Czech context]. In *Dva světy základní školy? Úskalí přechodu z 1. na 2. stupeň*, ed. Eliška Walterová et al., 16–51. Prague: Karolinum.

Wittmann, Erich Christian. 2001. Developing mathematics education in a systematic process. *Educational Studies in Mathematics* 48: 1–20.

Žlábková, Iva, and Lukáš Rokos. 2013. Pohledy na formativní a sumativní hodnocení žáka v českých publikacích [Formative and summative assessment in Czech publications]. *Pedagogika* 58 (3): 328–354.

Chapter 2
Traditions and Changes in the Teaching and Learning of Mathematics in Germany

Regina Bruder

Abstract This chapter investigates different effects on mathematics teaching in the Western and Eastern part of Germany after the fall of the Berlin Wall in 1989 and the reunification of the two German states. Striking developments and discussions on mathematics teaching are analyzed in their historical context, which is mainly affected by the different design of mathematics education and mathematics teacher training in the two German states [Federal Republic of Germany (FRG) and German Democratic Republic (GDR)]. Recurrent pendulum swing patterns can be observed in both the thematic emphases of mathematics teaching and the preferred approaches to teaching mathematical content. The aim of this chapter is to reproduce, in as fact-based a manner as possible, the recent history of German mathematics education, which is regarded as part of German social developments as a whole.

Keywords Mathematics education · German history · Cultural changes

1 Introduction

General mathematics teaching in Germany has an eventful history in terms of content and methodological orientation. These aspects were influenced by far-reaching historical events and social changes. After the Second World War (1945) Germany was divided into different occupation zones, see Fig. 2.1. In 1949 there was an administrative separation between the western part of Germany, named the Federal Republic of Germany (FRG) that contained the American, French, and British zones and the eastern part composed of the Soviet occupation zone, named the German Democratic Republic (GDR). The Western Allies (USA, Great Britain, and France) supported the construction of a Western democracy in the FRG. In the GDR, a

R. Bruder (✉)
Technical University of Darmstadt, Darmstadt, Germany
e-mail: bruder@mathematik.tu-darmstadt.de

Fig. 2.1 Occupation zones of Germany 1945

communist system similar to Poland, Hungary, and Czechoslovakia was established under Soviet influence.

In the FRG the individual federal states had educational sovereignty and were able to decide on the school types, timetables, and curricula themselves. Coordination was carried out by the Standing Conference of the Ministers of Education and Cultural Affairs of the States in the Federal Republic of Germany, known as the Kultusministerkonferenz (KMK). The KMK was a voluntary association and dealt with issues of cultural policy of superregional importance with the aim of forming a common opinion and will and the representation of common concerns (https://www.kmk.org/kmk/information-in-english.html).

In the GDR, a centralized state, a uniform educational system was established. In this system all children from the first grade up to the end of their eighth school year were taught together (Polytechnische Oberschule). In the Soviet occupation zone, professorships and courses in the "Methodology of Mathematics Teaching" were awarded and established as early as 1946 (Borneleit 2006). For more details on the different developments in education and mathematics education in divided Germany between 1945 and 1990, see works by Borneleit (2003), Bruder et al. (2013), Einsiedler (2015), Henning and Bender (2003), Jahnke et al. (2017), Neigenfind (1970), Schubring (2014), Wuschke (2018), and Zabel (2009) and the brief overview in Sect. 3.

The separation between the East and West of Germany manifested in 1961 with the construction of a massive border wall, which also ran through the capital, Berlin. This wall fell in 1989 as the result of a peaceful revolution in the GDR and led in 1990 to the reunification of Germany with the accession of the GDR by the FRG, see Fig. 2.2.

As a result of the rapid adaptation of the structures of the new federal states (the former GDR) to the educational system of the old federal states in 1990, there were

Fig. 2.2 The German federal states since reunification in 1991 and the neighboring countries of Germany. (Source: Panther Media GmbH/Alamy Stock Photo, JB50N7)

major changes to the framework for mathematics education and to the entire environment of education and teacher training in the new federal states (see also Sect. 4). However, educational leaders of the old federal states saw no need to make changes to their own mathematics education and teacher training.

The public perception and discussion about mathematics education in the entire country changed only a few years after reunification due to the weak German scores in some international studies, Trends in International Mathematics and Science Study (TIMSS) and Program for International Student Assessment (PISA).

> The participation in PISA was decided by the Conference of Education Ministers (Kultusministerkonferenz) in 1997 and then commissioned. …And so, with PISA 2000, a continuous performance evaluation of the German school system began through an international comparison, and the performance comparison of the federal states began. (Tillmann 2004, p. 478)[1]

This so-called "empirical turn" had a strong influence on the general conditions and the orientation of mathematics teaching and didactic research in united Germany. Not only were mathematical comparison tests launched for the diagnosis of basic competencies, but also large-scale projects for empirical educational research, such as the National Education Panel and technology-based testing (Buchhaas-Birkholz 2010). Output and competence orientation formed the main thread of these projects (for more details see Sect. 4).

This chapter will focus on the changes in German mathematics teaching after the fall of the Berlin Wall in 1989 and the reunification of the two German states.

Over the last 30 years there have been very few studies or publications dealing with the effects of the reunification of Germany on the teaching and learning of mathematics in the old and the new federal states (former GDR). The nature of the difficulty of conducting such studies is similar to what the Eastern German sociologist Hansgünter Meyer described 1 week after the GDR's accession to the FRG in his discipline:

> Whoever deals with the problem… is put into the role of a chronicler who speaks about a fact without historical-systematic processing, i.e. is very dependent on personal points of view. … The sociology that originated in the GDR still exists, the GDR no longer does. (Meyer 1991, p. 69)

These issues can be better understood and classified if differences in the social framework conditions and factors influencing mathematics teaching in both parts of Germany after 1945 are taken into account. These differences and factors are discussed in Sect. 3.

The influencing factors apply both to mathematics didactic research in the former GDR and to mathematics teaching itself, which was shaped by the specific socialization of teachers, pupils, and their parents in the new federal states. Some aspects of the current interior and exterior views of German mathematics teaching since 1990 are described in Sect. 4.

---

[1] All translations from German are by the author.

Mathematics didactic research results from the GDR period are briefly dealt with in Sect. 5. At the beginning of the 1990s there were many representatives from the old federal states, who were interested in a scientific exchange and the development of mathematics teaching and practice-oriented didactic research in the former GDR, according to the reports of the joint meetings in 1990 and 1996 in Henning and Bender (2003).

Sometimes, however, decisions on educational policy are also in reaction to developments in scientific discourse, even if they are conducted publicly. The eventful history of paradigms in educational science has always been heavily influenced by developments in the field of mathematics (let us recall New Math) and social changes.

Here, for example, interesting recurring patterns can be seen in the thematic emphases in mathematics teaching and the preferred approaches to learning mathematical content. In the next section, these patterns are described as pendulum movements. This section is positioned ahead of following sections, to provide a framework.

2 Pendulum Movements in the Main Focuses of Mathematics Teaching Since the Nineteenth Century—Using the Example of the German-Speaking World

The specialization of mathematics in the nineteenth century and the lack of practical relevance in pure mathematics prompted the establishment of technical colleges and the conception of "engineering mathematics." The development of mathematics education led, in turn, to the creation of "school mathematics." This area of teaching methodology was then named "Stoffdidaktik"—literally "subject matter didactics," a concept that combines mathematical theory, its application in real-world situations and the student's pedagogical and psychological state of development. Stoffdidaktik concentrated on the mathematical content of the subject matter to be taught (Sträßer 2019). It attempted to be as close as possible to the discipline of mathematics. A major aim was to make mathematics accessible and understandable to the learner:

> In the development of the didactics of mathematics as a professional field in Germany, subject-related approaches played an important role. Felix Klein created a model that has been referred to for a long time. A general goal was to develop approaches for representing mathematical concepts and knowledge in a way that corresponded to the cognitive abilities and personal experiences of the students while simultaneously simplifying the material without disturbing the mathematical substance. A fundamental claim was that such simplifications should be "intellectually honest" and "upwardly compatible" (Kirsch 1977). (as cited in Jahnke et al. 2017, p. 307)

Until the 1980s, content orientations in mathematics education were closely linked to developments in mathematics itself.

Picker (1991) discusses the effects of New Mathematics on mathematics lessons in elementary school and shows an interesting phenomenon in the choice of learning content, which dated back to the sixteenth century. He describes the following dispute about the right way of teaching lessons on calculating:

> Visual methodology or counting methodology …?
>   Quantity handling … or series formation …?
>   Cardinal number … or ordinal number …?
>
> This dualism is as old as Plato … and Aristotle …
>   They form no contrast, but are complementary mutual as are material and formal education. (Picker 1991, p. 335)

While Adam Riese advocated for material education around 1522, Pestalozzi advocated for formal education in 1801. In 1814, Harnisch called for a combination of material and formal education in schools.

In 1838, Diesterweg favored the viewing and recognition of point patterns without counting in primary school arithmetic lessons. While in 1888, Hartmann explicitly referred to counting as an intuitive access to quantity concepts. Kühnel (1916) recommended in his proposals for the "new building of the arithmetic lesson" to allow both: looking and counting.

While Wittmann, in 1929, proposed the concept of cardinal numbers and a set theoretical basis for initial instruction, Breidenbach, in 1947, advocated starting with ordinal numbers on the basis of children's first experience of counting off numbers. Here, too, 20 years later, a solomonic solution was found: Fricke recommended that both number aspects be taken into account from the outset: cardinal number and ordinal number.

The challenge behind these examples lies in a balanced use of the findings and proposals from personalities and various interest groups in educational policy. This applies to the selection of learning content and the associated objectives (e.g., formal and material education in the past and later around 1970 in the opposition between learning to think and learning to calculate). A balanced approach is also key to understanding different professional approaches and to partially contrary methodical ways of learning selected learning contents. Schneider writes:

> Between the poles of divinity and usefulness has always fluctuated the occupation with mathematics and thus also the teaching of arithmetic, space theory and mathematics. In the process, there have been and continue to be clubs of opinions up to the present day. (Schneider 1989, p. 7)

The history of alternative approaches to mathematical topics of (German) math lessons are characterized by pendulum swings.

The search for a better mathematical approach to subjects in schools played a central role in the discussion about mathematics education until the 1990s and repeatedly led to partially contradictory curricular changes. One could ask why these oscillations were not recognized early and why educators did not learn from them, but that is just a rhetorical question. A dialectical approach using multiple perspectives and taking into consideration the advantages and disadvantages of exposed technical approaches in regard to a particular goal or a particular method could (if done in conjunction with empirical evidence) perhaps have avoided extreme pendulum swings. Schneider, however, also gave some perspective:

> Are we questioning history to find answers for us today or are we just looking to history for confirmation of our current views? We can always find something there. The aim is probably only to track down progressive tendencies and to justify why they brought progress, and also, in the case of regressive views, to prove why they had an inhibitory effect, whatever their advocates subjectively wanted. (Schneider 1989, p. 4)

At the same time, teaching methodology in mathematics is still a very young academic discipline that is rooted in the field of mathematics. The new generation of professors who filled the few available places for teacher education in colleges and universities were mainly mathematicians—not all of whom had teaching experience in schools. The science of "teaching methodology"—known as *Fachdidaktik* in the western part of Germany and as *Fachmethodik* in the GDR—has only existed as an independent academic area since the 1960s. Teaching methodology in mathematics in the German-speaking world has its origins in the development of mathematics schooling. And for *Stoffdidaktik*, new challenges arose again. Jahnke et al. (2017) wrote about the resulting challenges to *Stoffdidaktik* and its current development:

> Concepts and explanations should be taught to students with sufficient mathematical rigor in a manner that connects with and expands their knowledge of the subject. For this reason, subject-matter didactics placed value on constructing viable and robust mental representations (Grundvorstellungen) to capture mathematical concepts and procedures as they are represented in the mental realm. In the 80s, views of the nature of learning as well as objects and methods of research in mathematics education changed and the perspective was widened and opened towards new directions and gave more attention to the learners' perspective. This shift of view issued new challenges to subject-related considerations that have been enhanced by the recent discussion about professional mathematical knowledge for teaching. (Jahnke et al. 2017, p. 307)

But the pattern of pendulum movements was repeated. Around the 1970s and 1980s, more attention was paid to the students' needs, own goals, and individual difficulties. That was a significant development, especially in the FRG. In the GDR, the centrally set goals for mathematics lessons were not up for debate, so teaching focused more on individual support, both to overcome deficits and to support endowments.

The growing interest in researching individual learning processes in the FRG had to do with various social developments. In addition to questions and new answers about the content of mathematics teaching (e.g., New Math and problem-solving), more questions were asked about the ways that classes were organized. The Sputnik shock and the 1968 movement spurred many of these questions (Schubring 2014, 2016) and led to the propagation of antiauthoritarian education. The institutionalized education of mathematics teachers at colleges and universities was influenced by developments in mathematics as well as education in both German states.

Parallel to the social developments that gave the individual and his or her needs more space, constructivist notions of learning received great attention and replaced previous behaviorist concepts. However, there were differences between developments in the FRG and in the GDR until 1990. In the GDR, the 1968 movement had little influence on schools, which were controlled by the state; as a result, the pendulum swing toward antiauthoritarian education was much less pronounced and behavioral ideas about learning were not thrown overboard so quickly. For example, in the new federal states the repetition of basic knowledge and basic skills has always been a common practice. So-called "daily exercises" or as we say now "mixed mental exercises" are very well suited to keep the basics alive. Such methods were hardly known in the FRG.

Some of the characteristic features (not only as a goal, but visible in the reality) of math classes in the eastern part of Germany were:

- application of mathematical knowledge
- close connection of mathematics with other subjects
- discovering theorems and proving them.

Of course, there was still great potential for development, and these aspects were subjects of teacher training. In the GDR, colleges of education and universities were solely responsible for teacher training, including practical phases. School-practical studies accompanied academic study.

In the old federal states, a two-phase teacher education was favored with a traineeship run by the state school administration after students finished university studies. However, there was also a pilot project for single-phase teacher training in Oldenburg and Osnabrück, which was comparable to the model in the GDR (Daxner et al. 1979). This model was not pursued for political reasons. The two-phase teacher education prevailed. There was no time for a discussion of alternatives. At the time of the accession of the GDR by the FRG, all structures of the old federal states were introduced in the east without examination.

The concept of single-phase teacher education was introduced at the same time, and a nationwide dispute over the introduction of comprehensive schooling began. The question was whether the schools should be differentiated—by age or by school type. The debate was also an example of the abovementioned pendulum swings. Now in united Germany there is a certain variety of schools, but, for the most part, only the names in the different federal states differ, not the concepts. The question of the "best" type of school and school system continues to smolder in the background and repeatedly comes up in political discussions.

At several universities and colleges for mathematics teacher training in the GDR, the idea of learning was shaped by the theory of social-constructivist-oriented activity as described by Vygotsky (1978), Luria, Davydow and Galperin—and later Kossakowski and Lompscher (Lompscher 1985; Giest and Lompscher 2006). After that, learners were understood as subjects rather than objects of instruction. However, the idea of the subject position of the learners did not lead to a methodological pluralism in the teaching of mathematics. In the teacher education a moderate constructivist approach to learning was associated with the idea that underdefined conditions there would be an optimal way to design a learning unit for all students so that everyone could benefit from it. Forms of internal differentiation were part of such concepts so as to overcome and avoid failures.

In the old federal states, constructivist learning concepts had to have many facets. Support was given to a constructivist view of learning through many qualitative analyses of the mathematical learning processes (Bruder et al. 2013; Schreiber et al. 2015). While mathematics teachers in the old federal states could learn to be aware of individual student difficulties with mathematics, the teachers in the GDR received more practical methodical solutions for dealing with the entire learning group. In the FRG curricula learning goals and content were given, but not any methods or concepts for mediation. The teachers had more freedom to design their lessons and less supervision by the school inspectorate than in the former GDR. How the teachers used these free spaces is another question. Of course, any description will be incomplete. It can only be a question of indicating general trends and not describing the situation of individual teachers or students.

Currently, constructivism is establishing individualized learning environments and discovery learning is regarded as the best instruction design. This current trend is visible in the expectations and requirements for the teaching of mathematics in the second phase of teacher education (traineeship) and in the questionnaires used in school inspection. Most of the federal states have school inspection now, but not as often, as strictly or as ideologically based as in the GDR. Inspections focus more on the exclusion of allegedly obsolete theoretical ideas of learning. Hamburg, for example, assessed the following for teaching, which clearly describes the expectations:

> The lesson observations should be about recording the quality of the lesson, not the assessment of the teaching teacher. When recording the quality, the superficial quality is meant. "Professionalism cannot play a role in recording the lesson with the lesson observation sheet." (Leist et al. 2016)

Item 15 of the lesson observation sheet:

The students are encouraged to actively participate in the lessons, or they actively participate in the lessons The students are given responsibility to actively participate in the lessons. The methodology is specified by the teacher, the content, however, determines the students	Core indicators • Students are responsible for the learning process of their classmates and the learning content, whereas the methodology is given in the form of cooperative learning forms by the teacher • The students change from a learning role to a learning facilitator role and practice their own teaching functions, such as communicating, supporting, and securing results
	Additional indicators • The teacher acts as a role model and conducts an activity related to the class during the student-centered lesson time. The teacher gives, for example, "tutoring" for students, which are weaker, corrects tests, gives private lessons for students who have missed something, etc. • The teacher refers to inquiries and requests for help of students first to classmates before they intervene • There is neither teacher-centered teaching nor individual work

Such concerns led to a strong pendulum swing of school practice in the direction of expected instructional design in the classroom, ignoring empirical knowledge about the conditions for successful discovery learning or inquiry-based learning (Hattie 2009; Bruder and Prescott 2013). It is well known from the research on teaching that enacting different learning goals effectively requires different methodological implementation (Weinert 1999; Bruder and Roth 2017).

The structural changes in the educational system combined with new instructional design had a lot of consequences, for example, for the kind and the understanding of kindergarten care as well as for the education of kindergarten teachers, and for the level and kind of learning in primary schools (visible in the new curricula) and also for the special schools with an advanced course of study in the area of mathematics and science. If there were no role models in the old federal states, it was very difficult to import specific GDR structures. In particular, this applies to schools for the gifted, especially in mathematics. A reconciliation between the sometimes opposite practices of the FRG and GDR had no chance in 1990.

Another consequence, founded in constructivist notions of learning, is noticeable today in textbooks, which often only contain tasks and dispense with coherent explanations. Other versions of textbook design are described in Sect. 4.

From these historical developments one can learn to separate but not to exclude different aspects of mathematical content, goals, and methods. A goal for further discussion regarding educational policy could be the linking of different aspects in dialectical consideration. We will come back to this phenomenon when we describe striking discourses and developments in mathematics education since 1990.

3 Different Framework Conditions and Developments in Teaching and Learning Mathematics in Eastern Germany from 1945 to 1990

This section presents aspects of the development of the school system of the former GDR, which are important for the understanding of the changes after 1990. The school system in the FRG after 1945 and its various influencing factors are already well documented in English (Schubring 2014).

The following remarks are a translation by the author of a piece from Porges (2017, p. 217):

> The development of the general school system of the Soviet occupation zone/GDR in retrospect reveals three phases (Köhler 2008). Social change after 1945 necessitated a structural change in the school system, specifically comprehensive denazification and new appointments. …The need for a school reform that would provide educational opportunities for all was expressed by representatives of various parties as early as 1945. Consequently, a new beginning of the school system developed and was called the antifascist democratic school reform. This change led to a law for the democratization of German Schools, which came into force on September 1, 1946. It regulated the objectives of the school reform and the

tasks of the school administration, and it defined the structure of the democratic unitary school. The law was based on the guiding principles of the uniform structure of the education and training system, the right to education at all educational institutions and the transfer of school affairs to the state. (Lang 1946)

In this context, private schools were facing dissolution, integration, or transformation to prevent "any offside education and segregation in youth education" (Lang 1946, p. 11). Students would attend an 8-year primary school designed as a democratic standard school for all, after which they could proceed to a 3-year vocational school or a 4-year secondary school. This scheme replaced the previous tripartism of the school system (parallel Hauptschule, Realschule and Gymnasium with selection after primary school, which was also the standard structure in the FRG). The aim of primary school teaching was "to overcome traditional popular education and to lead all children into the realm of educational opportunities which, from a proletarian point of view [...] were regarded as a privilege of social elites" (Geissler et al. 1996, p. 19). The secondary school consisted of a course system that was divided into new language, old language, and mathematical-scientific branches (Köhler 2008). In preparation for high school, grades seven and eight differed in core and course instruction. The introduction of 10-year schools in 1951 expanded the educational landscape. In this type of school, the instructors taught according to the curricula of the mathematics and science branch of the secondary schools (Köhler 2008). The aim was to shorten secondary education to 2 years and to provide direct access to technical schools. Only 2 years later this experiment was stopped. In 1955, a new form of the 10-year school system was introduced under the name "Mittelschule" (secondary school). The law regarding the socialist development of the school system in the GDR introduced the third structural change in December 1959 and required students to attend school for 10 years instead of eight in order to graduate (Köhler 2008).

A 1965 law on the uniform socialist education required structural changes which resulted in a 10-year general polytechnic secondary school (POS) dividing grades in the lower, one through four, and higher, five through ten. A 4-year extended secondary school (EOS), which led to the Abitur was developed (Rockstuhl 2011). These schools remained in existence until 1981.

Some further differences between GDR mathematics lessons and those in the FRG were:

- Centrally organized examinations: German, Mathematics, Russian, Science
- Systems of subject commissions created in every district and of circles (clubs) in each major subject in every school
- Supervisors for all subjects, regular visits and inspections of math lessons every 2 years at all schools
- Mandatory retraining for all teachers every 5 years in several fields: pedagogy, subject to be taught (say, mathematics), psychology, and philosophy.

For comparison here is an overview of the FRG school system before reunification (see also Fig. 2.3).

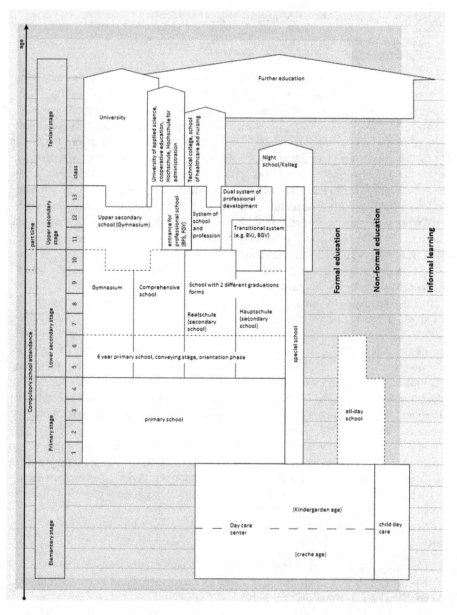

Fig. 2.3 Educational venues and learning worlds in Germany. (Source: Authors' group educational reporting—http://www.bildungsbericht.de/daten2012/vorspann_web2012.pdf)

The main organizational and structural differences between the school systems in East and West Germany before the reunification were as follows:

- Thirteen school years required for graduation in the FRG compared to 12 in the GDR

- Nationwide afternoon offerings and care in the Polytechnische Oberschule (GDR)
- Higher number of lessons in science, technology, engineering, and mathematics (STEM) subjects in the GDR
- Unitary school in the GDR, which enabled more than 8 years of joint learning
- Systematic talent promotion in the GDR with some compulsory participation in the first stage of the Mathematical Olympiad.

A collection of documents on regulations for teacher training in the GDR can be found in Richter (1972). Brislinger et al. (1998) published social science data from the GDR and the new federal states 1968–1996.

Porges (2017, p. 223) writes:

> In line with the curricula, teaching aids for teachers and textbooks for pupils were provided. Both were considered planning aids. Founded in 1945, the *Volk und Wissen* publishing house published all school textbooks without competition. In 1945 there was already a central picture and teaching material office for teaching materials. From 1948 onwards, the State Office for Teaching Materials and School Furniture acted as a sales organisation. In addition, the German Television Broadcasting Service (DFF) began broadcasting school television in 1964. In the year 1971/72 the total demand plan for teaching aids were published with the aim of creating unity between demand plan, curriculum, textbook and teaching aids.

Wuschke (2018) describes the development of the content of mathematics lessons in the Soviet occupation zone and early GDR from 1945 to 1959. Although there are no systematic overview studies for the later years up to 1990, reports on the school system and mathematics teaching in the GDR (Birnbaum 2003), on the conditions for school and science policy (Weber 2003) and on contributions to individual areas of mathematics teaching are available from the perspectives of those involved. These include a contribution by Sill (2018) to the didactics of geometry teaching and a contribution by Borneleit (2003) to curriculum and textbook development.

In fact, there have been some significant differences in the orientation of research in mathematics education. By the end of the 1980s in the FRG, interpretative teaching research had already become accepted as the prevailing method for study in the field. The preferred method was the case study and the explorative "small scale study," and interpretative methods were predominant. In a fundamental contribution to the development of the didactics of mathematics, Griesel (1975) describes the "development of practicable courses" as its most important task.

Some examples of performance assessments from 1990 until 2004 are comparative studies in German and English on mathematics teaching. Performance tests were also conducted in some cases.

Performance assessments of a single class or school or something larger always played a major role in the history of mathematics teaching in the GDR. The social system of the GDR defined itself as a performance-oriented society. School was to lay the foundations for a high level of performance and commitment from every citizen.

To inspect the performance of a school, a "comprehensive, discriminating system of centralised performance controls, analyses and control mechanisms, from giving grades, tests, centralised and local in-school extracurricular performance comparisons was created" (Döbert and Geißler 2000).

The teachers' performance was seen as the main factor in students' performance. Consequently, measures to improve the students' performance were targeted at the work of teachers. The conception of humanity was dominated by the ideal of a socialistic personality "which was attainable by all." Because of this perception, the cause of shortcomings in the knowledge and ability of pupils was seen as the result of the work of the teachers.

The empirical studies done in schools in the GDR were not isolated independent actions merely for the purposes of collecting data. They were invariably conducted with the aim of deducing necessary concrete changes to the school or to evaluate the effectiveness of measures already initiated. So, as a rule, these studies were linked to the introduction of new curricula and textbooks. Accordingly, the extensive empirical studies of the 1980s were referred to as stress tests for the new materials.

Overall, an almost diametrically opposed development of empirical research in mathematics teaching in the GDR and the FRG is evident. While large-scale empirical studies and field experiments were carried out in the GDR, focusing on the quantitative assessment of students' performance, empirical studies in the FRG concentrated increasingly on isolated, high-quality case studies.

Some of the personnel-related and social reasons for the development of empirical research in the FRG until the mid-1990s were:

- The roots of most didacticians lay in the fields of mathematics.
- The distinction in the 1960s and 1970s between practice and theory was based on separation of the academic and practical phases of teacher training.
- There was long-lasting trauma from the failed reform of mathematics teaching in the 1960s and 1970s.
- The educational administration had not, until that time, required the compiling of teaching results.
- Consequently, there were no regular, centralized performance reviews in most states.

The ignorance in both parts of Germany of the developments of the other side is regrettable. In scientific publications in the GDR dealing with mathematics learning, developments in the FRG were not discussed. And, even today, empirical studies which were done in the GDR are hardly known in the unified Germany.

Significant time had to elapse between historical developments in German education in order for authors to be able to present and describe the developments in Eastern and Western Germany without biases and prejudices. Thus, reflective contributions to mathematics teaching and research on teaching and learning mathematics in the two Germanies have only been available since about 2003. These sorts of publications began with the assessment on the proceedings of two meetings of East and West German didacticians held in 1996 (Henning and Bender 2003). This point in time seems connected with the weak results of the PISA study in 2000, which was particularly surprising to Germany's education leaders and triggered the so-called PISA shock. With PISA 2000 began a continuous performance evaluation of the German school system as well as international comparisons. It also began the performance comparison of the federal states (Tillmann 2004, p. 478). This completely new situation for Germany was met with great public interest which was strongly echoed in the media.

4 Aspects of the Current Interior and Exterior Views of German Mathematics Teaching since 1990

The period of change in 1990 was marked in the GDR by the dissolution and decentralization of existing structures. There were also profound changes in educational offerings both in content and in forms of mediation (Schneider 2003). Schneider reports that proposals for the reform of the East German educational structure were developed and discussed as early as the autumn of 1989. In it, the merits of the unitary school for the majority of children were to be combined with a stronger differentiation that would set in earlier. However, these ideas were neither discussed nor applied after 1990. All changes were politically motivated by the respective majorities; they were not organically grown, nor were they adequately prepared.

In the 1990s, mathematics instruction in the old federal states was still struggling with the aftereffects of "New Math," and discussions were held about comprehensive schools. The previous education system in the new federal states was adapted to the administrative structures and legal frameworks of the old federal states relatively silently. The decision-making positions within the new education administration were often filled by people from the old federal states. Schneider (2003) and others report on this phase of great uncertainty for both mathematics teachers at schools and representatives of didactics at universities and colleges.

The new federal states used their newly gained freedoms and leeway to set different priorities. While some introduced the 13th school year at Gymnasiums and, as in the state of Brandenburg, adapted curricular content and structure in a short amount of time due to their geographical proximity to Berlin, others remained with the 12-year school system (e.g., Saxony and Thuringia), but opened up to the variety of teaching and learning materials now available. The new federal states each had one partner from the old federal states whose influence was already evident in the structure of the new curricula (Schneider 2003, p. 260). After reunification, mathematics teachers in the new federal states were confronted with completely new textbook choices. With the publisher *Volk und Wissen* there was only one textbook in the GDR, the development and evaluation of which involved at least in some way all areas of methodology in the GDR. After 1990, all textbook publishers in the old federal states were producing special editions. *Volk und Wissen* also published a slightly modified and then revised edition of its textbook series. At a publishing house founded by former employees of the dissolved Academy of Pedagogical Sciences, a new textbook series was developed for secondary level I (e.g., Sill 2002). The aim was to develop new concepts as a result of analysis of textbooks from the old federal states and taking into account the experiences of the GDR. The structure of the book was not based on Lietzmann's suggestion for a methodical book but, rather, on a guideline suitable for pupils and a collection of tasks. The description of Lietzmann's book, *Das Wesen der Mathematik* (1949), provides essential background for understanding the new approach:

> The older mathematics textbooks began to answer the question of the nature of mathematics and its individual branches, and indeed to provide definitions of these terms. We have gotten away from it today. Rightly! For such a question does not belong to the beginning, but to the conclusion of the study of mathematics. Only when one already has learned something about mathematics, does it seem appropriate to get clear about the mathematical method, the structure of the teaching material and its basis. Numerous previous curricula for higher schools have moved a "repetitive structure of the number concept" into the upper classes. The Merano proposals and, according to them, other modern plans for materials have been taken as the conclusion of mathematical instruction: "Retrospect based on historical and philosophical considerations." The Prussian guidelines of 1925 emphasized philosophically-deepened retrospectives in both the methodological remarks and the curricula: "Logic and knowledge theory find a place in mathematics. Even the psychological foundations of mathematical thinking should touch the lesson. Individual questions such as numerical and spatial representations should, if possible, be deepened philosophically"—so it is said in them. The Marienauer proposals (1945), to name at least one of the new plans, demand: "Structure and basis of mathematics: development of the concept of number and function, axiomatic method of foundation by the example of geometry, prospects of logic and epistemology."

Another challenge for teachers were short-term changes to all curricula in some of the new Länder, which were in several ways due to changes in school structures. In order to adapt to the plans of the old federal states, content was deleted, such as explicit demands for evidence and derivations or for descriptive geometry, which was taught previously in the GDR. Also, new content was included, such as increased description of statistics and of probability theory, which were not included in the GDR curriculum. Since the existing system of mandatory teacher training in the GDR collapsed during this time, teachers were largely on their own when it came to implementing the new requirements.

The effects of these changes on the real practice of mathematics teaching in the new states have not yet been researched. This is partly due to the low level of textbook research in Germany. A search in the MathEduc database using the search terms "U20" (Textbooks. Analysis of textbooks, development and evaluation of textbooks. Textbook use in the classroom) and "German" returns only 82 contributions, of which only 3 deal with the comparative analyses of textbooks. Curricula are also only a marginal subject of didactic research (Sill 2018).

The international PISA tests and the German supplementary test in 2003 showed large differences between the federal states and the old and new states in particular. An analysis of the causes of these differences is difficult because the socioeconomic framework conditions varied in many ways. Nevertheless, it is clear that those new federal states that made only minor structural changes achieved significantly better results on the mathematics test than those federal states that had already been exposed to multiple structural changes—from new school types to educational transitions (say, changing the beginning of lower secondary school from grade 5 to grade 7) to changed timetables and the abolition of advanced courses in upper schools.

The different results of the old and new federal states offer an opportunity to deal with mathematics teaching in the GDR. Sill found that in the *Handbuch der Mathematikdidaktik* (Bruder et al. 2015), for example, only 18 of the 1700 bibliographical references are to teaching works from the GDR.

In 2012, almost 44,600 pupils from the ninth grades of all school types and from all federal states took part in a large learning status survey in mathematics and natural sciences on behalf of the KMK. The comparison between the federal states delivered such headlines in the media as here at SPIEGEL online 11/10/2013:

Maths and Natural Sciences
- The performance gap between students in the East and West is serious:

The East has model students: Saxony and Thuringia lead the nationwide school comparison in mathematics and science. Laggards are the city states and North Rhein-Westphalia. There, students are up to 2 years behind (Fig. 2.4).

This large-scale investigation shows that the performance of students in mathematics and natural sciences is very strongly dependent on the respective federal state in all four subjects examined.

The clear winners in the new study are the federal states of Saxony, Thuringia, Saxony-Anhalt, and with a few slight exceptions, such as Brandenburg, they are significantly above the German average in all four subjects.

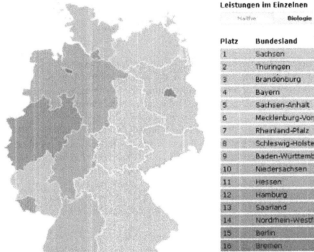

Fig. 2.4 School comparison of the federal states of Germany

Origin Makes a Difference in Performance of 3 Years
- Researchers are bothered about the extent to which educational success in Germany depends on the parents' home. ...Students from better-off families score an average of 82 points more in mathematics than children from weaker families, a difference of almost 3 school years. .... Children with two parents born abroad, regardless of their social background, had significantly worse results than classmates with only one or no immigrant parent.
- The first explanation given by education experts for the good performance of the East German states is the GDR's tradition of mathematics and science. Polytechnic secondary schools there focused on these subjects.
- Educational disadvantages based on immigration status, poverty and educational distance are the biggest problems that the authors of the study point out (Titz 2013)

Even after these media reports emerged, there were hardly any efforts to examine the phenomenon of the GDR school traditions more closely. Rather, the lower proportion of migration among pupils in the new states was identified as a reason for the differences in performance. This, however, overlooks some important experiences in the GDR school tradition, which could be helpful to solving current problems, for example, individual support for all-day schooling.

> Looking at the educational policy process according to PISA from the perspective of "output" (the decisions made and implemented), an ambivalent assessment is appropriate: on the one hand, meaningful and long overdue measures (e.g. more all-day schools) have finally gotten off the ground; on the other hand, all activities to reduce selectivity in secondary schools are excluded. And the measures that are most consistently implemented in primary and secondary education in all the states – setting standards and evaluation – are particularly controversial among educators. (Tillmann 2004, p. 483)

Another topic that has received a great deal of media coverage and is linked to the conclusions drawn from the PISA results is the transition between school and university with regard to mathematical fundamentals and the ability of school leavers to study:

> First of all, the point of departure at the interface between school and university is dominated by political framework conditions and changes to the education system. In our view (shared by numerous colleagues), the shortening of schooling from 13 to 12 years and the shortening of the number of hours in mathematics.... led to a decrease in the level of mathematics at school.
>
> The politically motivated increase in the number of university students additionally reduces the average mathematical competencies of new university students. As a result of all this, poor mathematical skills of students are diagnosed by those lecturing at universities, and administrations criticize high dropout rates in mathematics-intensive subjects. This applies, in particular, to engineering courses but also to mathematics courses for a teaching or a specialized bachelor's degree. This development is even more worrying now, as ministries and universities have started to move to financing concepts dependent on graduation rates.
>
> With a view to schools, the university side laments the fact that today's students mainly have shortcomings in the subjects taught primarily in the lower secondary level. This fact is evidenced by tests in the study entry phase, e.g. at bridge courses (Greefrath et al. 2018). In an open letter in 2017, the unsatisfactory situation concerning the gap between mathematics results in schools and the expectations at colleges was taken as an opportunity to criticize the introduction of educational standards and the orientation of teaching to competencies

instead of prescriptive knowledge (Baumann 2017). This public discussion on possible causes received a lot of media coverage. (DMV 2017)

> Based on the experiences of universities, a great number of school leavers show deficits in knowledge of fractional arithmetic and other calculation techniques; also, logical speech comprehension is often inadequately developed. This would indicate deficits in the sustainability of the transfer of knowledge at school. In addition, the university side has also noticed deficits in the general competencies of new university students, such as self-organization, self-assessment and a willingness to work hard. (Kramer 2010)

An interesting example of a constructive approach to educational policy with regard to student test results and empirical data on mathematics teaching is shown by the city-state of Hamburg (1.8 million inhabitants). Hamburg was not satisfied with the results of the comparative tests in mathematics and therefore used an Expert Commission of mathematics didacticians and human scientists to analyze the state of mathematics teaching. On the basis of extensive empirical data, the Expert Commission developed proposals for six kinds of action, ranging from the establishment of preschool programs to the further education and training of mathematics teachers working in Hamburg. The central message that runs through all fields of action is the need for subject-related and subject-didactic professionalization measures on the part of pedagogical staff (in day-care centers) and teachers (in schools) in order to increase mathematical competence on the part of children and young people. The report of the Expert Commission was published at the end of 2018 (Expert Commission Hamburg 2018).

The recommendations for action in Hamburg address both structural and organizational questions (all-day schools with support programs that are tailored to development of gifted students) as well as content and teaching-related methodological aspects. These include well-balanced goals for the curriculum that take both application and argumentation within mathematics seriously. But there must also be regular repetition and awakening of basic knowledge and basic skills.

If you look at these recommendations for action, there are some parallels to the differences between the FRG/GDR school systems (see Sect. 3) and to the idea of an individually supported mathematics lesson, which also gives sufficient space to specialized content and the general learning potential of students of mathematics.

From today's point of view, with an interval of about 30 years, some of the rapid changes and adaptations to structural conditions in the new federal states appear under a different light:

- In the kindergartens, the opportunities for all children included not only state-supported aspects, to build up the cult of personality related to the political leadership in the GDR, but also carefully thought-out learning and preparation for mathematical thinking and working in school. Today, the potential of early childhood education in the field of mathematics is (again) being considered in the old federal states as well (see also the recommendations of the Expert Commission on Mathematics Education in Hamburg 2018).

- In the afternoon care (after-school care) at the general education schools in the GDR, which was intensively used due to large number of working parents, there were a variety of supports offered both to pupils with learning difficulties and to high-performing pupils, as well as offers for leisure activities. After-school working groups in mathematics to promote gifted children were also a part of this but were not continued after joining the FRG due to the lack of a structural fit and ideological reservations.
- In some places there were private initiatives established with the support of associations [e.g., *Brandenburgischer Landesverein zur Förderung mathematisch-naturwissenschaftlich-technisch interessierter Schüler e.V* (BLIS) in the state of Brandenburg] that were able to continue to train students for the Mathematics Olympiad. The Mathematics Olympiads were continued mainly due to the activities of the mathematicians of the University of Rostock (and other colleagues, see Kugel 2019) and then extended to all federal states (Engel 1990). Since 1994 exist the Union of Mathematics Olympics e.V. (https://www.mathematik-olympiaden.de/). Only about 20 years later did the changed social framework conditions of the Federal Republic of Germany require more afternoon care at schools to answer the growing demand for a better work-life balance. New structures are now being set up, for which there were already possible role models with many years of experience from the GDR. However, these were not well enough known.

Since the 1990s, the promotion of gifted children was still little accepted in the old federal states from the point of view of educational policy. The special schools for mathematics and natural sciences working at a high technical level in the GDR had a hard time surviving with the FRG's concepts of holistic personal development of highly gifted children. In the meantime, the promotion of gifted students has received the status of an official educational goal of the United Nations. The PISA 2015 test results for mathematics in Germany showed that the proportion of students with very good results decreased and had been falling behind the rest of Europe for many years. These results were one of the reasons why the 2016 Standing Conference of the Ministers of Education and Cultural Affairs of the States and the Federal Government decided to launch a "Joint Initiative of the Federal Government and the States for the Promotion of Highly Efficient and Potentially Very Highly Efficient Pupils" focusing on mathematics, natural sciences, and languages. In this initiative, school development toward the direction of talent advancement was to be strongly supported.

After the results of the Third International Mathematics and Science Study (TIMSS) became known in Germany in 1997, a broad discussion about the quality of mathematics and science teaching began. Based on the disappointing TIMSS results, the *Bund-Länder-Kommission model programme* for increasing the efficiency of mathematics and science teaching (SINUS) was implemented at schools between 1998 and 2013 with a wide range of teacher training courses and material developments (see IPN 2003, among others). At the same time, binding educational standards for all federal states were adopted by the Conference of Ministers of

Education and Cultural Affairs in 2004 and 2005 for primary and lower secondary schools and in 2012 for upper secondary schools (KMK 2004a, b, 2005, 2012). Educational standards were described as performance standards, and performance was defined by competencies.

According to Weinert (2001, p. 27), competencies are "the cognitive abilities and skills available to or to be learned by individuals in order to solve certain problems." The focus should be on what pupils actually know and are able to do and not just what they are supposed to learn. Thus, educational standards constitute a pragmatic response to problems in traditional debates on education and curricula.

The first educational standards for mathematics were developed in a relatively short time by experts from the education ministries and were based on existing models, such as the American NCTM standards (2000), the Danish KOM project (Niss 2003), or on the international framework for PISA 2003 (OECD 2004).

The PISA studies and the development of standards in Germany were accompanied by clearly audible criticism within the didactic community (German Society for Didactics of Mathematics—GDM). Compare, for example, the critical analysis of educational standards in Sill (2008) and the German Mathematicians Association (DMV). On the one hand, there was agreement that mathematics teaching should be improved, but the path of controlling output represented by education policy through the standards and tests of PISA studies was ambivalently adopted by the scientific community. The criticism of the introduction of educational standards for mathematics teaching, combined with the new orientation toward competencies, reached a high point in March 2017 with an open letter that received a great deal of media coverage. This letter identified alarming symptoms of a crisis in mathematics education in schools. A central point of the criticism was formulated around the subject matter that mathematics in education had been thinned out to such an extent that the mathematical knowledge of many first-year university students was no longer sufficient for economic, mathematical or scientific-technical studies (see Baumann 2017). In the reactions to this letter the problems and issues were acknowledged but their explanation was questioned.

Dissatisfaction with the fact that experts from mathematics and mathematics teaching were not heard much regarding the implementation of educational standards at the state level led three associations, the DMV, GDM, and MNU (Association of the teachers for the mathematically scientific lessons/Verband der Lehrkräfte für den mathematisch-naturwissenschaftlichen Unterricht), to establish a joint Commission on Current Issues in Mathematical Education in the Transition from School to University in 2011. In particular, this Commission generates and communicates recommendations for the design of the school-university transition (Greefrath et al. 2018) and serves as a contact for the education administration. For activities and statements see http://www.mathematik-schule-hochschule.de/.

At the beginning of 2019, this Mathematics Commission presented the public with 19 recommendations for action to facilitate the transition from school to university. These recommendations for action are also a reaction to the urgent discussion in 2017, which called for, among other things, 4 h of mathematics lessons per week at each grade level, high-quality further training for teachers, a concretization of educational standards and a central examination section for the graduation examination (named *Abitur* in Germany or *Matura* in Austria and Switzerland) that is free of aids (Mathematik Kommission Schule-Hochschule 2019).

In the last 2 years, the quality debate in education has shown new developments, especially in mathematics. The so-called "empirical turnaround" in educational research initiated by the TIMSS and PISA studies increasingly led to the question of whether there was empirical evidence for proposed answers to the open questions—for example on the performance of different school types. Even if expectations cannot all be fulfilled at present, empirically working mathematics didacticians have been able to make themselves heard more. They are more in demand today when decisions on educational policy are to be made.

There are recurring discussions on the goals and content of mathematics as a teaching subject in general schools that are linked to social developments (Neubrand 2015). According to Heymann (1996, p. 50ff), the societal demands on general mathematics instruction are reflected in the "Seven Tasks of General Education Schools" at very different levels of quality: life preparation; foundation of cultural coherence; world orientation; guidance for the critical use of reason; development of willingness to take responsibility; practice in understanding and cooperation; strengthening the student's self.

However, such analytical categories must first be consciously associated with concrete subjects and situations in mathematics lessons. To this end, Winter (1995, p. 37f) formulated three "basic experiences" that share a consensus in Germany today:

> Mathematics teaching should aim to enable the following three basic experiences, which are interlinked in many ways:
> - to perceive and understand in a specific way phenomenon of the world around us that concern or should concern us all, from nature, society, and culture
> - to know and understand mathematical objects and facts (represented in language, symbols, images, and formulas) as spiritual creations, as a deductively ordered world of one's own kind
> - to acquire problem-solving skills beyond mathematics (heuristic skills) in dealing with tasks.

The finding of a harmonious balance between experience-based situational learning and systematic, cumulative knowledge acquisition proves to be a central problem of school-based learning, especially in mathematics:

> If one aims at cumulative knowledge acquisition within a specific field of knowledge, for example in mathematics or a scientific subject, the empirical findings prove the effectiveness of systematic, cognitively abstract learning: a well-organized knowledge base is the best prerequisite for subsequent learning within a domain.
>
> [...] If one rather aims at lateral transfer, at the transfer of what has been learned to parallel but distinct application situations, then situated learning proves its strength. In school, both perspectives of learning are important. The structural strength of the school undoubtedly lies in the organization of systematic, long-term knowledge acquisition processes.... The regulatory idea of school teaching is the long-term cumulative acquisition of knowledge using varying, and if possible also authentic, application situations, with a constantly new balance to be found between casuistry and systematics." (Bund Länder Kommision–BLK 1997, pp. 19–20)

This statement comes from the expert opinion of a Bund Länder Kommision (BLK 1997) project group in response to the results of the TIMSS study (Klieme and Baumert 1998), which founded the teacher training program SINUS.

At the end of the 1990s, the curriculum revisions of individual federal states focused more on student activities in connection with a more process-oriented and less product-oriented view of mathematics (Klieme et al. 2003, p. 45). There was a stronger orientation toward interdisciplinary action competence, including professional competence, methodological competence, personal competence, and social competence.

As an educational policy response to the public debate on the disappointing achievements in the international comparative studies, especially PISA 2000 (Baumert 2001), transnational educational standards were introduced via the *Kultusministerkonferenz* (KMK). Among other things, the "expertise for development of national educational standards" (Klieme et al. 2003) recommended the description of minimum standards and the development of core curricula. But the KMK did not follow all aspects of this recommendation—now we have so-called *Regelstandards* (rule standards) in Germany.

> However, schools will need additional guidance, support and counseling to be able to deal productively with educational standards.
> It follows from this that teacher training, curriculum work, school supervision, and other instances of educational administration must take up the impulses of educational standards and assume new functions. (Klieme et al. 2003, p. 90)

5 Aspects of Dealing with Mathematics Didactic Research and Development Results from the GDR Period

Already in 1990 there was a meeting of didactics representatives from East and West Germany in Ohrbeck for an exchange about didactic research in both parts of Germany, organized by Hans-Georg Steiner (IDM, Bielefeld) (Bruder and Winkelmann 1991). In 1996 there were two follow-up meetings in Osnabrück and Magdeburg, for which proceedings were not published until 2003 (Henning and Bender 2003).

It is a fact that until today East German research results have found only little entry into the all-German research on the development of mathematics teaching. There are still noticeable prejudices due to the undeniable proximity of research and development in mathematics teaching in the GDR to the failed East German political system. However, there are other reasons for the poor reception of the results to date. As Sill (2018) describes, after 1990 there were very few East German didactics professors who were still active and could continue or communicate previous research traditions. The greatest difference between the teaching and researching of subject didacticians from the old and new federal states are visible in their practical relevance. A (theory-based) practical relevance in the GDR was much stronger. This discrepancy led to a different degree of appreciation of the subject of didactics in the

1990s in the East and in the West of Germany. (This assessment is based on personal experience of the author during many years of teaching and in-service trainings in both the East and West.)

The lack of access to East German dissertations made the reception of East German research and developments in mathematics teaching more difficult; they were not published due to scarce resources and were essentially only available at specific research locations through individual specimen copies. Until 1990 there was also no domain-specific research journal in the GDR like the *Zentralblatt for Didactics of Mathematics* (ZDM) and the *Journal of Mathematics Didactics* (JMD) in the FRG. There were, however, special editions and interdisciplinary scientific university journals in small print runs at some universities and colleges. Unfortunately, these articles and materials are still not digitally available today. Sill writes further on the possible causes for the widespread disregard of the results of mathematical methodology in the GDR:

> The most interesting results for today are not to be found mainly in the journal 'Mathematik in der Schule' or the teaching aids, but in the dissertations, the scientific journals of the institutions and the so-called gray literature. The qualification work is available in only a few copies in the libraries of the institutions or in the German National Library. (Sill 2018, p. 5)

Two international conferences in recent years on teaching and learning mathematics, which were organized in Germany, opened up the possibility of presenting developments in mathematics education and its research from both the old and the new federal states. This opportunity was used at the Psychology in Mathematics Education conference in Kiel (Bruder et al. 2013) and at the ICME 2016 in Hamburg (Bruder and Schmitt 2016).

Jahnke et al. (2017) described and analyzed developments that have taken place in German mathematics education research during the last 40 years. The 16 authors are experts and identify eight themes, which "were characteristic for the discussion on how Germany was influenced by and how it interacted with the international community" (Jahnke et al. 2017, p. 305). The authors show:

> the profound changes that have taken place in German-speaking mathematics education research during the last 40 years. The development comes near to a sort of revolution—not very typical for Germany. The only themes that could have appeared in the program of the Karlsruhe Congress in 1976 are subject-matter didactics and, with qualifications, design science and *Allgemeinbildung*. All other topics, especially modelling, theory traditions, classroom studies, and empirical research represent for Germany completely new fields of activity. Today, they define the stage on which German mathematics educators have to act. (Jahnke et al. 2017, p. 317)

But theoretical traditions (e.g., the development of subject didactical theories) concerning typical teaching and learning situations (Steinhöfel et al. 1978) as well as for structuring math lessons, classroom studies (e.g., about using hand held computers, see Fanghänel 1985) and empirical research about initial differentiation or problem-solving (Bruder et al. 2013) were very well established in the GDR, but not in the old states. And that hardly has changed. Since most of the professorships for specialized didactics in mathematics now come from the old federal states and naturally bring along their own training culture, there is a danger that theoretical foundations and research results of the former GDR will become even less visible in the future.

Hopefully, future subject-didactic German-language research on teaching and learning mathematics will:

- become aware of historical oscillations and provide for a timely necessary balance to extreme positions by keeping a memory of research results from more than 30 years ago
- be, above all, responsible for its service function in the further development of real mathematics education and
- engage in open discourse with reference disciplines and without ideological reservations, valuing and taking part in all forms of insight gained at various research locations.

Acknowledgments Many thanks to Hans-Dieter Sill, Rostock, and Axel Brückner, Potsdam, for their support in the discussion of this chapter.

References

Baumann, Astrid. 2017. Open letter. https://www.tagesspiegel.de/downloads/19549926/2/offener-brief.pdf. Accessed.

Baumert, Jürgen, ed. 2001. *PISA 2000 – Basiskompetenzen von Schülerinnen und Schülern im internationalen Vergleich* [PISA 2000 – Basic competences of pupils and pupils in an international comparison]. Wiesbaden: Verlag für Sozialwissenschaften.

Birnbaum, Peter. 2003. Schulsystem und Mathematikunterricht in der DDR [School system and mathematics lessons in the GDR]. In *Didaktik der Mathematik in den alten Bundesländern – Methodology of Mathematics Teaching in the GDR*. Report on a double conference on the joint reappraisal of a separate story, eds. Henning, Herbert and Peter Bender, 13–25. Magdeburg.

Borneleit, Peter. 2003. Lehrplan und Lehrplanerarbeitung, Schulbuchentwicklung und -verwendung in der DDR [Curriculum and curriculum development, textbook develop-ment and use in the GDR]. *Zentralblatt für Didaktik der Mathematik* 35 (4): 134–145.

———. 2006. Zur Etablierung der Methodik des Mathematikunterrichts an Universitäten und Hochschulen in der Sowjetischen Besatzungszone (SBZ) 1946–49 [On establishing the methodology of mathematics teaching at universities and colleges in the soviet occupation zone (SBZ) 1946–49]. In *Beiträge zum Mathematikunterricht*, 139–142. Hildesheim and Berlin: Franzbecker.

Brislinger, Evelyn, Brigitte Hausstein, and Eberhard Riedel. 1998. Sozialwissenschaftliche Daten aus der DDR und den neuen Bundesländern – 1968 bis 1996 [Social science data from the GDR and the new federal states – 1968 to 1996]. In *Materials for the study of the GDR society*, eds. Gesellschaft Sozialwissenschaftlicher Infrastruktureinrichtungen e.V. (GESIS). Wiesbaden: Verlag für Sozialwissenschaften.

Bruder, Regina, Bärbel Barzel, Michael Neubrand, Silke Ruwisch, Gert Schubring, Hans-Dieter Sill, and Rudolf Sträßer. 2013. On German research into the didactics of mathematics across the life span—National Presentation at PME 37. In *Proceedings of the 37th conference of the international group for the psychology of mathematics education*, ed. Anke M. Lindmeier and Aiso Heinze, 233–276. Kiel.

Bruder, Regina, Lisa Hefendehl-Hebeker, Barbara Schmidt-Thieme, and Hans-Georg Weigand, eds. 2015. *Handbuch der Mathematikdidaktik* [Handbook of mathematics didactics]. Berlin, Heidelberg: Springer.

Bruder, Regina, and Anne Prescott. 2013. [Research evidence on the benefits of IBL]. ZDM. *International Journal of Science and Mathematics Education* 45: 811–822.

Bruder, Regina, and Jürgen Roth. 2017. Welche Methode passt? [Which method fits? Matching methods to lesson objectives in typical teaching-learning situations]. *mathematik lehren* 205: 2–9. Seelze: Friedrich.

Bruder, Regina, and Oliver Schmitt. 2016. Joachim Lompscher and his activity theory approach focusing on the concept of learning activity and how it influences contemporary research in Germany. In *Theories in and of mathematics education. Theory strands in German speaking countries. ICME-13 topical surveys*, ed. Angelika Bikner-Ahsbas and Andreas Vohns, 13–20. Berlin, Heidelberg: Springer.

Bruder, Regina, and Bernard Winkelmann. 1991. Bericht über das "Symposium zur Förderung der wissenschaftlichen Zusammenarbeit in der Mathematikdidaktik in Deutschland" [Report on the symposium for the promotion of scientific cooperation in mathematical didactics in Germany]. ZDM. *International Journal of Science and Mathematics Education*, 23(4): 148–155.

Buchhaas-Birkholz, Dorothee. 2010. Die "empirische Wende" in Bildungspolitik und Bildungsforschung: Zum Paradigmenwechsel des BMBF in der Forschungsförderung [The "empirical turning point" in educational policy and research. On the BMBF's paradigm shift in research funding]. *DIE Zeitschrift für Erwachsenenbildung* 4: 30–33.

Bund-Länder-Kommission für Bildungsplanung und Forschungsförderung (BLK). 1997. Gutachten zur Vorbereitung des Programms "Steigerung der Effizienz des mathematischnaturwissenschaftlichen Unterrichts". Materialien zur Bildungsplanung und Forschungsförderung [Opinion on the preparation of the program "Increasing efficiency of mathematical-natural science teaching". Materials for educational planning and for research funding], no. 60. Bonn.

Daxner, Michael, Marianne Kriszio, and Ulrich Steinbrink. 1979. *Modellversuch einphasige Lehrerausbildung: Erfahrungen mit der Durchführung eines Reformmodells in Oldenburg uund Osnabrück* [Pilot project one-phase teacher training: Experiences with the implementation of a reform model in Oldenburg and Osnabrück]. Zentrum für Pädagogische Berufspraxis: Oldenburg.

DMV (Deutsche Mathematikervereinigung). 2017. DMV-Blog. https://www.mathematik.de/dmv-blog/1464-ein-brandbrief-und-seine-folgen. Accessed.

Döbert, Hans, and Gert Geißler. 2000. *Schulleistung in der DDR: Das System der Leistungsentwicklung, Leistungssicherung und Leistungsmessung* [School performance in the GDR: The system of performance development, performance assurance and performance measurement]. Frankfurt: Peter-Lang-Verlagsgruppe.

Einsiedler, Wolfgang. 2015. *Geschichte der Grundschulpädagogik. Entwicklungen in Westdeutschland und in der DDR* [History of primary school education. Developments in West Germany and the GDR]. Bad Heilbrunn: Klinkhardt.

Engel, Wolfgang. 1990. Entdeckung und Förderung mathematischer Begabungen in der DDR [Discovery and promotion of mathematical gifts in the GDR]. *Zentralblatt für Didaktik der Mathematik* 1 (1990): 23–34.

Expert Commission Hamburg. 2018. Report. https://www.hamburg.de/contentblob/11904704/a80cee49fc0febd76d810b6514f1c108/data/mathegutachten.pdf. Accessed.

Fanghänel, Günter. 1985. Über die Einführung von elektronischen Taschenrechnern im Unterricht der allgemeinen Polytechnischen Oberschule [On the introduction of electronic calculators in the teaching of general polytechnic high school]. *Mathematik in der Schule*: 151–174.

Geissler, Gert, Falk Blask, and Thomas Scholze. 1996. Geschichte, Struktur und Funktionsweise der DDR Volksbildung [History, structure and functioning of GDR folk education]. Bd. 1. In *Schule: Streng vertraulich! Die Volksbildung der DDR in Dokumenten*. Berlin: Basisdruck.

Giest, Hartmut, and Joachim Lompscher. 2006. *Lerntätigkeit – Lernen aus kultur-historischer Perspektive* [Learning activity – learning from a cultural-historical perspective]. Berlin: Lehmanns Media—LOB.de.

Greefrath, Gilbert, Max Hoffmann, and Wolfram Koepf. Eds. 2018. Mathematik in Schule und Hochschule – Wie groß ist die Lücke und wie gehen wir mit ihr um? [Mathematics in schools and colleges – How big is the gap and how do we deal with it?] *Der Mathematikunterricht* 64(5). Seelze: Friedrich.

Griesel, Heinz. 1975. Stand und Tendenzen der Fachdidaktik Mathematik in der Bundesrepublik Deutschland [Status and tendencies of subject didactics mathematics in the Federal Republic of Germany]. *Zeitschrift für Pädagogik* 21 (1): 19–31.

Hattie, John. 2009. *Visible learning*. London, New York: Routledge.

Henning, Herbert, and Peter Bender. Eds. 2003. *Didaktik der Mathematik in den alten Bundesländern – Methodik des Mathematikunterrichts in der DDR – Aufarbeitung einer getrennten Geschichte* [Didactics of mathematics in the old federal states – Methodology of mathematics teaching in the GDR]. Report on a double conference on the joint reappraisal of a separate story. Magdeburg: University of Magdeburg.

Heymann, Hans-Werner. 1996. *Allgemeinbildung und Mathematik* [General Education and Mathematics]. Beltz: Weinheim.

IPN (Institut für Pädagogik der Naturwissenschaften). 2003. *Steigerung der Effizienz des mathematisch-naturwissenschaftlichen Unterrichts. Abschlussbericht des BLK-Modellversuchsprogramms* [Increasing the efficiency of mathematics and science teaching. Final report of the BLK model test programme]. Kiel: IPN.

Jahnke, Hans-Nils, Rolf Biehler, Angelika Bikner-Ahsbahs, Uwe Gellert, Gilbert Greefrath, Lisa Hefendehl-Hebeker, Götz Krummheuer, Timo Leuders, Marcus Nührenbörger, Andreas Obersteiner, Kristina Reiss, Bettina Rösken-Winter, Andreas Schulz, Andreas Vohns, Hofe vom Rudolf, and Katrin Vorhölter. 2017. *German-speaking traditions in mathematics education research*. New York: Springer.

Kirsch, Arnold. 1977. [Aspects of simplification in Mathematics teaching]. In *Proceedings of the third international congress on Mathematical education* . ZDM, 98–120. Karlsruhe.

Klieme, Eckhard, Hermann Avenarius, Blum Werner, Peter Döbrich, Hans Gruber, Manfred Prenzel, Kristina Reiss, Kurt Riquarts, Jürgen Rost, Heinz-Elmar Tenorth, and Helmut J. Vollmer. 2003. *Expertise zur Entwicklung nationaler Bildungsstandards* [To the development of national educational standards – An expertise]. Berlin: Bundesministerium für Bildung und Forschung (BMBF).

Klieme, Eckhard, and Jürgen Baumert, eds. 1998. *TIMSS – Impulse für Schule und Unterricht. Forschungsbefunde, Reforminitiativen, Praxisberichte und Video-Dokumente* [TIMSS – Impulses for school and lessons. Research findings, reform initiatives, practice reports and video documents]. Bonn: Bundesministerium für Bildung und Forschung (BMBF).

Köhler, Helmut. 2008. *Datenhandbuch zur deutschen Bildungsgeschichte. Band IX: Schulen und Hochschulen in der Deutschen Demokratischen Republik 1949–1989* [Data manual on the history of German education. Bd. 9. Schools and colleges in the German Democratic Republic 1949–1989]. Göttingen: Vandenhoeck & Ruprecht.

Kramer, Jürg. 2010. Die Perspektive der Universität [The perspective of the university]. In *Interface School – University. Beiträge zum Mathematikunterricht, GDM München*, ed. Hans-Wolfgang Henn, Regina Bruder, Jürgen Elschenbroich, Gilbert Greefrath, Jürg Kramer, and Guido Pinkernell, 75–82. Münster: WTM-Verlag.

Kugel, Manuela. 2019. Databank. https://www.olympiade-mathematik.de. Accessed.

Kühnel, Johannes. 1916. *Neubau des Rechenunterrichts* [Arithmetic lesson]. Leipzig: Klinkhardt.

Kultusministerkonferenz. 2004a. *Bildungsstandards im Fach Mathematik für den Hauptschulabschluss* [Educational standards in mathematics for lower secondary school leaving certificates (grade 9)]. Darmstadt: Luchterhand. https://www.kmk.org/fileadmin/veroeffentlichungen_beschluesse/2004/2004_10_15-Bildungsstandards-Mathe-Haupt.pdf. Accessed.

———. 2004b. *Bildungsstandards im Fach Mathematik für den Mittleren Schulabschluss* [Educational standards in mathematics for secondary school leaving certificates]. Darmstadt: Luchterhand. http://www.kmk.org/fileadmin/veroeffentlichungen_resolutions/2003_2003_12_04-Bildungsstandards-Mathe-Mittleren-SA.pdf. Accessed.

———. 2005. *Bildungsstandards im Fach Mathematik für den Primarbereich* [Educational standards in mathematics for the primary school (grade 4)]. Darmstadt: Luchterhand. http://www.kmk.org/fileadmin/veroeffentlichungen_resolutions/2004/2004_10_15-Bildungsstandards-Mathe-Primar.pdf. Accessed.

———. 2012. *Bildungsstandards im Fach Mathematik für die Allgemeine Hochschulreife* [Educational standards in mathematics for the general higher education entrance qualification]. Cologne, Wolters Kluwer. https://www.kmk.org/fileadmin/veroeffentlichungen_beschluesse/2012/2012_10_18-Bildungsstandards-Mathe-Abi.pdf. Accessed.

Lang, Johannes. 1946. *Die Demokratische Einheitsschule* [Democratic Unity School]. Leipzig: Volk und Buch.

Leist, Sebastian, Tom Töpfer, Stefan Bardowiecks, Marcus Pietsch, and Simone Tosana. 2016. *Handbuch zum Unterrichtsbeobachtungsbogen der Schulinspektion Hamburg* [Handbook for the lesson observation sheet of the School Inspectorate Hamburg]. Hamburg: Institut für Bildungsmonitoring und Qualitätsentwicklung. https://www.hamburg.de/contentblob/4017978/897e411bb701eefc20277b4b0ec77f33/data/pdf-instrumente-handbuch-zur-unterrichts-beobachtung.pdf. Accessed.

Lietzmann, Walter. 1949. *Das Wesen der Mathematik* [The essence of mathematics]. Braunschweig: Vieweg & Sohn.

Lompscher, Joachim, ed. 1985. *Persönlichkeitsentwicklung in der Lerntätigkeit. Ein Lehrbuch für pädagogische Psychologie an Instituten für Lehrerbildung.* [Personality development in the learning activity. A textbook for educational psychology at institutes for teacher training]. Berlin: Volk und Wissen.

Mathematik Kommission Schule-Hochschule. 2019. Handlungsempfehlung von DMV, GDM und MNU für einen leichteren Übergang von der Schule an die Hochschule [Recommendations for an easier transition from school to higher education]. https://www.mathematik.de/presse/2497-handlungsempfehlung-von-dmv,-gdm-und-mnu-f%C3%BCr-einen-leichteren-%C3%BCbergang-von-der-schule-an-die-hochschule. Accessed.

Meyer, Hansgünter. 1991. Die Soziologie in der DDR im Prozeß der Vereinigung der beiden deutschen Staaten [The sociology in the GDR in the process of the union of both German states]. In *The modernisation of modern societies: Negotiations of the 25th German sociologist's day in Frankfurt am Main 1990*, ed. Wolfgang Zapf, 69–86. Frankfurt: Campus.

National Council of Teachers of Mathematics. 2000. *Principles and standards for school mathematics*. Reston: National Council of Teachers of Mathematics.

Neigenfind, Fritz. 1970. Die Entwicklung des Mathematikunterrichts in der DDR [The development of mathematics teaching in the GDR]. *Mathematik in der Schule* 8 (3): 166–183. Berlin: Volk und Wissen.

Neubrand, Michael. 2015. Bildungstheoretische Grundlagen des Mathematikunterrichts (Kap. I.3) [Educational Theoretical Foundations of mathematics education]. In *Handbook of mathematics didactics*, ed. Regina Bruder, Lisa Hefendehl-Hebeker, Barbara Schmidt-Thieme, and Hans-Georg Weigand, 51–73. Berlin, Heidelberg: Springer.

Niss, Mogens. 2003. The Danish KOM project and possible consequences for teacher education. In *Educating for the future. Proceedings of an international symposium on mathematics teacher education*, ed. Rudolf Strässer, Gerd Brandell, Barbro Grevholm, and Ola Helenius, 178–192. Gothenburg: Royal Swedish Academy of Science.

OECD. 2004. Learning for tomorrow's world – First results from PISA 2003. http://www.oecd.org/education/school/programmeforinternationalstudentassessmentpisa/34474315.pdf. Accessed.

Picker, Bernold. 1991. 25 Jahre Neue Mathematik in der Grundschule [25 years New Math in primary schools – A balance]. In *Beiträge zum Mathematikunterricht*, 393–396. Bad Salzdetfurth: Franzbecker.

Porges, Karl. 2017. Evolution and school in the SBZ/DDR. In *Negotiations on the history and theory of biology*, vol. 20, 215–242. Berlin: Volk und Wissen.

Richter, Wolfgang. 1972. *Lehrerausbildung in der DDR. Eine Sammlung der wichtigsten Dokumente und gesetzlichen Bestimmungen für die Ausbildung von Lehrern, Erziehern und*

Kindergärtnern. [Teacher training in the GDR. A collection of the most important documents and legal provisions for the training of teachers, educators and kindergarten teachers]. Berlin: Volk und Wissen.
Rockstuhl, Harald, ed. 2011. *School law of the GDR 1965–1991/92. Law on the uniform socialist education system of the GDR of 25 February 1965.* Bad Langensalza: Rockstuhl.
Schneider, Siegfried. 1989. *Mathematikunterricht in Geschichte und Gegenwart. Vortrag zur Eröffnung der 9. Tage des Wissenschaftlichen Rates* [Mathematics lessons in history and present. Lecture for the opening of the 9th Days of the Scientific Council]. Dresden: Pädagogische Hochschule Dresden.
———. 2003. Entwicklung des Mathematikunterrichts und der Mathematikdidaktik nach der Wiedervereinigung in den neuen Bundesländern [Development of mathematics lessons and didactics of mathematics after reunification in the new federal states]. In: *Didaktik der Mathematik in den alten Bundesländern – Methodology of mathematics teaching in the GDR.* Report on a double conference on the joint reappraisal of a separate story, eds. Henning, Herbert, and Peter Bender, 257–265. Magdeburg.
Schreiber, Christoph, Marcus Schütte, and Götz Krummheuer. 2015. Qualitative mathematikdidaktische Forschung: Das Zusammenspiel von Theorieentwicklung und Anpassung von Forschungsmethoden [Qualitative mathematics didactic research: The interplay between theory development and adaptation of research methods]. In *Handbuch der Mathematikdidaktik*, ed. Regina Bruder, Lisa Hefendehl-Hebeker, Barbara Schmidt-Thieme, and Hans-Georg Weigand, 591–612. Heidelberg: Springer.
Schubring, Gert. 2014. Mathematics education in Germany (modern times). In *Handbook on the history of mathematics education*, ed. Alexander Karp and Gert Schubring, 241–256. New York: Springer.
———. 2016. Die Entwicklung der Mathematikdidaktik in Deutschland [The development of mathematics didactics in Germany]. *Mathematische Semesterberichte* 63 (1): 3–18.
Sill, Hans-Dieter, ed. 2002. *Mathematik. Textbook for grade 5.* Mecklenburg-Vorpommern, Berlin: PAETEC Verlag für Bildungsmedien.
———. 2008. PISA and the educational standards. In *PISA & Co. Critique of a program*, ed. Thomas Jahnke, 391–431. Hildesheim: Franzbecker.
———. 2018. On the didactics of geometry teaching in the GDR. In *Geometrie mit Tiefe. Arbeitskreis Geometrie 2017. Geometry with Depth. Working Group Geometry 2017*, 3–42. Hildesheim: Franzbecker. https://www.math.uni-sb.de/servive/lehramt/AKGeometrie/AK_Geometrie_Tagungsband_2017.pdf. Accessed.
Steinhöfel, Wolfgang, Klaus Reichold, and Lothar Frenzel. 1978. Zur Gestaltung typischer Unterrichtssituationen im Mathematikunterricht [To design typical teaching situations in mathematics lessons]. In *Lehrmaterial zur Ausbildung von Diplomlehrern: Mathematik* [Teaching material for the training of diploma teachers: Mathematics]. Berlin: Hauptabteilung Lehrerbildung des Ministeriums für Volksbildung.
Sträßer, Rudolf. 2019. The German speaking didactical tradition. In *European traditions in didactics of mathematics, ICME-13 Monographs*, ed. Werner Blum, Michele Artigue, Maria Alessandra Mariotti, Rudolf Sträßer, and Marja Van den Heuvel-Panhuizen, 123–152. Berlin: Springer.
Tillmann, Klaus-Jürgen. 2004. Was ist neu an PISA? Zum Verhältnis von Bildungsforschung, öffentlicher Diskussion und Bildungspolitik [What's new about PISA? On the relationship between educational research, public discussion and educational policy]. *Neue Sammlung* 44 (4): 477–486.
Titz, Christoph. 2013. Mathe und Naturwissenschaften: Leistungsgefälle zwischen Schülern in Ost und West ist gravierend [Maths and science: Differences in performance between students in East and West is serious]. https://www.spiegel.de/lebenundlernen/schule/laendervergleich-ostdeutsche-schueler-in-mathe-besser-als-westdeutsche-a-927216.html. Accessed.
Vygotsky, Lev S. 1978. *Mind in society. The development of higher psychological processes.* Cambridge, MA: Harvard University Press.

Weber, Karl-Heinz. 2003. Mathematikunterricht und mathematikmethodische Forschung in der DDR – wesentliche schul- und wissenschaftspolitische Rahmenbedingungen [Mathematics education and subject didactical research in the GDR – Essential school and science policy framework conditions]. In: *Didaktik der Mathematik in den alten Bundesländern* [Methodology of mathematics teaching in the GDR]. Report on a double conference on the joint reappraisal of a separate story. eds. Henning, Herbert, and Peter Bender, 1–12. Magdeburg.

Weinert, Franz-Emanuel. 1999. Die fünf Fehler der Schulreformer. Welche Lehrer, welchen Unterricht braucht das Land? [The five mistakes of school reformers. Which teachers, what lessons does the country need?]. *Psychologie Heute* 26 (7): 28–34.

Weinert, Franz Emanuel, ed. 2001. *Leistungsmessung in Schulen* [Performance measurements in schools]. Weinheim and Basel: Beltz.

Winter, Heinrich. 1995. Mathematikunterricht und Allgemeinbildung [Mathematics education and general education]. *Mitteilungen der GDM* 61: 37–46.

Wuschke, Holger. 2018. Entwicklung von Inhalten im Mathematikunterricht der SBZ und der frühen DDR (1945–1959) [Development of contents in mathematics lessons of the SBZ and early GDR (1945–1959)]. In *Suggestions from the mathematics methodology of the GDR*, eds. Lambert, Anselm, and Hans-Dieter Sill. *Der Mathematikunterricht* 64(6): 5–14.

Zabel, Nicole. 2009. Zur Geschichte des Deutschen Pädagogischen Zentralinstituts der DDR. Eine Studie zur Geschichte des Instituts. (Dissertation) [On the history of the German Pedagogical Central Institute of the GDR. A study of the history of the institute]. Technische Universität Chemnitz. http://monarch.qucosa.de/api/qucosa%3A19263/attachment/ATT-0/. Accessed.

Chapter 3
The Traditions and Contemporary Characteristics of Mathematics Education in Hungary in the Post-Socialist Era

János Gordon Győri, Katalin Fried, Gabriella Köves, Vera Oláh, and Józsefné Pálfalvi

Abstract The traditions of mathematics education in Hungary date back over two centuries. This chapter attempts to give an analysis of the essential features of mathematics education in the post-communist era. To provide context, the authors also describe some characteristics of Hungarian mathematics education after World War II. We put an emphasis on the reforms of the 1970s, specifically the Complex Mathematics Education Experiment (CMEE) led by Tamás Varga, since this had an immense influence on today's mathematics education.

The social and economic developments after 1989 changed the centralized structure of education and caused new types of schools and centers to be founded. The textbook market was liberated and new textbooks appeared. The educational content of the curriculum changed and so did the assessments. Changes to teacher training followed the changes in the school system and the changes to the curriculum. Talent management programs were extended to all age groups. We follow these in our study and discuss how the educational policy of the government influenced mathematics education.

Finally, we mention the difficulties our education system had to face in recent years and will have to face in the future.

Keywords Mathematics education in Hungary · Teacher education · Tamás Varga · Mathematics competitions · Gifted education · National curricula

---

J. Gordon Győri (✉) · K. Fried · J. Pálfalvi
Eötvös Loránd University, Budapest, Hungary
e-mail: janos.gyori@ppk.elte.hu; kfried@cs.elte.hu

G. Köves
Károli Gáspár University of the Reformed Church in Hungary, Budapest, Hungary
e-mail: koves.gabriella@kre.hu

V. Oláh
János Bolyai Mathematical Society, Budapest, Hungary

Abbreviations

CMEE	Complex Mathematics Education Experiment
IMOF	International Mathematical Olympiad Foundation
KöMaL	Mathematical Journal for Secondary Schools; later: Mathematical and Physical Journal for Secondary Schools (Középiskolai Matmatikai Lapok) (All translations from Hungarian in the chapter were done by the authors)
NCC	National Core Curriculum (Nemzeti Alaptanterv)
OPI	National Institute of Education (Országos Pedagógiai Intézet)
PISA	Programme for International Student Assessment
TIMSS	Trends in International Mathematics and Science Study
TIT	Society for Dissemination of Scientific Knowledge (Tudományos Ismeretterjesztő Társulat)

1 Introduction

The chapter below presents the main features of mathematics teaching in Hungary following the changes to the economic and political systems. Although that change occurred in 1989, it is imperative to discuss the roots, traditions, and main features of mathematics education in the socialist era at the beginning of this chapter because they will be referred to repeatedly later on. In the socialist era, one crucial question in mathematics education was what from the relevant and highly valuable traditions of the past could be preserved and to what extent it could be preserved. It was quite clear that the new, socialist approach to universal education represented enormous challenges for teachers in the field. It was also clear that innovation and preservation had to be accomplished together, and that traditions and new elements had to be integrated into meaningful, coherent and efficient new teaching patterns after World War II. In many respects, the same scenario recurred after the political change in 1989. The socialist era brought many achievements in mathematics education that needed to be carried forward, but many of its elements became outdated and superfluous in the interim. After 1989, the main challenge, therefore, was again to determine what to preserve and/or innovate and what and how to integrate in an environment where politicians were apparently intent to start a brand-new era in education, mathematics included, after each change of government.

This chapter aims to depict the past and current fluctuations in the field under study. What occurred was driven by the wish to preserve and develop mathematical sciences and culture and, first and foremost, the interests of the children, who are the future.

2 Historical Background of Hungary

Throughout the history of the last 500 years, Hungary had hardly ever been independent with the exception of a few years. The Ottoman Empire invaded Hungary in the beginning of the sixteenth century and stayed for over 150 years until it was defeated by the Habsburgs in the late seventeenth century. Afterward, the Habsburgs took over, ruling Hungary and stayed until the late nineteenth century. In 1848, Hungary started its Revolution and War of Independence against Austria, which Hungary lost in 1849. As a result of the Austro-Hungarian Compromise of 1867, Hungary became a part of a dual Austro-Hungarian Monarchy. Hungary entered World War I on the side of the Germans, and lost, and the situation was the same for World War II.

After World War II, there was a short period of democracy with a multi-party system, but in 1948 the communist party seized political power with the support of the Soviet Army, stationed in Hungary. The communist party nationalized both church-owned and private property, factories, companies, and schools; it abolished private ownership and declared a dictatorship of the proletariat.

The aim of the revolution and uprising for freedom in 1956 was to restore democracy and regain independence for the country. However, the Soviet Army crushed the revolution. The new first secretary of the Hungarian Socialist Labor Party, János Kádár, seized power. After years of retaliation, the dictatorship loosened. Employment was compulsory and involved practically everybody. The standard of living was at about the same level for everybody; however, it was a fraction of the average level for people in North America and Western Europe. In 1963, following a wave of political amnesty for past revolutionaries, a number of democracy-related regulations were introduced, although the one-party system was kept. This made Hungary a much more livable country than others in the socialist block.

In the mid-1980s, the socialist system weakened and opposition organizations that wanted social and political reforms were set up. In 1988–1989, a peaceful change to the form of the state took place, and a new government took over political power through free elections. Although 1989 is considered the year that the social and political system changed, the signs of change were in fact already seen in the early 1980s (when the political elite recognized the need to change the management guidelines for economic growth). After 1989, new institutions of parliamentary democracy were founded and the market economy took the place of socialist ownership. In addition to positive results, the change also had many negative consequences. Many people lost their jobs, the unemployment rate rose, especially in the former industrial centers of the country, and inflation was above 30%. A new and much richer compared to earlier times layer of society appeared, and, at the same time, the amount of poverty greatly increased. Governments that came to power in the 4-year parliamentary elections generally failed to introduce long-term reforms, and the ruling governments repeatedly attacked the actions of the previous governments. This caused major problems in the fields of health and education. The problems persist in all areas of education. The laws concerning education and curriculums were greatly changed from one governmental term to another, reducing educational effectiveness. To put this issue into more context, we will discuss in detail the topic of curriculum changes.

3 The History of Mathematics, Education, Mathematics Education, and Mathematics Education for Gifted Students in Hungary Before the Political Changes in 1989

3.1 From the Early Years Until 1945

Until the eighteenth century, schooling in Hungary was governed by churches. The language of education in the schools that correspond to today's secondary schools was Latin. The first textbook of mathematics in the Hungarian language was published in 1743 by a teacher of Debrecen College (Maróthi 1743).

The first institutions of higher education (university level) also operated within churches. The first Hungarian university was founded by King Louis the Great in 1367 with the permission of Pope Urban V. During the Reformation, Protestant academies and colleges founded in Transylvanian cities became of great importance. They later served as the base of famous universities in the nineteenth and twentieth centuries such as the one in Kolozsvár (Cluj). The legal predecessor of Eötvös Loránd University in Budapest (the biggest and most prestigious university currently in Hungary)—originally named Pázmány Péter University—was founded by Jesuits in Nagyszombat (Trnava) in 1635 (Németh 2001).

The eighteenth century Habsburg rulers, especially Maria Theresa, tried to incorporate education under state supervision while maintaining the denominational nature of the schools, from elementary school to university, through regulations known as the *Ratio Educationis* in 1777. In the eighteenth and nineteenth centuries, apart from the study of classical humanities, the teaching of natural science subjects, including mathematics, became more and more widespread in the formerly denominational institutions at the secondary and university levels.

There were outstanding scholars in the history of Hungarian mathematics from the first half of the nineteenth century. The greatest Hungarian mathematician, János Bolyai, lived and worked between 1802 and 1860. In 1831, he showed (at the same time as Carl Friedrich Gauss and Nikolai Ivanovich Lobachevsky) that the parallel postulate is independent of the other Euclidean postulates and created a non-Euclidean axiomatic geometry. However, János Bolyai pursued his activities in isolation in a context where mass education did not exist and mathematics as a science was not yet an organized social activity (Hersh and John-Steiner 1993).

Also from the first half of the nineteenth century, a number of events took place that had great impact on the development of Hungarian mathematics teaching. The Hungarian Academy of Sciences was founded in 1831 (the intention to found it originated in 1825), and a (still functioning) association for popularizing science, called the Society for Dissemination of Scientific Knowledge (Tudományos Ismeretterjesztő Társaság or TIT), was founded in 1840.

After losing the 1848–1849 Revolution and War of Independence against Austria, the laws of the winner, Austria, shaped Hungarian education. Secondary school education and the final examination at the end of it (graduation or matriculation) was regulated by the so-called *Entwurf: Entwurf der Organisation der Gymnasien und Realschulen in Oesterreich* (Rules for the Organization of Secondary Schools in Austria) (Ministerium des Cultus und Unterrichts 1849).

On the basis of the Austro-Hungarian Compromise in 1867, the Austro-Hungarian Monarchy (1867–1919) made it possible to enforce the Hungarian aspirations to establish an independent and modern Hungarian education system, which, with minor modifications, was preserved until the end of World War II. Baron József Eötvös, Minister of Religion and Public Education, is associated with the 1868 Folk High School Act, which introduced the compulsory education of 6 to 12-year-old children, established a six-grade school, and a four-class "civil school" based on it. The emergence and unfolding of an academically oriented secondary grammar school took place at this time in Hungary. It provided a classical education shaped according to the German pattern in which natural sciences and mathematics received major roles. The law ensured state control over education and introduced a system of supervision. Within a few years, the secondary school law was also established: regulating the school leaving examination, the operation of eight-grade secondary schools, and the so-called *real school*. The law set the framework for vocational training and made it possible for girls to go to school. Institutes for teacher education were founded, which in addition to university faculties led the education of prospective secondary school teacher candidates.

A wide-ranging system of identification and development for mathematics talent also emerged in the era of the Austro-Hungarian Monarchy. Its starting points can be tied to Baron József Eötvös and his son Loránd Eötvös (Frank 2012); the foundation of the Association of Mathematics and Physics (Mathematikai és Physikai Társulat) in 1891 (Gosztonyi 2016) was due to the ministerial position of Loránd Eötvös. The Association of Mathematics and Physics became an effective organizer of research as well as public activities in the fields of natural sciences and mathematics. During these intellectually vibrant decades of the Austro-Hungarian Monarchy, a number of Hungarian scientific journals were founded and published; among them was a journal for students titled *Középiskolai Mathematikai Lapok*, which was launched in 1893. The journal's name means the *Mathematical Journal for Secondary Schools* or as it was called later, *Mathematical and Physical Journal for Secondary Schools*, which we are going to refer as *KöMaL* for short, after its Hungarian title. *KöMaL* has been playing a pivotal role through generations ever since by identifying and developing talent in mathematics nationwide. At least partly, the high level of mathematics teaching in the secondary grammar schools led to the appearance of such scientists as János Neumann (Johann von Neumann), Jenő Wigner (Eugene Wigner), and Ede Teller (Edward Teller). Noticeably, they all were students of the Fasori Lutheran Comprehensive Grammar School (Fasori Evangélikus Gimnázium), where they could develop under the guidance of the internationally renowned mathematics teacher László Rátz (Frank 2011) and his colleagues. It was through Rátz's teaching activities and editing *KöMaL* by himself for approximately one and a half decades that it became possible that even primary school students who happened to live in the smallest villages in the countryside of Hungary could send in their solutions to tasks published for the competitions organized by *KöMaL*. Students could become acclaimed for their achievements in a broad circle in Hungary. By this system, a countrywide competition unfolded, in which students gifted in mathematics could compete against one another, no matter where they lived.

The salient fact about the Austro-Hungarian Monarchy period in terms of Hungarian gifted education is that it provided the context in which an organized system of mathematics competitions among schools could be launched for the first time, even internationally, in order to identify and develop students gifted in mathematics. This is the reason that the International Mathematical Olympiad Foundation (IMOF) begins its presentation of the history of international mathematics competitions as follows: "Mathematics competitions began as inter-school competitions in the Austro-Hungarian Empire in the nineteenth century." (International Mathematical Olympiad Foundation 2019).

The first Hungarian nationwide competition in mathematics for students graduating secondary school was launched by the Hungarian Mathematical and Physical Society in 1894 during the Austro-Hungarian Monarchy. This particular competition was so successful that a large collection of the mathematics problems from it were published in 1929 in Hungarian and later in English (in four volumes, Kurschak 1963a, b; Hungarian problem book III 2001; Barrington Leigh and Liu 2011). This competition is still running; today it is called the József Kürschák competition (Suppa 2007).

Some years after the collapse of the Austro-Hungarian Monarchy, another important innovation in Hungary was the organizing of the first Hungarian Students' Competition in Mathematics in 1923, which was followed by a competition in physics in 1927. From the middle of the century, the competition was organized annually in several other subjects as well for students aged 17–19 in their last 2 years of secondary school.

Between the two world wars, mathematics education reached a very high level in some schools thanks to some excellent teachers with outstanding knowledge. We already mentioned László Rátz and the Fasori Lutheran Comprehensive Grammar School in Budapest. Other secondary grammar schools (gimnáziums), mainly attended by the children of middle-class families, also continued to be places for the identification and development of gifted students in mathematics. The educational policy of the time placed great emphasis on the development of primary-level public education; some special scholarships even became available for gifted students from poor family backgrounds. The most important one, the Nicolas Horthy Scholarship was established by the ideologically and politically highly controversial government of Admiral Horthy in 1937 (Horthy 2000; Pornói 2011; Ujváry 2017).

3.2 The Socialist Period: From 1945 to 1989

3.2.1 New Ways in Education and Mathematics Education After World War II

After World War II, a new educational structure was built according to the educational statutes of the Ministry of Religion and Education in 1945. A 4-year secondary education was planned based on the primary education of grades one through eight. The structure of grades one to four (lower primary) was based on the former public-school system, with one teacher per class. An upper primary education for grades five to eight was introduced. Compulsory education lasted from ages 6 to 14. The purpose of primary school for children aged 6–14 was to provide a unified, basic education that a vocational or secondary education could build on, from 1946 onwards. According to Act 33 of the 1948 law, all non-state schools were secularized. This educational structure was different from the Soviet school structure, which had different age groupings for students than did the Hungarian school system. The Soviet influence at that time was primarily manifested in the rigorous and centralized management of education policy and its unified political orientation.

The new educational system was struggling with many difficulties due to at least two main reasons: a change in the structure of education and the lack of teachers for the age group of 10–14. Previously, only about 15% of all children went to school beyond the age of 10 (Pukánszky and Németh 1996), so there were no textbooks for students aged 10–14. Teacher education did not prepare enough teachers for this age group either. Moreover, many teachers had died during the war. New needed teaching materials, such as textbooks, had to be prepared in a very short period of time. The pre-1945 textbooks had to be revised. New textbooks were written and approved by the government for the use of all children. With this, the "single-textbook" era began in which only one mathematical textbook existed for each grade in the entire country.

In 1947, the former Mathematical and Physics Society split up into two parts: the physics and the mathematical societies. Ever since, the mathematical society part has organized annual meetings for teachers that are named after László Rátz (a summer school, which is also considered as in-service training for teachers), as well as the Varga Tamás Methodological Days, since 1989. The members of the János Bolyai Mathematical Society consisted of mathematicians and teachers of mathematics, who hand-in-hand tried to modernize and reinvent mathematics education. The society took over the publication of *KöMaL*. Already during this time, a number of excellent mathematicians began to work on improving Hungarian mathematics teaching as well. They renewed and further developed the already well-functioning mathematical talent training system, which was aimed primarily at secondary school students, especially those in general, non-vocational secondary schools, called "gimnázium." (We will discuss the training of gifted children in detail later in this chapter.) Well-known Hungarian mathematicians of the mid-twentieth century (Alfréd Rényi, Rózsa Péter, Tibor Gallai, János Surányi, László Kalmár, György

Alexits, etc.) also helped to revive the active professional public life of mathematics educators and to make mathematics teaching more effective. One of the most significant works in this period was the secondary school textbook, volumes I and II (for the first 2 years) written by Rózsa Péter and Tibor Gallai in 1949–1950. This textbook did not go beyond the curriculum developed between the two world wars, but instead of the formal, dry statements for procedures, definitions, and theorems used in the previous secondary school textbooks, it tried to make the reader discover new knowledge by revealing reasoning with arguments and explanations. Imre Hajnal wrote of the books, "The authors worked using a heuristic method from the beginning to the end, taking all the inconvenience that goes with it, the lengthy deductive thoughts, which made the book long and in many cases required repetition" (1984). Most teachers were not prepared at this time to teach the new methods the book would have required them to. Due to public pressure, the following edition and the books for the subsequent grades had to be revised thoroughly and appeared from other authors like Jenő Tolnai, Endre Hódi, and Piroska Szabó.

During these years, the goal of primary education was to improve numeracy and to help students to solve practical tasks directly applicable to everyday life. In secondary schools, a performance style was the dominant manner of teaching. The frontal form of work (i.e., work with the entire class) prevailed; education in terms of content did not go further than classical algebraic and geometric issues. The mathematical teaching methods used were based on the Hungarian translation of the teaching books used in the Soviet Union, for example, Chichigin's *The Methodology of Arithmetic* (Csicsigin 1951) and Bradis's *The Methodology of Teaching Mathematics in Secondary School* (Bragyisz 1951). These books, having reliable mathematical content and occasionally acceptable methodological suggestions, had actually helped teachers. At the same time, they contributed to consolidating the rigorous consistency of the centralized teaching style. The most significant of the Russian books related to mathematics teaching was Laricsev's *Problem Book*, which was translated into Hungarian by Tamás Varga (Laricsev 1952). For two decades, this collection of problems was the fundamental document for mathematics teaching in secondary schools. It was the basis of curricula and textbooks, and teachers built their lessons on sets of its tasks. This problem book is the foundation of the generally accepted teaching style in Hungary, which still has effect. The essence of this style is that the backbone of each lesson is a series of tasks selected by the teacher.

The 1960s and 1970s can be called the Great Reform Era for Hungarian mathematics education. In 1957, partly because of the so-called "Sputnik shock," there were worldwide movements aimed at improving mathematics education, the modernization of curriculums, and introduction of new goals for mathematics in schools. On the other hand, they urged the application of new findings in psychology and pedagogy research in mathematics education to replace the formerly used and ossified formal teaching methods.

These intentions had an effect on Hungarian mathematics education. In the beginning of the 1960s, the János Bolyai Mathematical Society invited Zoltán Dienes to give a 2-week course for teachers, presenting his principles of mathematics education and introducing his methods and tools. As a result, Tamás Varga started his first training experiment in September 1961. It was such a major event that in 1962 UNESCO organized its symposium on mathematics education in Budapest, where "Soviet, Polish, Czechoslovakian, Romanian, Hungarian, Italian, Swiss, French, Belgian, Dutch, Danish, Swedish, American, Australian, and Japanese delegates unanimously made the decisions (Servais and Varga 1965) that can be regarded as the statutes of the reform movement" (Varga 1970).

In the 1962–1963 academic year—under the guidance of Ferenc Lénárd—a longitudinal and complex psychological experiment began within the National Institute of Pedagogy (OPI, for short). The study of lower primary mathematics was named "variational mathematics teaching" (Demeter 1990), while that of the higher primary classes was named "individual mathematics in the classroom" (Forrai 1972).

These events greatly contributed to the start of the CMEE in 1963 under the leadership of Tamás Varga in Hungary. Through one and a half decades of continuous work, Tamás Varga and his colleagues developed a single, effective pedagogical system for eight classes of primary school, in which they achieved a harmony of sophisticated mathematical content, methods, and tools that took into account age-specific characteristics. On this basis, in 1978, a new primary school mathematics curriculum was introduced which was completely different from the previous ones in all aspects. Instead of the formal, dissected teaching style of the old arithmetic-geometry structure, it laid down criteria for a modern way of teaching mathematics from the very beginning in a spirally structured way. It began from the very beginning of school with a system for modern mathematics teaching.

Introduced in 1978, the content of the new curriculum was spiral-structured with five main topics elaborated in the complex experimental work throughout the 8-year curriculum: sets and logic; arithmetic, algebra; relations, functions, sequences; geometry, measuring; and combinatorics, probability, statistics. In addition to content issues, this curriculum gave a detailed description of methodological principles. It outlined a long process of conceptualization based on the discovery and activity of children, including adapting to individual differences, aspiring to get children to like mathematics, developing thinking and creativity, freeing students to make mistakes and argue, and meaningfully involving tools and games. A new set of textbooks was published based on the new curriculum for grades one through eight, the authors of which were participants and supporters of the CMEE. The experiment relied on some of the basic ideas of the Western European and American New Math movement, but fortunately avoided the pitfalls of those, such as early abstraction, the formal use of notations, definitions, and the self-serving use of modern work forms and tools. In the 1950s and 1960s, the influence of the world movement for modernizing international mathematics teaching also appeared in mathematics education in secondary schools.

In the meantime, the secondary school curriculum changed as well in 1965. Apart from classical algebraic content, it included sets, vectors, combinatorics, and probability theory. It became important to develop the functional approach and the use of geometric transformations in problem solving. Based on this curriculum, a textbook for secondary school classes written by Pálmay–Horvay was published and used for over two decades. The book was accompanied by excellent teacher handbooks and problem books.

The changes in teaching methodology, however, did not follow contextual changes. In secondary school, frontal teaching based on lectures and explanations remained crucial. There were only a handful of teachers trying to focus on activity-based teaching methods (Somfai 2009).

3.2.2 Gifted Education in Mathematics in Hungary During the Socialist Era

Following World War II, the socialist era was a scene of activities carried out not only by significant mathematicians but also by teachers of mathematics who were considered outstanding internationally. Tamás Varga was the most salient figure of this group. He developed his new mathematics teaching methodology using a thinking skills development scheme applied in the eight-grade primary school (Gosztonyi 2015; Halmos and Varga 1978; Karp 2014). Although it was the genuine intention of Tamás Varga to provide a methodological tool that would be useful and efficient for any student's development, in practice, it turned out to work more efficiently in the case of gifted students (Pálfalvi 2018).

In 1962, a special mathematical class (the so-called *specmat* class) for the development of gifted students was launched within Fazekas Gimnázium in Budapest (Fazekas Mihály Fővárosi Gyakorló Általános Iskola és Gimnázium) (Gordon Győri 2018). It was practically independent of the ongoing educational reforms of this time in Hungary, but acted on Soviet influence (Karp 2009). Ten mathematics lessons were offered in this class weekly. A number of internationally renowned mathematicians and teachers of gifted mathematics education emerged from the former students of this class. The "specmat" classes continued to be taught the next academic year in the Fazekas Gimnázium and were launched in some other schools in Budapest and around the country.

At the end of the 1970s, classes specializing in some other subjects (e.g., biology–chemistry, Hungarian literature, or a language) were introduced in secondary schools, where students had more lessons in the given subject than the students of non-specialized classes. In mathematics, there were several levels of education: basic or "normal," type A, type B, and the special classes mentioned above. Normal classes had fewer lessons a week than type A classes, which had fewer than type B classes. There were 9–10 lessons a week (out of at most 36) for special mathematics classes. The curriculum of these classes explicitly applied differentiation. Apart from the different number of lessons, it tried to take into account the needs of the students concerning their further studies. Also, it handled talented students distinctively, especially in type B classes. Later, the Ministry of Education ordered the publishing of new textbooks for all these types of classes, and they were published by the official textbook publishing house (Tankönykiadó). From the early 1980s, personal computers spread in Hungary. In 1983, schools started to acquire computers (Szentgyorgyi 1999) and computer literacy education began. These lessons were taught mostly by enthusiastic mathematics teachers focusing on algorithmic thinking, mathematical modeling, and developing computer literacy, such as computer programming (contrary to today's practice when children learn only the applications of computer programs).

3.2.3 Mathematics Teacher Education During the Socialist Period

Teacher education institutions for the teachers of lower primary classes existed in Hungary even in the nineteenth century. The teachers of further school years traditionally were trained at a university, and usually they earned a degree in some sort of a science (mathematics, physics, or chemistry).

After World War II, the loss of professionals showed in every area of life; there was a lack of teachers as well. In the 1950s, teacher preparation was subdivided into three parts according to the age grouping of the school system: lower primary, which already had a history of teacher preparation (grades one to four, and even to grade six with some additional studies), upper primary (grades five to eight, even ten if needed) and secondary, where students earned science knowledge rather than methodological and didactical comprehension (grades nine to twelve or five to eight as well, if needed). The communist government sought a way to quickly (quicker than it takes to earn a university degree) educate teachers for this system. Teacher education colleges were established where teachers got a license to teach children between the ages of 10 and 14 (grades five to eight, upper primary). It was satisfactory then, since the age limit for compulsory education was 14 years. Quality in this period was not the most important issue. Actually, a person's occupation was not an individual decision; often it was decided by the government and depended on which areas needed more workers. Many young, quickly trained people—even non-trained ones—were forced to start teaching, despite the fact that their abilities were not satisfactory. So, these factors had a cumulative effect and led to the decline of the quality of mathematics education in Hungary in the 1950s.

3.2.4 Educational Reforms of the Interim (Transitional) Period of the 1980s

From the mid-1980s, significant changes and new phenomena took place in many areas of social and political public life. Significant changes also occurred in education compared to the systematic, centrally controlled system of socialism from previous decades. Ferenc Gazsó, Deputy Minister of Education, expressed the purpose of the 1985 Education Act in an interview as follows:

> The education system should be freed from direct political control and from state administrative control and direction. The aim should be to create a public educational system which is autonomous or, to be precise, works within relatively autonomous frames with all its consequences. (Báthori 2000, p. 3–5)

This law ensured the independence of schools according to teaching content and determined the rights and duties of the participating groups in education (pupils, teachers, parents, and the local community). It also created favorable conditions for the professional independence and autonomy of schools; it provided the opportunity for the appearance of an alternative school system, changing the 8 + 4 grades system into a 6 + 6 and 4 + 8 grades system. At the same time, the development of the new core curriculum began with these three structural systems. In 1989, a ministerial decision was made for preparation of the National Core Curriculum (NCC for short, NAT in Hungarian). The NCC is a law, and, thus it is an obligatory ruling. It gives a frame of knowledge without concrete description of the content compulsory for all schools, giving general descriptions of knowledge required for children to acquire by the end of specific grades.

In this era, there were also some attempts to correct the curriculum of 1978. First of all, a number of studies had been undertaken to reveal the reasons for problems and improve the situation (Klein 1980; Hajdu and Novák 1985; Hajdu 1989). Based on his small-scale impact assessment of 1980, Sándor Klein found that there was a greater difference in student performance in the class he studied than there was before: the number of creative solutions increased, but the number of those who could not solve the problems increased as well. That is, the difference of performance or the "performance scissors" opened wider (Klein 1980). In this political and social situation, the revisions of the 1978–1979 reform curriculum appeared.

As a result of these studies, a new series of textbooks were published starting in 1989 (edited by Sándor Hajdu). The so-called Hajdu-type textbooks became very popular in mathematics education and gave more choice in the textbook market. The two different textbooks coexisted and were based on the same ideas of the CMEE, but there were significant differences in them as well.

In the experimental textbooks of Varga's team, which were based on the active involvement of children, the concepts were built through interpreting and experiencing them. Using these textbooks required a lot of work from teachers, as the interrelationships between concepts were presumably discovered through solving tasks. Contrary to this, the so-called Hajdu textbooks, though preserving much of the content of the 1978 curriculum, were less built on the creative work of teachers and the independent activity of students. The concepts of those textbooks were introduced at the given level and the acquired knowledge was reinforced by repeated practice. As a result, according to an increasingly articulated opinion among teachers, this series of textbooks helped prepare and give lessons in a more direct way than before. Compared to the 1978 reform-textbooks, however, Hajdu's textbooks were more formal; they contained more notations and made statements as precise as possible far too early to lead to lengthy verbalizations of concepts and connections. Hajdu's textbooks also promoted differentiation in upper primary schools in mathematics education (following the example of secondary schools): they separated the basic material and the extra-curricular materials. All this resulted in broadening the spread of mathematical content differences in primary schools that was already there. While teachers of schools in some of the districts with more disadvantaged pupils had difficulties teaching even the basic requirements, students of outstanding schools had already acquired considerable mathematical literacy by the age of 10–14.

4 The Post-Socialist Period: After 1989

4.1 Changes in the Structure of Education

The 1990 XXIII (1990) law amending Act I of the 1985 Law on Education loosened the structure of the educational system. It restored the right to free schooling, giving back the free choice of schools to parents, contrary to the practice of the socialist era, when—with some exceptions—children had to attend district schools. The Act made it possible for churches and other legal entities to set up and maintain schools. Eight-grade secondary school training (grades five to twelve) started, first in Budapest at the László Németh Secondary School, and then at the Fasori Lutheran Comprehensive Grammar School, in which, prior to World War II, László Rátz (Frank 2011) was the mathematics teacher. After completing the fourth grade of primary school, children could enter the eight-grade secondary school. Many schools followed the example of László Németh Secondary School, and most of them had developed their own mathematics teaching materials which did not exist before.

In the eight- and six-grade secondary schools (grades five to twelve and seven to twelve), students were usually enrolled on the basis of entrance examinations required by the school. In the beginning, the above-mentioned new forms of education spread rapidly, but later their number decreased as it turned out that not all schools had the appropriate staff and conditions for a new training structure.

The 1993 LXXIX Public Education Act gave public education a unified framework. It secured freedom of education, abolished the state monopoly on school maintenance, and made the final year of kindergarten compulsory. It preserved a flexible starting age for school depending on the child's developmental level; that is, some children had to repeat the final kindergarten year before starting elementary education. Previously, after completing eight grades of primary classes, children could enter a 4-year secondary school. According to the 1993 act (though still based on an 8-year primary education), it was possible for children finishing grade four, six, or eight to take an entrance examination to get into secondary school in their fourth, sixth, or eighth year. Vocational training was regulated by Act LXXVI of the 1993 law. The same law regulated the training of ethnic minorities and children with disabilities.

In the field of public education, we can also observe the emergence of a large number of organizations operating outside the public administration but with management functions of their own from the mid-1990s.

The changes to the age of compulsory schooling through time are shown in Table 3.1. This also shows a reduction in required education age in 2013.

Table 3.1 Changes to the age limits for compulsory education (Pukánszky and Németh 1996)

Year	1845–1912	1912–1945	1945–1961	1961–1985	1985–1996	1996–2013	2013 to present
Age	6–10/12[a]	6–14[b]	6–14	6–16	6–16/18	5/6–18	5/6–16

[a]10 for girls, 12 for boys
[b]Did not come into effect

According to government rule 134/2016 (VI. 10) in 2016, public schools lost their right to decision-making and their financial independence and found themselves under the maintenance of a state supervised center called the Klebelsberg Center. This proved to be a step back in the autonomy of the schools.

4.2 Changes in the Curriculum and National Curriculum

4.2.1 The Birth of the National Core Curriculum (NCC), Its Changes, and Its Impact on Mathematics Teaching

Already at the end of the 1980s, following political change, educational policy makers and educational researchers immediately proposed to prepare a modern, comprehensive curriculum covering all of education.[1] Thus, at the beginning of the 1990s, professional committees were formed and the preparation work started for the National Core Curriculum (NCC) compulsory for all schools. According to their ideas, a compulsory minimum curriculum should have been developed, which then could have been expanded and increased using a locally developed curriculum according to the needs of the given school, as opposed to the uniform centralized curriculum of the former socialist era. The leaders and participants of the professional committee in mathematics came from the supporters of the previous 1960–1970s Tamás Varga's reform.[2] The idea was to give a less specific description of the material and to give guidelines only in the core curriculum, leaving schools to decide the details. But the mathematical framing of the content was practically the same as before. The first version of the NCC (1991) was sent to schools for opinion polling. The responses it received were rather divided in many respects and reflected the presence of different elements in thinking of Hungarian pedagogical society. There was a broad consensus that "the majority of teachers did not like the idea of general requirements in the curriculum, but would have liked to see exact instructions for the level of requirements instead" (Liskó 1991).

---

[1]The research group led by Zoltán Báthory at the National Institute of Public Education was commissioned by Ferenc Glatz (1989–1990) to prepare the National Core Curriculum.

[2]The chairman of the committee was Lóránt Pálmay, who was one of the authors of a popular, high-quality mathematics textbook series for secondary schools related to the curriculum of 1965.

However, this response was partly due to the concerns of the pedagogical community, which was accustomed to regulation in the former socialist times. From another point of view, this concern was also legitimate, since in order to make the new approach work in practice, a number of infrastructural changes should have been done. For example, an NCC-compliant examination system, a suitable selection of textbooks, and, most importantly, teachers needed to have been prepared for the new approach.

In 1995, the then socialist-liberal government finalized and introduced the NCC (Government Decree 1995), which had passed through several changes by that time, taking into consideration the results of the polling. The government ordered all schools in Hungary to meet the common requirements for education and training. The school curriculum was subdivided into different areas of education and the requirements for completing grades four, six, eight, and ten. The NCC also outlined the approximate proportions for local curricula in educational areas.

The NCC subdivided the content into cultural domains and subordered subjects in these domains. However, besides being a subject, mathematics was also considered a cultural domain in itself, being a part of other subjects. Its share in the first 4 years of education was 19–23%, in grades 5 and 6 it was 16–20%, and 10–14% in grades 7–10. The NCC set a minimum requirement for everyone but also gave more freedom of choice to schools and teachers compared to previous curricula. Schools had to develop their pedagogical programs and their local curriculum for each subject based on the guidelines of the NCC. This caused difficulties for teachers in general, since this form of NCC was unusual for them and differed from the traditional curricula characteristics of the socialist system.

Following the socialist-liberal government, the conservative right-wing government of 1999 reinforced the 8 + 4 grade educational system with an amendment to the law (which allowed the 4 + 8 and 6 + 6 grade school structures to continue to operate). NCC-based frame curricula were published, which, in contrast to the NCC, detailed the purpose of education and training, the subject requirements, and the number of lessons needed. It regulated the knowledge requirements of the first eight grades and even grades nine to twelve. These new regulations were a significant step back from the original NCC that represented the decentralization of education from the centralized structure known from socialism, as well as from free teacher curriculum development and back toward a central curriculum. It is understandable that a significant number of teachers welcomed these changes (Eszterág 2010). Schools have since been able to build their local curricula based on framed curriculum plans. According to the questionnaires returned to institutions and advocacy organizations at that time, teachers welcomed the changes because they eliminated some of the problems in connection with the NCC and dispelled existential uncertainty through the rearrangement (Eszterág 2010).

4.2.2 Mathematics in NCCs

The First NCC

The impact of the CMEE and the primary school curriculum from 1978 that was built on it can be found in the mathematical content and spirit of the NCC. In the systematization of its detailed requirements, one can see the five topics[3] developed there. The topics of sets, logic, combinatorics, and graphs became part of a newly detailed unit, called "thinking methods." This topic contains mathematical knowledge that is processed by discussing problems from topics that are related to it.

All further sections of the NCC preserved the previous (1978) titles as well (Arithmetic and algebra; Relations, functions, sequences; Geometry and measuring; Probability and statistics).

> Development-centricity and the spiral structure are also an important feature of NCC. The five topics listed (with minor title changes) appear on all four levels – in the 1–4, 5–6, 7–8, and 9–10 grade requirements. One can recognize the manner of developing concepts suggested by the CMEE as well: starting from direct experience; age-dependent, playful activities; leaving a long maturation time for developing concepts; the active participation of students in acquiring knowledge; and the forming of concepts. (Pálfalvi 1997, p. 8)

NCC 2003

The beginning of the 2000s brought another change of government as well as changes in education policy. As a result of the activities of the socialist-liberal coalition that was re-elected at that time, a revision of the NCC appeared in 2003, including the mathematical content of it. This NCC terminated the compulsory nature of the framework plans. It did not change the three-tier system of content regulation, but it changed the role of each level:

> The emphasis was placed on preparing for "lifelong learning" and the policy of opportunities, which were defined by the *keycompetences* suggested by the concept of EU (Table 3.2). Detailed requirements were omitted as well as canonical knowledge; development tasks were not defined in relation to the literary content. (Eszterág 2010, p. 89)

Instead, they set up a series of activities based on the logical development for each area of competence.

Joining the European Union (in 2004) had a significant impact on Hungarian educational policy, including corrections to the NCC, which further strengthened the role of competence-based education. Accredited, framed curricula based on the concept of the NCC were a great help in preparing local curricula.

Textbook publishers competed with each other to adjust their textbooks to key competencies articulated in the new regulations. Some leading publishing houses

---

[3] The five topics covered by the 1978 curriculum are (1) sets and logic, (2) arithmetic and algebra, (3) relations, functions, and sequences, (4) geometry and measurements, and (5) combinatorics, probability, and statistics.

Table 3.2 Competencies in mathematics education (Fábián et al. 2008)

Skills	Thinking abilities	Communicational abilities	Knowledge-acquiring abilities	Learning abilities
Counting, numeracy, quantitative thinking, estimation, measuring, interchanging measurement units, solving word problems	Systematizing, combinativity, deductive reasoning, inductive reasoning, probabilistic reasoning, reasoning, arguing, proving	Relation vocabulary, literacy, text interpretation, spatial vision, spatial relations, depicting, presentation	Problem sensibility (questions), representation of problems, originality, creativity, problem solving, metacognition	Attention, perceiving whole and part, memory, keeping mind on the problem, speed of problem solving

had their own competence-based, framed curricula accredited for this work, and tried to find the best professionals to give in-house training courses for teachers willing to use their textbooks.

The OECD PISA measurements (since 2001) have had a significant impact on education policy, including mathematics, and as a result, new competencies were introduced. The government supported the development of competence-based education. Multiple program packages were developed for five priority areas of competency for all types of schools and for all age groups; several program plans were accredited. The task of the program developmental center the Educational Social Service Nonprofit Kft. (successor of the suliNova Kht.) between 2004 and 2008 was to accomplish the central program of the National Development Plan (Human Resources Development Operational Program, Measure 3.1). Within the framework of this measure, new programs for mentor and teacher education were to be developed (Lannert 2009). They had been tested in practice within a wide-ranging national school network. Further general extension and the sustainability of new software packages "were not realized due to the lack of continuous production of teaching tools" (Trencsényi 2015). The schools did not receive any feedback on the processing and results of the tests. The experiment was then discontinued in the educational policy of the next government in 2008.

All of these events contributed to the modification of the system of final examination. The government introduced a two-tier graduation system in 2005 and at the same time abolished the system of university entrance examinations inherited from the socialist era. In mathematics, the intentions of the EU and other international trends had become more and more prevalent: the reduction of lexical knowledge, the increase of practical applications, and the spread of modern methods and forms of work, such as different forms of cooperative methods, project methods, and differentiation.

NCC 2012

The change of government in 2010, during which the conservative right-wing coalition came to power again, brought about a much stronger change in Hungarian education than all other changes before. This change was based on the Education Laws from 2010 and 2011 (Government Decree 2010, 2012), which are still in effect at the time of the writing of this chapter. According to the new laws, schools maintained by the former local councils became state owned, the rights of school principals were greatly reduced, and schools were subordinated to the control and supervision of a state-organized centralized institution. The government prescribed a system of career development for teachers which created unjust circumstances for the development of teachers. Within a few years, textbook publishers were nationalized, with only a couple of sets of textbooks remaining per subject. Many elements of the new system were introduced without any substantive professional consultation and preparation, thus causing a lot of damage to the Hungarian education.

In the 2012 version of the NCC, the predominant viewpoint of global educational goals and tasks phrased by earlier versions of the NCC were replaced by ones with tendencies to raise awareness and preserve national values by the government. At the same time, however, it retained several features of earlier versions, such as the fact that development requirements were still based on European key competences:

> Mathematical competence is the ability to develop and apply mathematical thinking, abstraction, and logical conclusion. It also means that we use mathematical knowledge and methods in solving everyday problems. Knowledge and activities play an important role in the development of mathematical competence. (NCC 2012, pp. 10653–10654)

Mathematics remains a separate cultural domain (besides being a subject itself). The percentages it takes up of all lessons at different stages are 15–20% in grades 1–4, 14–18% in grades 5–6; 10–15% in grades 7–8; 10–15% in grades 9–10; and 10% in grades 11–12.

The developmental tasks are presented in tables (see Tables 3.3 and 3.4) according to age groups and are more transparent than before. The five main topics of the old (1978) reform re-appear in the public education content and a sixth topic is added. These topics appear in all grades from grades 1 to 12 and are divided into three stages: grades 1–4, 5–8, and 9–12.

1. Thinking methods, sets, mathematical logic, combinatorics, and graphs.
2. Number theory and algebra.
3. Geometry.

Table 3.3 Example of table of developmental tasks for grade 1

(1) Knowledge	(2) Developmental requirements	(3) Connection points
(4) Sorting some elements by trial	(5) Fine motor coordination: Stacking small objects	(6) Physical education and sports: grouping according to different properties

Table 3.4 Example of table of developmental tasks for grade 5

(1) Knowledge	(2) Developmental requirements	(3) Connection points
(4) The concept of fractions	(5) Illustrating, interpreting fractions according to the two types of meanings (fraction of a unit and part of the numerator), recognizing fractions in text environments	(6) Music: the connection between fractions and the value of music notes

4. Functions and elements of analysis.
5. Statistics and probability.
6. History of science and mathematics and renowned mathematicians.

The explanation of the content in this version is more specific and more detailed than in previous versions of the NCC.

4.2.3 The So-Called Frame Curricula

The frame curricula of 2012 declared that the aim of the cultural area of mathematics education is to reveal the world of mathematics and mathematical thinking by combining different topics. Also, as it states, the development of concepts, connections, and thinking requires an ever-increasing spiral structure—according to age, individual developmental, and interest characteristics; complex knowledge; and abstraction ability (Sinka et al. 2014).

The frame curricula in general and the one from 2012 specifically represent thematic units constructed according to the NCC. The main topics according to the spiral way of structuring are as follows:

1. Thinking methods, sets, mathematical logic, combinatorics, and graphs
2. Number theory and algebra
3. Geometry
4. Functions and elements of analysis
5. Statistics and probability—these go through the 12-year curriculum.

The sixth topic of the NCC is content related to interesting and important facts about the works of national and international scientists. This is not an independent topic itself, but, rather, an important requirement for students to get acquainted with. Topics regarding mathematics history can help raise the interest of students who are less interested in theoretical mathematics. Another important motivational factor is mentioned in the framework curriculum: "Positive attitude to mathematics can be greatly enhanced by games with mathematical content and interesting problems and puzzles related to mathematics" (Oktatáskutató és Fejlesztő Intézet 2016).

According to the NCC 2012, schools should prepare their local curricula on the basis of a frame curriculum, and they are free to decide about 10% of the content. The frame curriculum defines the weekly number of lessons. In all four grades of lower primary schools, the number of lessons a week in mathematics is four. In the upper primary (grades five through eight), the minimum number of lessons is four,

Table 3.5 Example of developmental requirements in specialized mathematics classes

Knowledge and development requirements	Connection points
Observing random processes. Rolling dice, flipping coins—teamwork	Informatics: simulation of random processes

three, three, and three, respectively, according to the basic level (called level A) and for the higher level (called level B) it is higher: four, four, four, and three, respectively. In secondary schools, the number of mathematics lessons a week is three in all 4 years for the basic level and for the higher level (level A or B) it is more; for level B, it is four, three, three, and four.

The frame curriculum based on the NCC 2012 contains more prescribed detailed knowledge in mathematics compared to the frame curriculum based on the previous NCC (2003), and many people believe that there is too much material contained in it.[4] Apart from this, the lessons teachers are free to use make it possible for them to provide a higher level of knowledge. A higher level is provided in outstanding schools, if there is an appropriate, highly qualified teacher at the school. There usually is one per region where there is a high social demand for a higher level of mathematics education.

The requirements of this thoroughly detailed curriculum are organized in a three-column table below and the related components appear in a row:

Example from grade 1:

Example from grade 5:

More detailed content arranged in two columns was offered in the secondary school frame curriculum. Specialized mathematics classes can apply curricula and textbooks, which prescribe special instruction as well as the appropriate number of lessons that apply to this form of training. For example (Table 3.5):

4.3 Textbooks

As mentioned briefly earlier, under the 2011 law, the state centralized and nationalized textbook publication, practically abolishing the free selection of textbooks in the textbook market. The production of a series of new textbooks was ordered. Some of the series of textbooks that were used earlier and had become successful in the past years went under state control and were permitted (for a while) to be chosen by schools.

Some other popular textbooks used earlier did not become state owned but were corrected according to the new NCC and framework curriculum, and were temporarily—for some years—permitted by the government. Their content was not much

---

[4] "It strongly builds on the declarative knowledge and the operation of knowledge (procedural knowledge); its applications (knowledge transfer) are held back" (Trencsényi 2012).

different from that of the state textbooks although they differed in their style and use. The content of some textbooks that were suitable for classes with highly motivated children exceeded the requirements of the framework curriculum, while newly designed mathematics textbooks were suitable for classes covering the basic curriculum. Some previously used textbooks clung (sometimes more than necessary) to a strict scientific precision. Others tried to attract pupils with colorful content closely relating to their everyday life and tried to raise student interest with curiosities. The previously used textbooks were also different from a teachers' point of view: some required independent creative thinking from the teacher in order to apply the book effectively, while others practically built up the content of the curriculum by a series of lessons to help teachers prepare. Teachers' handbooks were also published along with the textbooks. They contained methodological recommendations and the solutions to the problems in the book to help the teachers with their work. After the trial period of the new, experimental state textbooks, a study (Dárdai et al. 2015), ordered by the EU and conducted by Ágnes Dárdai, concluded that the textbooks needed serious revisions and corrections. However, this never happened due to a lack of funding. Teachers also pointed out a number of issues that would be worth taking into consideration, but the government did not call for a revision.

4.4 Measurement and Assessment

4.4.1 Examinations

One of the strongest traditions in Hungarian mathematics education is teaching based on solving well-selected problems. Many teachers design their lesson plans with the aid of problem books. This tendency is reinforced by the effect of the system of matriculation. Before 1973, all students finishing secondary school had to take a matriculation examination. Those who wanted to enter university had to sit for a test in given subjects. This two-tier system of examination was made easier in 1973 for those who wanted to take a college or university examination in mathematics. This was the first year in Hungary that students had the possibility to choose between two types of examinations in mathematics: taking either the matriculation examination of their secondary school or taking the entrance test of a university and having that considered as their matriculation test as well. This was only available for those who wanted to enter a university where mathematics was an entrance subject. For the rest of the students finishing secondary school, the mathematics matriculation examination was compulsory. With the scarce exception of a few years and a few types of secondary schools, mathematics has always been a compulsory subject for matriculation. This system, with minor changes, was in force until 2005. From the 1990s onward, a professional graduation committee with a ministerial commission worked to amend the graduation process to adjust it to the NCC and the frame curricula. In 2005, the university entrance examinations were abolished and a new two-tier graduation system was born, covering all subjects. The original intention of the matriculation committee was to prescribe an advanced level of examination for those entering a university. This was not realized in practice; some universities required the advanced level of the matriculation test, but usually entering a university could be achieved by taking the intermediate level matriculation test. The intermediate written examination was based on the knowledge contained in the basic curriculum. Tasks with relatively simple mathematical content appear in a context of problems chosen from everyday life. The content of the advanced level test is much more difficult, and the knowledge required to solve it goes beyond the requirements of the basic curriculum. In addition to the written examination, an oral examination has to be taken with a centrally appointed committee too.

There are no other centralized examinations apart from the final examination (matriculation or school exit examination). This is the first official measurement and assessment of knowledge in the Hungarian school system and is conducted at the end of the 12th grade. In Hungary, the final examination is a state examination and the certificate for it is an official document. The requirements of the exam (defined at two levels—intermediate and advanced) are given separately for each subtopic. Topics are the standard five: Thinking methods, sets, logic, combinatorics, graphs; Number theory and algebra; Functions and elements of analysis; Geometry, coordinate geometry, and trigonometry; Probability and statistics (Ministry of Human Capacities 2017).

Here are some sample problems from the intermediate-level matriculation examination in 2018:

- The fifth term of an arithmetic progression is 7, the eighth term is 1. Give the difference of this progression.
- Given that $32\log_8 x = \log_2 32$, is it true that $x > 32\,000$? Explain your answer.
- Calculate the perimeter and area of the six-sided polygon shown in the diagram. The lengths of the edges of the rectangular prism shown below are $AB = 63$ cm, $BC = 16$ cm, $BF = 72$ cm. Calculate the angle between the solid diagonal CE and the face $ABCD$.

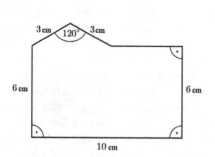
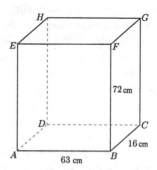

- Roll a fair dice twice and record the results (in the order of rolls), thereby obtaining a two-digit number. What is the probability that the number obtained will be divisible by 7? Show your work.

Sample problems from the advanced-level matriculation examination in 2018:

- (a) The angles of a planar quadrilateral (in degrees) form consecutive terms of a geometric progression. The common ratio of the progression is 3. Give the angles of this quadrilateral.
- (b) The angles of a convex polygon (in degrees) form consecutive terms of an arithmetic progression. The first term of the progression is 143, and the common difference is 2. Give the number of sides of this polygon.
- The bottom part of a circus tent is a prism with a regular 12-gon as a base. The top part is a pyramid, also with a regular 12-gon base, that fits onto the top of the prism. The length of the base edges is 5 m, the height of the bottom prism is 8 m, the height of the top pyramid is 3 m. During winter, the tent is heated with a number of (identical) heaters, each of which is rated to heat 200 m^3.

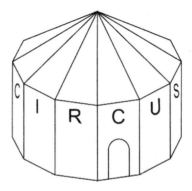

- What is the minimum number of such heaters required? Titi and Jeromos are two jugglers, working for the circus. At one point in their show they are tossing clubs to each other. Both jugglers are very skillful; they only miss 3 clubs out of 1000 on average (this could also be interpreted as a probability: the probability of missing a club is 0.003). In their new show, the two jugglers plan to catch clubs a total 72 times.
- What is the probability that they will miss no more than one club during their show? Round your answer to two decimal places.
- Ottó is arranging a class lottery game. Five numbers will be selected out of 1, 2, 3, 4, 5, 6, 7, 8, 9 and, accordingly, 5 numbers should be marked on each lottery ticket, too. (Shown below is a blank ticket and also one with a possible way of filling it.)

(a) András would like to get at least three of the winning numbers right with as few tickets as possible. What is the minimum number of tickets he should play with, in order to get at least three of the winning numbers right on at least one of his tickets for certain?
(b) Dóra and Zoli both fill one (valid) ticket each, randomly. What is the probability that exactly four numbers on their tickets will be the same?
(c) How many different ways are there to fill in this class lottery ticket, so that the product of the five numbers marked is divisible by 3780?

4.4.2 School Assessments

Following World War II, a summative assessment on a five-point scale (1–5) was typically used in Hungarian public education. The five-point summative assessment is still prevalent today in Hungarian education. The assessment of pupils' knowledge of mathematics is mostly done by means of written assignments or test sheets. Teachers select the problems from the teachers' handbooks or from their own collection.

Some types of schools have written/oral examinations, according to their own, pre-published examination requirements (valid at their own institution only) for certain subjects such as eighth and tenth grade mathematics, for example. The aim is to help children experience a test situation, learn how to relieve stress, improve self-esteem, and to help them decide on a career.

4.4.3 Entrance Examinations to Higher Primary and Secondary Education

In Hungary, children are required to attend kindergarten from the age of 5. Under legislation which took effect in 2015, at age 6 or 7 they must be enrolled in a primary school in their districts. However, it is also possible for parents to choose another school, for example, one that runs classes under a special or alternative pedagogical program. Primary schools that focus on sports, music, or language classes are particularly popular. For enrollment, schools use their own measuring tools to select suitable preschoolers.

Since 2004, secondary schooling has been organized centrally by the Educational Authority (Oktatási Hivatal). Some selected schools require tests for those entering grade nine. These students have to sit for a unified written entrance examination. Mathematics is a compulsory subject for all admission examinations:

> Students taking part in the entrance examination must solve a series of ten problems in 45 minutes. The primary objective of the written admission examination is to assess the basic skills and competences required for effective secondary school performance. Therefore, the worksheets are not the typical ones students are used to for the given subject. In addition to the application of well-practiced knowledge gained in the educational process, some problems are presented in a new context, in an unusual way, opening up space

for the mobilization of gained knowledge and the ability to apply imagination and creativity. (Oktatási Hivatal 2016a, preamble)

It is also common for selective schools to arrange entrance examinations for pupils in grades four and six (who will enter grades five and seven). These tests are also organized centrally by the Educational Authority.

4.4.4 Teachers' Assessment

During the socialist era, teachers regularly took part in in-service training and their work was observed by members of a supervisory board. The board often made their decisions from a political perspective. As the power of the socialist party weakened in the 1980s, this supervisory board lost its power and was eventually abolished. This also meant that the quality of teaching was not directly checked any longer, which resulted in in-service training losing its importance. In 2013, the government decided that schools formerly maintained by local governments would now be maintained by the state and, at the same time, it set up a career model for teachers (Eurydice 2013). The law mandated that a person had to hold a special license in order to teach. It also described the possible progress one could make as a teacher, as well as the potential salary. Teachers formerly were employed as civil servants, and with this law they lost that status.

4.4.5 Creating and Developing the Measuring and Evaluating Culture of Teachers

Basic ideas, different possibilities, and methods of measurement and assessment are being taught more and more as part of teacher education. Most of the professional knowledge required by the higher levels of the pedagogical career model can be achieved and practiced by becoming a measurement and evaluation expert. In addition to traditional summative evaluation, more and more people are using pre-knowledge or basic knowledge diagnostic evaluation, which allows comparison with the outcome of a new measurement following the completion of the learning unit. Developmental assessment built into the learning process is common; it can be a regular source of feedback from students on the progress they have achieved. In the first two grades of all primary schools, a developmental evaluation is used instead of the summative evaluation. The same evaluation is used in the first 4 years in some types of primary schools, where the end-of-year certificate also includes a personalized written assessment. The purpose of the developmental evaluation is similar to that of the formative evaluation, which is to give feedback based on the information collected multiple times during the learning process (Csépe 2018).

4.4.6 Some New Forms of Assessment

In this section, a few new forms of assessment in mathematics education are discussed.

Electronic Diagnostic Measurement System (eDia)

The Educational Theoretical Research Group of Szeged Scientific University launched a project called Development of Diagnostic Measurements in the early 2010s. This allows pupils and schools to try and to practice computer-based testing. The eDia system (Electronic Diagnostic Measurement System) contains approximately 20,000 multimedia-supported tasks in the fields of mathematics, science, and reading (Molnár and Csapó 2019), and the tests of the system allow pupils to prepare for future online measurements (www.edia.hu) without any pressure. This is important because PISA measurements are already done online and because a system for both the National Competency Measurement and for the final examination to go online is being developed. (The work is currently in an experimental stage; therefore, there is no publication available on this topic at the moment.)

eDia already studies the readiness of pre-school children for school. Its main goal is to evaluate pupils' development several times a year in three major fields of education (mathematics, science, and reading). To do this, experts have developed an electronic measurement and evaluation system, which measures children's skills and abilities in grades one through six, tracks their development individually, and reveals their learning difficulties. The implemented online task bank is enriched with images, sounds, animation, video, and varying types of responses (selection, clicking, coloring, moving, or rearranging) that make tasks more lifelike and enjoyable.

Alternative Assessment in Teaching Mathematics

Presentations and talks on alternative assessment methods are popular at national conferences and forums organized for mathematics teachers (such as the László Rátz Meetings, the Varga Tamás Methodology Days, and the online forum of mathematics teachers). One of the most popular topics of these presentations and talks is assessment based on gamification: "Gamification is the use of game design elements in non-game contexts." The idea comes from Deterding and his colleagues (Deterding et al. 2011) who, based on the results of a form filled in by hundreds of school students, pointed out that the anxiety triggered by the usual forms of assessment (writing, answering, marking) can hinder understanding and favoring the subject of mathematics. An evaluation method based on including playful activities, where different levels of performance can be accomplished with the help of online games (homework assignments, online searches for concepts, definitions, joint projects, etc.) appeared to be successful in a Hungarian experiment and attracted great interest among teachers (Barbarics 2018).

4.4.7 Large-Scale National and International Assessments Connected to Mathematics

There have been three regular, recurrent evaluations used in Hungarian education since the change in 1989. These evaluations cover the whole country and also measure the effectiveness of mathematics education: the National Competence Measurement, PISA, and TIMSS, which we will discuss in detail later in this subsection. However, since neither the tasks of the National Competency Measurement nor those of international assessment directly meet the requirements of NCC, these measurements cannot be taken as an official school evaluation of the students. In spite of this, the results of these national measurements provide more important information to schools and teachers (and not least of all for parents), compared to those "input measurements" (like the optional entrance examinations we mentioned before), for example, which have increased in the last one or two decades. All three measurements are organized by the Educational Authority.

National Competency Measurement in Mathematics Since 2001

In response to the PISA test, a national competence-based assessment was launched in 2001 for grades six, eight, and ten that was connected to the previously detailed frame curricula to inform authorities about the state of education. The areas of education measured are literacy and skill in using mathematical tools. The aim was to measure both pupils' ability to apply their knowledge in life and to use it for acquiring further knowledge. As well, the goal was to check whether they had the knowledge of the necessary tools essential to successful progress in their studies. The questions in the competency tests are not necessarily about meeting curriculum requirements, but about solving real-world problems. The aim of the measurement was to develop a common scale, independent of grades, along which the results of the pupils could be evaluated. Therefore, the tests for different grades included common tasks. These common tasks, or benchmarks, connect the scale and provide continuity through the grades. The National Competency Measurement provides important information on the state of education for the Educational Authority only. Teachers, pupils and parents are not informed of the results.

Students of grades five and nine of all schools took a National Competency Measurement test in 2001 for the first time in Hungary. It was given at the same time and under the same conditions. Students in grades six and eight were included in 2002, and students in grade ten joined in 2004 with a literacy and mathematics test.

The mathematics component of the National Competency Measurement fit into the international measurement evaluation trends, as they were based on a detailed content framework. Somewhat similarly to PISA tests, they measure how students can apply the knowledge they acquire in school in real, everyday life situations and in real-life contexts. The mathematical tools of the pupils are measured through lifelike problems in four content areas:

- amounts, measurements, numbers, and mathematical operations
- relations and functions
- shapes and coordination
- statistical indicators and probability.

For each content area, Hungarian experts identified three categories of thinking operations:

- factual knowledge and simple operations
- application and integration
- complex solutions and evaluation.

The worksheets were categorized on the basis of the test matrix for content areas and thinking operations: 55–65% of the tasks are multiple-choice and 35–45% are open-ended tasks (requiring a short answer or multi-step calculations with longer explanations). Tasks consist of one to four questions in an increasing order of difficulty, and the situations presented in the exercise are lifelike for the most part. The results are evaluated according to seven levels of mathematical skills students have acquired (Balázsi et al. 2014). These levels refer to the mathematical skills and representational level of the children. For example, level one refers to the ability to carry out basic operations, while level seven refers to abstract skills, such as the ability to reason.

Schools whose pupils achieve good results in the National Competence Measurement have higher prestige and have more student candidates to choose from.

The PISA Test

The achievement of our children in the 3-year PISA test has been at the center of interest of the Hungarian public since Hungary has first taken part in the PISA tests (Oktatási Hivatal 2017). In 2003 and 2012, mathematics was the focus of the measurement. Already by 2003, the test results of Hungarian students had caused an outrage in the press, and the mathematical performance, which was below the OECD average, gave the public reason to question the educational system. Changes in educational policy did not bring the long-wanted performance improvement in PISA tests: in 2012 and 2015 Hungary continued to score (477 points) significantly below the OECD average (500 points). And, even worse, this was 13 points below our 2009 average. (Between 2003 and 2009, the average result of Hungarian students did not change.)

Although there is no direct proportional relation between PISA results and a country's educational expenditures, the criticism is that the Hungarian budget had not spent enough money on education for years (Totyik 2016). One of the reasons for the weak scores of the Hungarian children lies in the great difference between students' level of knowledge. In 2012, 28.1% of Hungarian 15-year-olds did not reach level two out of the six, and the ratio of children not reaching level two in year 2015 was 27.7%. There were no significant changes in the proportion of those who performed at an excellent level between 2009 and 2012. Although the Hungarians were still proud of their excellent mathematicians, those (8.1%) who reached level five or six in 2015 were below the 10.7% OECD average (Balázsi et al. 2012, 2014).

The analysis of the results of PISA 2012 has highlighted the fact that our school system does not address the differences arising from the socio-cultural backgrounds of students. These differences greatly determine the differences between pupils' mathematical performance (Balázsi et al. 2012).

The PISA 2015 data also showed that in Hungary, students' performances were more homogeneous within schools than in OECD countries in general, but differences between different types of schools were higher. The 4, 6 and 8-year secondary schools had pupils of similar—more advantageous—socio-cultural backgrounds, while vocational secondary schools and vocational training schools enrolled students with less advantageous (even disadvantageous) socio-cultural backgrounds. There is an average of 90–95 points of difference in natural scientific, literacy, and mathematical knowledge between an 8-year secondary school student and a student from a vocational secondary school in Hungary. According to the different tests of the PISA study, the same difference can be found between the similar skills of vocational secondary school students and vocational training school students (Balázsi et al. 2014).

TIMSS

Hungarian students' results on the 4-year TIMSS tests organized by the IEA (International Association for the Evaluation of Educational Achievement) since 1995 are recorded and the results for the years 2007 to the present can be found on the webpage of the Educational Authority (Oktatási Hivatal 2016b). The TIMSS results of Hungarian students are much better than their corresponding results for PISA. In the 2015 TIMSS study, fourth graders (scoring 529 points) scored significantly above the 500 point average and eighth graders (scoring 514) were above the average as well (TIMSS 2016).

The difference between the results of these two measurements can be accounted for by the fact that the TIMSS mathematics tasks measure the acquisition of acquired knowledge and not its application in real life. However, Hungarian mathematics education traditionally supports the study of the exact structure of the subject and the pure theoretical solution of problems, devoting less attention to applications. The tasks of TIMSS are those that Hungarian children are well acquainted with from their lessons, whereas the types of tasks on the PISA are not practiced at their schools. The preparation of Hungarian mathematics teachers, the national curricula, the frame curricula, and the system for final examination require different competencies than the PISA (Balázsi et al. 2012).

It is worth examining the fourth-grade TIMSS results of 2015. Three content areas—numbers, geometric shapes, and data representation—as well as cognitive areas such as knowledge, application, and interpretation, have been the developmental scope of the four-level evaluation. The average score on geometric shapes for Hungarian pupils of grade four is 536 and the average score for numbers is 531 (meaning, they scored 7 and 2 points above the 529 average), while the data representation, with an average score of 513 points, is 17 points below the average—showing that the latter topic is much less represented in Hungarian mathematics lessons than the otherwise strongly represented geometry. A similar interesting fact can be observed in the study of cognitive areas. The score of knowledge is 532, that of interpretation is 529, but the score for application is only 526. So at this age, the children still possess the knowledge covered by the curriculum, and they even understand it, but their application of the knowledge is less successful.

Unfortunately, the good results of TIMSS have less of an impact in the Hungarian media than the weak scores on PISA. These weak results have greatly damaged the public perception of mathematics education in Hungary over the past two decades. There is a general belief that teachers are incompetent and that the content of the curriculum is not good enough or is too much. It is considered unnecessary to stuff the heads of children with a great deal of mathematics (HVG 2016; Balla 2016; Lannert 2014; Tutiblog 2016; Prókai 2013).

4.5 Traditions and Innovative Elements in Gifted Education and Mathematics Competitions in Hungary After the Political Changes

The traditions of Hungarian mathematical talent management dating back over a hundred years are still present. Actually, the most successful branch of mathematics teaching in Hungary has always been gifted education (Vásárhelyi 2002). György Pólya once stated in an interview to Wieschenberg Arvai (1984) that "...mathematics is the cheapest science" (quoted by Frank 2011), which can at least partly explain why mathematics could gain such a momentum in its development, which led to results that were influential even internationally. And this occurred in the context of the predominantly agricultural and poor Hungary of the nineteenth and even of the first half of the twentieth century (Csapó 1991).

The social and economic changes of 1980s also helped to expand the frameworks of mathematical talent development. Again the shifts in gifted education were not absolute, and many old forms of work are still in use. For example, *KöMaL* still plays a very important role in the gifted student identification and development processes. Also, a number of special classes for mathematics in secondary education (in gimnáziums) function very effectively in and out of the capital of the country. Extracurricular programs for the gifted and highly motivated students and many different types of competitions for each age are still very common in Hungary (Connelly 2010a; Connelly Stockton 2010).

4.5.1 Some Methods and Programs of Gifted Mathematics Education Within School Frameworks

Since 1962, when the very first class with a special curriculum for mathematically gifted students was launched in Hungary, the system has become established in a relatively wide portion of the country. There are approximately a dozen of these special classes in Hungary today. The number of these special classes seems to be enough if we consider how many talented students in mathematics should be served by the special gifted education program in Hungary. It also explains the fact that in Hungary—because of demographic reasons—it was the system of specialized curriculum classes and not specialized schools that became prevalent, as opposed to what happened in this respect in the Soviet Union or the US.

It is worth mentioning that the teachers of the special classes have long been acting in compliance with a tradition that was established simultaneously with the launch of the first special classes. These teachers are intensively involved in the maintenance and nurturing of a nationwide professional network with one another and their university colleagues; they teach one another, share their experiences, and join common development and research projects. However, even though the students involved in these special classes obtain a certain part of university-level curriculum during their secondary school years, they are not permitted to gain university credits at this age for their achievements. Besides, it is worth mentioning that, unlike in certain East Asian countries (Sun and Tian 2016; Kim 2016), there is no established institutionalized preparation for the Mathematical Olympiads. However, a number of the students attending the special mathematics classes do participate in preparation workshops organized at the national level. It is not only the special mathematics classes where students can study mathematics at an advanced level and with a higher than average number of lessons. There are also a relatively high number of secondary school classes specializing in mathematics and physics, or another natural science subject (Ádám et al. 2008) where the students receive advanced and intensive mathematics education in five to six lessons a week as well as an intensive education in the science. In a number of secondary schools, the last two grades, the 11th and the 12th, give their students opportunities to specialize in mathematics with a view to preparing for entering university programs in the natural sciences or engineering. Certain schools indulge in innovative, alternative methodology experiments. For example, students of the Radnóti Miklós Secondary School in Budapest (ages 12–18) study mathematics in varying depths and ways. Those who are successful in competitions and are aged 12–16 can study in extracurricular workshops where participation requires a personal invitation based on achievements. Those aged 16–18 can deepen their knowledge within the "Táltos" ("Miraculous") mathematics specialization, an elective subject, developed to suit the needs of the highly gifted.

The prevalent form of support for the gifted within the school context is the system of extracurricular courses, which is present in nearly every primary and secondary school in Hungary (Páskuné Kiss 2014). These are free opportunities to attend classes after the compulsory lessons: almost every school offers such opportunities for their gifted students. One of their main advantages lies in their extracurricular nature, because they allow more freedom both in terms of their content and character than compulsory lessons can ever allow. Furthermore, as opposed to the strictly age-dependent grade system of regular Hungarian education, gifted and interested students from a wide age range may gather in an extracurricular course (Connelly 2010a; Connelly Stockton 2010). The requirements students must meet in order to gain entrance in these extracurricular courses can vary based on the particular class, but it is a common feature of all of them that the students do not get marks for their performance, hence the relatively stress-free atmosphere the students can learn and develop in.

Along with the above, the organization of lessons based on the individual capacities of students is a common form of nurturing talent within the school framework (Balogh 2012; Radnóti 2004). In most teachers' resource books and within course book tasks, there are ones that are marked as intended "for those interested" (Csahóczi et al. 2013), which are meant to help develop the students of a higher-than-average capacity and/or performance in the given subject.

4.5.2 Out-of-School Gifted Education

The KöMaL: Competition and Journal

KöMaL is one of the most important means of recognizing and educating talented children even today. The volumes of the journal, founded in 1893, have appeared in every month of every year—apart from the years during the two wars—posing very interesting problems (often outside the regular school material) to solve, and then publishing the best solutions of the contestants in later volumes. The creative activity of problem solving and solution writing and the reading of mathematical articles that appear in the journal inspired many students who later became famous Hungarian mathematicians. It has encouraged mathematical research for the last 125 years.

A problem from 1893:
Solve the following system of linear equations

$$x + y + z = 1 \tag{3.1}$$

$$ax + by + cz = h \tag{3.2}$$

$$a^2 x + b^2 y + c^2 z = h^2 \tag{3.3}$$

A problem from 2019:
After the elections in Nowhereland, there are $50 < n < 100$ representatives in the parliament, all from a single party called the Blue Party. (The Blue Party has a single president.) According to the law, a party in the parliament may be divided into two parties as long as the following conditions are met:

- The president of the old party is not allowed to become a member of the newly formed parties. His or her parliament mandate will terminate, thereby reducing the total number of representatives.

- Every other member may decide which new party to join.
- Each of the new parties must have at least one member among the representatives.
- Each of the new parties must elect a president from their representatives.

If at least one such splitting of a party results in all parties in the parliament having the same number of members, the parliament will be dissolved. What should be the value of n so that this could never happen?

The mathematics competition and the journal of *KöMaL*, which both started at the end of the nineteenth century, served as the basis for the education of gifted students in mathematics. Several other competitions and journals have been launched since. *Abacus*, for example, is a journal similar to *KöMaL* for upper primary pupils. During the socialist era, the operation of the journal *KöMaL* was financed by the Ministry of Education. For the last 30 years, the journal has been funded by a form of a foundation that is supported by companies. Also, since 2008, grants that are earned, with varying success, from the 20-year-long National Talent Program (Parliamentary Resolution 2008) help the journal to be published regularly. *KöMaL* is still very popular, though the number of published copies dropped from 4000 to 2000 (partly for demographic reasons). Teachers and parents who were contestants of the journal find its development of today's pupils very important. Since 1989, the journal has been publishing articles and problems to solve on the topics of mathematics, physics, and information science for several age groups and at different levels of difficulties. Today there are 22 categories of contests for children to measure up their knowledge. The solutions are typically not hand-written or sent to the editors by "snail mail," but are sent online via an electronic workbook. After checking the solutions, the individual results and points received are available on the webpage of *KöMaL*, where the problems are published. Thus, anybody can take part in the competitions for free. This is considered a great achievement since most competitions in Hungary require a registration fee.

The journal's corresponding problem-solving contests motivate a high number of gifted secondary school students today. *KöMaL* helps to identify gifted young people early, and they are motivated to undertake further studies. The journal forms an essential part of nurturing the gifted in mathematics in Hungary.

4.5.3 Mathematics Competitions

Competitions in general play a pivotal role in supporting gifted students in Hungary (Connelly 2010a). Hungary has been participating in the International Mathematical Olympiads since the beginning, which was during the socialist era of the country, and it has never ceased to take part since then.

After many social and political changes, new types of competitions emerged, especially for primary school students, including several for which participation fees had to be paid. Some better-off schools took over the fees for their students.

The finals of the multi-round, national competitions were often organized in a multi-day camp with many activities prepared for the participants. Not only the winners, but also their teachers and schools would be generously rewarded. The teachers of students who achieved good results at competitions have in the past won the recognition of the professional public and have received professional awards (the László Rátz award is one of these; it was founded in 2000 by three international and Hungarian enterprises: Ericsson Hungary Ltd., Graphisoft Ltd., and Richter Gedeon Plc. For more information on the award see Alapítvány a Magyar Természettudományos Oktatásért 2019).

There are more than a hundred mathematical competitions in Hungary, most of them are local, school-organized, regional, and countywide competitions. About 10 of these competitions are nationwide ones (Ács et al. 2010).

The competition tasks are usually compiled by excellent and enthusiastic teachers as well as volunteering mathematicians who not only strive to measure higher-than-average knowledge in competitions, but who also try to create problems not necessarily connected to a curriculum. At the competitions, one can find problems that require unusual types of approaches, have interesting wording, and are of types that are less likely to come up in school mathematics (such as combinatorics, graph theory, number theory, or spatial geometry). Competitions have had a great impact on education as well. Problems from competitions are widespread. They enrich not only the knowledge of students, but also teachers. Textbooks and problem books have incorporated several problems from previous competitions.

4.5.4 Mathematical Camps for the Gifted

Some of the essential and most effective parts of gifted education in Hungary are the activities for nurturing talent that are carried out in mathematical camps. Although there were mathematical camps for the gifted back in socialist Hungary (Connelly 2010b), which were held when schools dismissed for the summer or other holidays, the launch date of the most important such camps coincided with the year of the commencement of the regime change, 1989. The most important camps can be tied to one person, Lajos Pósa, who was also a student of the very first special mathematical class of Fazekas Gymnasium. At the age of 14, he studied there while he was co-authoring works with Pál Erdős, the world-famous mathematician (Gordon Győri and Juhász 2017). Nearly each gifted student with notable national or even international achievements in mathematics has participated in his camps since 1989; these camps are truly pivotal to gifted education. Through the decades, Pósa has developed an exceptionally complex and effective gifted education pedagogical system to nurture talent, which, unfortunately, cannot be detailed within the framework of this chapter (see Gordon Győri and Juhász 2017). The method of teaching in the camps is under further development even today; for example, more and more attention is being paid to the scientific research and developmental work. Lajos Pósa and Péter Juhász—of the Alfréd Rényi Institute of Mathematics operating within the Hungarian Academy of Sciences ("MTA Rényi Alfréd Matematikai Kutatóintézet")—are the leaders of this work.

4.5.5 Medve Matek—Bear Mathematics: A Camp, Competition, and Miscellaneous Program

One of the most important initiatives of recent years is the "Medve Matek" ("Bear Mathematics") (Fried and Szabó 2018; Mécs 2015).

The story of Medve Matek goes back to 1999, meaning that its history started right after major political changes (Mathematics Connects Us 2019). In 1999, Gábor Kis, an excellent secondary school teacher, organized the very first outdoor mathematics competition (in a program called "Challenge") in Hungary. In 2013, a group of his former students—former Medve Matek program members—founded a non-profit association called the Mathematics Connects Us Association and through this organization they upgraded the program to a much higher scale.

The program contains three types of subprograms:

- Medve Outdoor Mathematics Challenge
- Medve Summer Camps and Mathematics Weekends
- Medve Experience Days.

As its name also reflects, Medve Outdoor Mathematics Challenge programs are organized in parks, low mountains, and other spots in nature where the organizers set up stations with challenging mathematical tasks. The children who take part in the program are organized into small groups, and the groups walk from station to station where they try to solve the given tasks. Their routes are algorithmically organized: if a group figures out the proper answer to a mathematical problem, it can continue on their way in a certain direction while another group which fails continues in a different direction, so that they reach a different station.

The Medve Summer Camps offer a special social space for the children with its own fictitious time zone, currency, and other traits. The most important parts of the camps are their laid-back mathematics classes and games. There are hikes and other physical and social activities as well.

The Medve Experience Days offer special programs for a mixed population: for teachers who are not among the organizers of Medve programs but are interested in them, or for parents who want to learn more about their children's Medve programs. In Hungary, it is quite unusual that specialists (the Medve team members), other teachers (non-Medve team members), parents, and children work together on an activity and can learn from each other, share ideas, and get involved in the activities of the program.

While the organizers of Medve programs try to involve as many students between the ages of 10 and 18 as they can, during the programs they also try to identify students who are the most interested in mathematics and who are able to work at the highest level of mathematical problem solving. The programs are held throughout the whole year and devote special attention to developing children from disadvantaged family backgrounds and/or who are living in places situated on the periphery in terms of access to gifted education. The organizers help these children participate in mathematically gifted education programs for free or for a reduced fee (Medve Matek 2019). The outdoor programs that offer a unique combination of open-air activities and hiking with solving mathematics tasks are incredibly popular. For example, in 2017, more than 10,000 children (and teachers and parents) were involved in these programs (Mathematics Connects Us 2019).

Medve Matek has a remarkable ideology and practice: they view mathematics as an intellectual and social experience. They view it as something in which the participants can develop intellectually in a healthy environment in nature. Here senior experts in gifted education in mathematics, interested teachers, parents, and children can stimulate and develop each other intellectually, emotionally, and socially.

4.5.6 Erdős Pál Camp and Gifted Support Program

One of the prominent events of mathematics in the post-socialist era was the launch of the Pál Erdős Gifted Support Program in 2001 (named after Pál Erdős, the world-famous Hungarian mathematician) (Connelly 2010b). The support center was set up within one of the universities situated in the country, the Technical and Informatics Faculty of the University of Pannonia in Veszprém. It was set up because of the founders "realizing the fact that decrease in interest towards the technical and natural sciences imposes a serious threat to the supply of researchers, highly qualified engineers, and educators of outstanding qualifications" in Hungary (Erdős Pál Matematikai Tehetséggondozó Program 2019).

The program—in a rather unusual manner given its Hungarian context—included children below the primary school age while also undertaking the development of scientists in mathematics as well. The program covers all age groups between small children and adults, but focuses mostly on students between 14 and 18 years of age. In 1 year, altogether 200–250 students attended the courses of the program, and most of them were not from Budapest, the capital of the country. The core operations of the program take the form of weekend training sessions five times a year. One class of the program takes 90 min, and in one weekend the Erdős program offers seven classes for each child. So a student can attend altogether 35 classes over the course of five weekends and in this way throughout the whole year he or she can spend more than 3000 min on extra education in mathematics.

Most of the teachers of these programs run short-term courses, so the students can meet with a wide variety of mathematical topics during the programs (Connelly 2010b), from functions to number theory, from algebra to geometry. However, mathematical problem-solving strategies are at the core of the programs along with creative problem solving and critical thinking development. Also, the teachers prepare their students for different types of national and international contests in mathematics. From a methodological point of view, it is notable that beyond typical whole-group work and individual practice, the teachers offer different types of small group work to the students. In course of this training, apart from an intensive mathematical training and intensive competition preparation, the participants receive personality and co-operation skills development with career counseling as well. As part of career counseling, the participants have opportunities to meet successful scientists and other professionals who share details of their lives and careers with the participants.

4.5.7 Opportunities in Gifted Education in the Future

Although young people gifted in mathematics today have several opportunities to develop their talent, whether in special mathematics classes, extracurricular workshop courses, mathematics camps, weekend trainings, competitions, or other forms of education, there are certain areas in which Hungarian gifted mathematics education should strengthen itself. We wish to highlight a few of them here.

First, even though there have been encouraging initiatives in this field, for example, the MaTech competitions launched in 2018 under the patronage of the world famous mathematician László Lovász (MaTech 2019), and the complex competitions combining mathematics, physics, and informatics organized in several schools, the ties between mathematics and other related sciences should still be strengthened in line with the main trend present in Hungary today.

Second, the development of the gifted in mathematics could be merged with real-life professional issues and the challenges of today's Hungary. It is similarly essential to find ways and methods that could help the youth of the language/ethnic minorities of Hungary, especially those of the Roma/Gypsy population, which accounts for 8–10% of the Hungarian population. They need assistance in order to have better chances to become high achievers in terms of mathematics performance so that they can have the chance to gain access to gifted mathematics education (Vecseiné Munkácsy 2011). This is a very complex and difficult task, in which nearly no success has been achieved to date, whether considering the socialist era or the pursuant decades afterward. As much as there are several social-historical differences, the Hungarian gifted mathematics education could learn a lot in this respect from other countries (see, for example, Walker 2007 or the Tribal Mensa Nurturing Programme established by Narayan Raghvendra Desai from India (Roy 2017; Teklovics and Gordon Győri 2016)). The relative underdevelopment of mentoring programs in mathematics and other fields alike, despite certain initiatives such as the János Arany Gifted Education Program (Gyarmathy 2013), cannot be viewed independently from the above. The number of properly qualified and trained experts is very low in this field.

Finally, the opportunities available for girls to unfold their talent in mathematics has never been low, neither before, nor after the change of regime; however, the situation has not shown significant improvement either. The higher the achievement segments we observe, the lower the ratio of girls/women we see, and there has not been any significant positive shift for a long time. Gifted mathematics education and the society as a whole certainly have a lot to do to address this issue and achieve real improvement in the ratio of girls among the highest achievers, no matter how successful those Hungarian girls were who participated in the European Girls' Mathematical Olympiad of 2012 (EGMO 2019).

4.6 Teacher Education and the International Connections of Teachers and Teacher Education Institutions

4.6.1 Teacher Education from 1989 to Today

As previously mentioned, Hungarian teacher education was subdivided into three parts in the 1950s according to the age grouping of the school system: lower primary (elementary), upper primary, and secondary (grades 9–12, or, additionally, 5–8 as well, if needed). Despite the fact that the social system of Hungary was changed in 1989, the essential changes in teacher education started only in 2003. This was due to monetary restrictions, because after the change in 1989, there was an economic crisis in Hungary, and the government tried to find ways to reduce expenses. The parallel nature of the training of upper primary teachers at teacher education colleges and university teacher education was discussed and understood; however, the significant differences were not taken into account, so teacher education colleges were closed down in 2003. This was an unfortunate decision that has an effect on teacher education today and will have more effects in the future. Thus, a huge change in the system of teacher training began with this decision.

The institutions for lower primary teacher education were merged into universities as independent faculties. They were allowed to keep their previous forms of training but were forced to reduce the content of their teaching.

The departments of the upper primary teacher education colleges were forced to work together with the corresponding departments for training teachers for secondary schools. In some subjects, it was not such a hard task, since upper primary training was not much different from secondary training. However, in mathematics, due to the difference between how 10–14 and 14–18-year-old pupils perceive mathematics, teacher education has to prepare teachers for different mathematical knowledge and different teaching techniques. It took quite a lot of time and a lot of effort to try to adjust the two levels of training, so much so, that the curriculum of teacher education is still under revision. Another serious problem was (and still is) that the three substances of teacher education were split; the mathematical content was taught by theoretical mathematicians who were not acquainted with education, the didactical background was taught by psychologists who were not acquainted with mathematics, and the didactical content was left to be taught independently of both mathematical and psychological content.

The Bologna agreement from 2005 deteriorated teacher education even further. It subdivided the training into two levels; however, teacher education was one unit and could not be terminated after 3 years of study. Those who only finished 3 years of teaching did not receive any kind of a degree and most certainly did not receive permission to teach, which they would have needed 5 years of study for. The new, unfamiliar structure discouraged secondary school students from becoming teachers, and the number of students entering into teacher education dramatically dropped.

The Curriculum for Teacher Training

The curriculum for lower and upper primary teacher education followed the changes made after 1978 and, as a matter of fact, many of the educators of teacher education colleges took part one way or another in the CMEE of Varga. Contrary to the secondary school teacher education system, which still based its teacher education mostly on scientific content, the curriculum of both lower and upper primary teacher education colleges contained not only a relatively high level of mathematics, but also stressed understandability and teachability. This type of education became popular and was acknowledged outside Hungary as well.

The Structure of Teacher Training

Before 2003, teacher education students had compulsory subjects and had the chance to pick some optional subjects. The subjects were generally taught utilizing both lecture and practice lessons. Teaching practice (pre-service) training has always been included in the teacher education program at all levels, and it usually takes place at one of the schools assigned to the university. Teachers of lower primary classes teach all subjects, while teachers of upper elementary and secondary classes usually have a degree for teaching two subjects.

The credit system was introduced in Hungary around 2000 (only at the university level and only gradually from university to university). The credit system itself did not bring a significant change and only some of the compulsory subjects became optional as a result of its implementation. The structure of teacher education was changed drastically with the introduction of the Bologna agreement, which caused the number of class hours to drop significantly.

In 2013, the government agreed to abolish the Bologna system in teacher training, but upper primary training did not regain its independence.

The Program for Lower Primary Teacher Education

Teachers of grades one to four were allowed to teach up to grade six if they completed some additional courses. They were obliged to pick at least one subject for additional preparation. The training takes 4 years (eight semesters) now, including the one semester of preparation for the field practice and a 6-week field practice. The number of clock hours for the programs have dropped by approximately 20% over the last 30 years.

The Program Before 2008

The compulsory courses as of 2006 were Elementary Mathematics (problem-solving techniques); Foundations of Mathematics (two semesters); Elementary Algebra (two semesters); Pedagogy of Mathematics Education (four semesters);

Geometry (two semesters); Combinatorics, Probability, and Statistics; and Functions. Optional courses concerned Soroban, Developmental Mathematics, Difficulties in Concept Forming, Tools in Mathematics Education, Introduction into the Practice of Teaching, and Interesting Mathematical Historical Facts.

The Program in 2018

The compulsory courses offered in 2018 were Foundations of Mathematics (two semesters), Teaching Number Sets and their expansion, Thinking Methods, Geometry and the Pedagogy of Its Teaching.

Those who want to get an additional degree in order to teach in grades five and six take extra compulsory courses in mathematics. This education includes subjects like Foundations of Mathematical Concepts, Elementary Mathematics (a problem-solving course), Functions, Synthesizing Seminar, Number Theory, Elementary Algebra, Geometry (two semesters), Teaching of Number Sets and Operations (two semesters), Teaching Thinking Methods (two semesters), Combinatorics, and Probability and Statistics.

Apart from the compulsory courses (mentioned above), the following optional courses are offered to them: Teaching Practice, Mathematics and Its Pedagogy, Foundations of Mathematical Concepts, Teaching Number Sets and Their expansion, and Pedagogy of Teaching Geometry.

The content of this program did not change much, but the structure did. The two-level training (bachelor and master levels) was never introduced for lower primary teacher education.

Upper Primary Teacher Education Until 2003

This type of education started in the 1950s and kept its form until 2003. The education lasted for eight semesters (4 years) and included practice in a school assigned to the university. Upper primary teachers had permission to teach up to grade ten; however, it did not often happen. Students in this education program had to learn somewhat less mathematical content than those in secondary training, but with the same mathematical precision.

The compulsory subjects in mathematics for the training in 2003 were Mathematical Analysis (four semesters); Geometry (three semesters); Algebra and Number Theory (three semesters); Probability, Elementary Mathematics, and Methodology (six semesters, practical lessons only); Numerical Methods (two semesters, practical lessons only); Teaching Mathematics (two semesters); and Logic and Set Theory. The students had to pick four optional subjects, which could change from year to year, for example, Mathematics in English, Studies of Curricula and Tools of Teaching, and History of Mathematics.

Upper Primary and Secondary Teacher Education from 2003

A committee studied the similarities between upper primary and secondary training, and while there were a lot of similarities, especially in the names of the subjects, they did not find significant differences between the two types of education. The duration was different (ten semesters for secondary teachers and eight for upper elementary teachers), and the contents of the courses were significantly different.

Due partly to budget cuts and partly to political battles, they decided that the two systems of trainings should be combined in 2003, at the time when the credit system was introduced.

The two subjects that students wanted to teach were categorized as their major and minor. They studied the subjects connected to their major in their BA course and those connecting to their minor during their MA course. The BA training was practically parallel to the training of theoretical and practical mathematicians; it was almost like a BSc program.

The mathematical subjects were taught at the faculty of sciences and pedagogy- and psychology-related subjects at the faculty of education and psychology. The methodology content (what to teach at a given age group and how to teach it) was drastically reduced. For those who took mathematics as their major, the compulsory courses of the six semesters of BA/BSc (in 2003) included Algebra (two semesters), Mathematical Analysis (three semesters), Geometry (three semesters), Number Theory, Combinatorial Mathematics, Introduction into School Practice (in the sixth and seventh semesters), Elementary Mathematics (problem solving) (three semesters), Foundations of Mathematics, Numerical Analysis, and Probability. Optional courses were Algebra (third and fourth semesters), Geometry (fourth semester), Experimental Mathematics on Computers, and Complex Functions.

The duration of the BA program was 3 years (six semesters) and the duration of the MA program was 2 years.

The program failed for several reasons. The most important reason was that at the BA/BSc level, the mathematics courses were taught by distinguished mathematicians, sometimes members of the Hungarian Academy of Sciences. This level might have been right for prospective mathematicians, but for students, who otherwise would have become high-quality teachers, it was far too high and many students failed. As a consequence, Hungary today suffers from a lack of teachers.

In 2015, teacher education was partially separated from the education of mathematicians, the two-level (BA/MA) Bologna-style education system was abolished, and teacher education was partially split for prospective teachers of the upper primary and secondary levels.

At the moment, the training for upper primary teachers takes 10 semesters; secondary teachers must study for 12 semesters, and an additional 1 year of practice has to follow. The system is still changing because there continue to be serious discussions about it.

The compulsory courses for the two training systems are Mathematical Analysis (five/seven semesters), Algebra and Number Theory (four semesters), Geometry (five/six semesters), Combinatorial Mathematics (three/four semesters), Problem

Solving and Elementary Mathematics (five/six semesters), Informatics, Probability, and Statistics (one/two semesters), Teaching Mathematics (four semesters), and Seminar on the Practice of Teaching (three semesters).

4.6.2 International Connections

In the years following the change in 1989, international relations were greatly expanded, which also had an impact on education. A number of grants and projects from many Western countries (TEMPUS, SOCRATES, ERASMUS, etc.) helped teachers to gain experience abroad and to make it possible for students to have more experience by traveling. These grants also contributed to the development of the assets of educational institutions. From the second half of the 1990s, education management launched a number of different development projects and application opportunities to spread the use of IT tools in schools.

The Teacher Education College of Budapest had connections with American and Western European institutions and showed its education system to several visitors from these places. One of the projects of the Teacher Education College of Budapest METE (Mathematics Education Traditions of Europe) was based on such connections. The METE research was a comparative study (Andrews et al. 2004) of mathematics education involving five countries (Great Britain, Belgium, Finland, Hungary, and Spain). The study compared the teaching practice of five different countries in several ways; the teaching of the same topic to the same age group was observed and compared. Despite the fact that Finnish data was not available, it can be stated—and this confirms the methodological ideas of Tamás Varga's experiments—that there was significantly more interactions (exploring, coaching, sharing, questioning, explaining) going on in Hungarian lessons than in the lessons of the other four countries (Török 2006).

4.7 The Popularity of Mathematics and Mathematics Education as a Subject in Hungarian Society in the Post-Socialist Era

Looking back some 200 years, in Hungary, mathematics traditionally was highly respected among school subjects. Despite the scarce research on the issue, the authors of the present chapter believe that there are many signs suggesting the deterioration of the position of mathematics, both the subject and discipline, in Hungarian society since the socialist period. During the socialist era, before the change of the political system, high-level mathematical knowledge confirmed by a degree provided the opportunity for social mobility. Today, this type of respect has been reduced because students can see several examples of successful people who have little mathematical achievements. However, public opinion still recognizes the respect of the scientific community that is acquired by Hungarian mathematicians, and that has an influence on school mathematics as well.

All surveys conducted in Hungary after the social and political systems change of 1989 indicate that the situation of mathematics as a school subject is rather troubled. Data from IEA surveys is available even from the socialist era. In the early 1970s, mathematics was among the moderately popular subjects (Ballér 1973; quoted in Csapó 2000). Measurements dating from after 1989 show the continuous decline of its popularity among students as they progress through the grades (Bánfi 1999; Csapó 2000). Children like mathematics lessons in the early years of school. However, this positive feeling diminishes more and more in the upper classes. In many cases, it is replaced by a fear of mathematics and mathematics lessons (Nótin et al. 2012). As established by Csapó based on the TIMSS data, this is in line with international trends, although this phenomenon is stronger in Hungary than in other countries. Csapó's own analyses revealed that the popularity of mathematics was below the popularity average of all other subjects for students aged between 10 and 18, with girls and boys showing the same patterns (Csapó 2000). However, the growing unpopularity of mathematics is not a unique phenomenon: it correlates with similar trends in physics, chemistry, biology, and geography. The alarming point is that the unfavorable situation of mathematics and the natural science subjects is concurrent in the past one to two decades with the drastic decrease in the number of students applying for teacher education programs in these fields. Almost a decade after Csapó's nationally representative research, regional research by Csíkos (2012) found identical trends in the popularity of mathematics among students. This is partly explained by the emergence of a multitude of financially more remunerative professional options in Hungary after the systems change of 1989—such as in IT in the first place.

The data of Szénay (2009a) indicates that, although the popularity of mathematics as a subject is not satisfactory in general, there are great differences by school type (e.g., whether a secondary school student studies at a general or a vocational secondary school), and the subject-specific performance of students. Somfai (2009), on the other hand, found that mathematics teachers are of the opinion that parents consider mathematics much more important than the students themselves, so there is also a generation gap present. Laczkovich (2016) concluded that the structural problems of mathematics curricula may have contributed to the problems in mathematics teaching and the lack of popularity of the subject. He noted the non-spiral subject structure, the "too early, too high" level of abstraction, and inadequate solutions to making mathematics teaching continuous between the specific school levels. Szász (2006) found that differentiation in mathematics classes is also a problem. The high number of primary and secondary school students in need of shadow education (for-profit private supplementary tutoring) in mathematics points to both the importance of the subject and the problems of teaching it (Szénay 2009b).

Although no relevant research data is available in Hungary, the authors of this chapter are of the opinion that the general social perception of mathematics has deteriorated since the political system changed. In the socialist era, mathematics education was an essential component of the pedagogical and social ideal of the cultured individual. Today, social and financial success seems much more attractive to young people. Under socialism, mathematics was one of the few escape routes toward the West from a country open exclusively toward the eastern bloc normally. Today, in our new political system, the situation is, of course, radically different. In 1966, in the finals of the competition "What are you good at?" (Ki miben tudós?) broadcast on Hungarian television, 18-year-old László Lovász and Lajos Pósa tested their knowledge against each other (both have been mentioned in this chapter). Millions watched the program in a country of ten million, and the tasks of the quiz show were also published in a popular book (Fried et al. 1968). True, there was only a single TV channel in Hungary at the time. However, today in the age of commercial TV and the internet, such social interest would be inconceivable in Hungary.

Interestingly, when mathematics is a topic of social discourse, whether in the context of scientific achievements or any other regard, in papers or in chat rooms, the issue of mathematics teaching is always raised. People typically discuss where they studied, who their mathematics teachers were, and whether learning mathematics at school was a positive or a negative experience. This further underlines the importance of mathematics teaching at school and its relevant responsibilities.

4.8 Deterministic Features of Hungarian School Mathematics—A Long-Lasting Dilemma of Teaching Theoretical and Applicable Mathematics

There are, of course, several reasons behind the trend of declining popularity of mathematics as a science and as a school subject, but as the authors of this chapter believe, this is crucially related to the content of mathematics teaching.

Hungarian mathematics education has been struggling for a long time to decide how it can make the material digestible and even lovable to those who are less interested, are less skilled, or have fewer mathematical abilities without the loss of precise mathematical content. Hungarian mathematics educators believe that the problems of the PISA test are not well-formulated—mathematical content is not sufficiently precise. The struggle for and against the precision and "purity" of mathematics over the last 30 years can be observed very well in the changes of the NCCs. Even today, the old debate about the appearance (and the ratio) of theoretical to applicable mathematics is still going on. The PISA measurements, launched in 2001, brought this problem to the forefront again. The application types of problems that arise from everyday situations that the PISA test uses are less precise than what Hungarian students are used to. Unfortunately, a very high percentage of participating Hungarian students achieve very poor results (Balázsi et al. 2012, 2014).

Assessments have shown that the great difference in performance is related to the social background of students—poor results were found in poorer, economically less-developed regions of the country. After the political change, with the emergence of a market economy, many people lost their previous livelihoods, unemployment increased, and some areas lagged economically. In such places, the quality of education deteriorated, the demand for knowledge dwindled, and students who wanted to continue their studies left and went to better schools in the cities. The gap between poor and wealthy people increased a great deal and, at the same time, a growing discrepancy between school performance and mathematical knowledge level was found. This problem, of course, cannot be solved by mathematics education methods only. Also, a student's mathematical self-efficacy in Hungary has a significant relationship to their family SES (Csüllög et al. 2014).

The PISA measurements and the resulting national competency measurements helped to emphasize the importance of applied mathematics in the NCC and in the frame curriculum. The contradiction between pure mathematics and applied mathematics in school mathematics education is not only a problem related to curriculum development but arises also from the nature of mathematics itself. This is an important question for mathematics but also has been a crucial point of argument in the society of mathematicians and mathematics teachers in Hungary. The field of applications for mathematics is very wide but understanding what kind of mathematical knowledge can be applied in a given area usually requires a very deep and thorough mathematical knowledge that can only be acquired at a higher level of study. Of course, basic mathematical knowledge may be sufficient for simple everyday computational problems (shopping, small measurements, etc.). Therefore, it is easy to teach the children of lower grades solve problems they might meet in everyday life and in their own environment. The children of higher grades are interested in the phenomena of the modern world around us, even in the most serious issues, such as the use of the internet, space exploration, artificial intelligence, the problems that need to be solved in medicine, future research, sustainable development, and so on. Surely they would be interested in mathematics if they could solve problems in connection with these topics by applying school-level mathematics. Researchers of school mathematics have long been trying to find answers to these teaching challenges.

In the twentieth century, it was thought that teaching relatively sophisticated mathematics was primarily the task of secondary school; children in lower grades should only be prepared for simple everyday calculation and measurement problems. This approach changed in the second half of the twentieth century as a result of international and national modernization reforms. In the NCC, based on the 1960–1970 reform, the content of mathematical literacy is based on "real" mathematical topics from the first grade to the twelfth.

Today, it is only natural that the classical algebraic and geometric content of the traditional twentieth century school system has been supplemented with modern chapters on mathematical science that have dropped the obsolete, less-used topics. In the rapidly expanding world of pocket calculators and various digital devices, it is no longer necessary to devote so much time to practicing basic operations, complicated algebraic transformations, rules of logarithmic operations, knowledge of trigonometric identities, and complex trigonometric equation-solving tricks. Instead, there are more and more possibilities for the development of independent problem solving and creative thinking by getting to know the methods of thinking, combinatorics, the basic elements of graphs, and the integration of mathematical games into the curriculum. This position is expressed in the literature and official documents.

> Pupils should regularly perform tasks independently and participate actively in the teaching and learning process. Through solving problems, pupils can gain the ability to do precise, persistent, and disciplined work. The need for self-checking and respect for other's opinions can grow in them. To achieve these, we must strive to ensure the positive motivation of pupils and to develop their autonomy during the course of teaching. (Oktatáskutató és Fejlesztő Intézet 2016, p. 2)

The development of these skills is essential for all those who want to be prepared by school for the intelligent and effective application of mathematics. Getting to know sets and learning elements of mathematical logic also contribute to the development of a unified mathematical approach, even in the teaching of small children. The probability and statistics topic, which is also included in the curriculum from the beginning of schooling, can help students to understand the possibilities for applying mathematics. Teaching practice is of particular importance concerning the teaching of mathematics application. Teachers' aversion could be explained partly by their inadequate preparedness and partly because this topic needs flexible lesson planning, the use of teamwork, and other modern forms of work and methods rather than usual teaching styles. The introduction of the new graduation system (2005) contributed to the strengthening of probability and statistics teaching, because such problems are almost always present in final examinations.

Compared to previous school content, the more modern approach to the teaching of geometry also prevailed. Instead of the mosaic-like descriptive style, the dynamic viewpoint of building on vectors and geometric transformations came into force. Also, the amount of material was reduced; for example, on the topic of analytic geometry, detailed discussion of conic sections was left out. As a result of the decreased number of lessons, some classical Euclidean geometric theorems were omitted from the requirements of the NCC. Compared to the curriculum of 1978, there is less emphasis on constructing and proving tasks (like the central angle theorem proof or the proof of the angle bisector theorem), but more and more schools use dynamic and geometric software.

Applicable mathematical knowledge in most areas requires mathematical analysis, like applications in physics, chemistry, or other natural science topics. In the twentieth-century curriculum modifications, the teaching of analysis in secondary school was an almost constant subject of debate. Some modifications sought to make mathematical analysis an obligatory topic, some intended to omit it based on the argument that it was impossible to require sufficient conceptual accuracy at school. As a compromise, the curriculum with fewer school lessons did not contain analysis; only curriculum for specialized classes included the topic. The NCC and the frame curricula have chosen this solution as well; analysis appears only in curricula for specialized classes and it is in accordance with the requirements of the final examination. Analysis is not included in the intermediate-level final examination, while it is present in the advanced level of national matriculation.

The changes to the curriculum also made it necessary to modernize methodological principles. The idea that mathematical conceptualization takes a long time, from the beginning of school going through the whole period of education until graduation, has become accepted. It has also become a common idea that children should be involved in the process of learning and participate actively in a form appropriate for their age, always keeping in mind developmental methods according to their different abilities. These aspects also apply to the development principles of the NCC and frame curricula. In the application of modern methods in mathematics education, the opportunity for differentiation, developing individual skills, and learning effective collaborative techniques is important. The NCC and frame curricula encourage the application of methods which promote the development of a correct image of mathematics and the development of a positive attitude toward mathematics. "With diverse problems and tasks, one can point out the benefits of problem-solving skills in everyday life. Mathematics education plays a key role in the development of financial-economic competencies" (Oktatáskutató és Fejlesztő Intézet 2016, p. 2). The rapid development of digital techniques makes it necessary to seek the application of mathematics in their use. "The use of subject-based computer software for educational, science popularizing, and development purposes should, as far as possible, be incorporated into the school's pedagogical program and the local curriculum" (Oktatáskutató és Fejlesztő Intézet 2016, p. 2).

More and more teachers, schools, foundations, parents' groups, and private companies have started to develop teaching materials, methods, and forms of work that involve joyful experiences, games, other subjects and arts, and the use of digital techniques to try to improve the efficiency of mathematics education.

5 Summary

Let's conclude this chapter with an overview of the main features of and changes to Hungarian mathematics teaching since the end of the socialist era.

It is worth repeating that in the socialist era the many features and characteristics of the mathematics education were saved from the bygone and recent past by professionals of mathematics teaching. Concurrently, efforts were made to meet the requirements of universal primary education, as well as for advanced support of mathematics talent in secondary education. Tamás Varga was the most prominent figure of this period; his Complex Mathematics educational model had an impact on mathematics teaching the world over. And the new social system has preserved many elements of the past, such as the system of special mathematics classes, the *KöMaL*, and several national and lower-level mathematics competitions.

The 1989 changes led to the competency-based National Core Curriculum (NCC), which has been subject to many significant modifications since then. Centralized socialist education management, under which mathematics teaching was also highly content-oriented, became first decentralized and competency-centered and then swayed back, a little like a pendulum, to being centralized and knowledge-centered again around the 2000s. The latter was promoted by the framework curricula and many other legislative frameworks instituted at the time.

The same scenario applied to mathematics textbooks: the ones used in the socialist era were first replaced by a highly diversified supply of new textbooks, then their range was again narrowed strongly by law in recent years. The same steps are discernible in mathematics teacher education: first, the system for teacher education, and specifically mathematics teacher education within it, changed to the BA/MA Bologna System, with scientific education taking place at the BA level and teacher education at the MA level, but the former uniform 5-year training system has been restored for now.

In talent support education—which seems to be traditionally one of the strongest parts of Hungarian mathematics education—many previous practices were preserved, like specialized mathematics classes for the gifted, the Hungarian Olympiad Team's activity and good results, extracurricular programs, and many others. However, many new initiatives also emerged in the past decades, among them professional, high-quality programs with fees.

After 1989, the range and nature of national and local mathematics competitions became significantly richer and were modernized: today many require teamwork, creativity, and complex problem-solving skills. Weekend and summer camps in many new forms became an important part of talent support. However, now some of the competitions and development programs are for-profit ones. Obviously, it reduces access to young talent coming from underprivileged groups, despite the many kinds of support available to them under mathematics programs.

The content of mathematics teaching also changed after 1989. The Mathematics NCC continues to rely on the reform curriculum prepared by Tamás Varga before the political system changed for specifics concerning relevant knowledge and development requirements. The material of the 12 grades is built up spirally along the main topic headings that have been retained. Probability and statistics became stronger, the topics of sets, logic, combinatorics, and graphs were given more emphasis, and classical algebra and geometry content were increasingly relegated to the background. Applications to real-world practice are increasing and proof-based problems are losing importance. Importantly, optional advanced-level curricula and syllabi exist already in primary school, and there are several advanced-level curricula at the secondary level. Intermediate-level graduation in mathematics requires no proofs and most written assignments start out from some type of practical problem, whereas the material for advanced-level graduation is much deeper and wider including analysis chapters, for example.

The modern methods advocated in the NCC and the framework curricula have not yet become general practice at the national level, but minor, local centers have launched many new initiatives to improve mathematics teaching and to boost interest in mathematics. For example, the use of digital devices in school lessons is spreading.

All in all, the most widespread classroom methods have changed little since the socialist era. It must be admitted that the methods applied by today's mathematics teachers in the classroom are in many aspects the same or similar to the ones that had been typical in the socialist era or even earlier.

Performance measurement by subject has become much more diversified than it used to be, and one must not underestimate the considerable impact both internationally and in Hungary of TIMSS and PISA.

Today's youth sometimes prefer knowledge fields and jobs promising high earnings over such sciences as mathematics. It is difficult to show them in and outside school that IT, bio-technology, pharmaceuticals sciences, and other fields that are progressing by leaps and bounds could not develop without advanced mathematics and, that the latter have opened up many attractive and brand new fields of mathematical research and development. As shown by the data, in Hungarian higher elementary and secondary schools, mathematics is among the less popular subjects today. It is challenging to identify pragmatic contexts where mathematics could seem important and attractive to the student population of today or to show the attractive and eternal beauties of mathematics as a form of art to schoolchildren.

References

1990 XXIII. 1990. *Law amending Act I of 1985 on education.* https://mkogy.jogtar.hu/jogszabaly?docid=99000023.TV (in Hungarian). Accessed 29 Apr 2019.

Ács, Katalin, József Kosztolányi, and Józsefné Lajos. 2010. *Cserepek a magyarországi matematikai tehetséggondozó műhelyekből* [Shards from the Hungarian talent management workshops]. Hungary: Bolyai János Matematikai Társulat. http://www.mategye.hu/download/cserepek/cserepek.pdf. Accessed 21 Apr 2019.

Ádám, Péter, József Baranyai, Sándor Bán, László Csorba, János Kertész, Katalin Radnóti, and Luca Szalay. 2008. *A természettudományos közoktatás helyzete Magyarországon: Az OKNT-bizottság jelentése* [The state of the education of natural sciences in Hungary: Report prepared by OKNT-committee]. Manuscript.

Alapítvány a Magyar Természettudományos Oktatásért. 2019. *Rátz Tanár Úr Életműdíj* [Rátz Lifetime Achievement Award for Teachers]. http://www.ratztanarurdij.hu/. Accessed 12 Mar 2019.

Andrews, Paul, Jose Carillo, Fabrice Clement, Erik De Corte, Fien Depaepe, Katalin Fried, Gillian Hatch, George Malaty, Peter Op't Eynde, Sári Pálfalvi, Judy Sayers, T Sorvali, Éva Szeredi, Judit Török, and Lieven Verschaffel. 2004. International comparisons of mathematics teaching: Searching for consensus in describing opportunities for learning. Conference representation. In *Discussion group 11: International comparisons in mathematics education, the tenth international congress on mathematics education (ICME-10)*.

Balázsi, Ildikó, Péter Balkányi, Lászkó Ostorics, Ildikó Palincsár, Annamária Rábainé Szabó, Ildikó Szepesi, Judit Szipőcsné Krolopp, and Csaba Vadász. 2014. *Az Országos kompetenciamérés tartalmi keretei: Szövegértés, matematika, háttérkérdőívek* [The contextual frames of the national competency test: Comprehension, mathematics and background questionnaires]. https://www.oktatas.hu/pub_bin/dload/kozoktatas/meresek/unios_tanulmanyok/AzOKMtartalmikeretei.pdf. Accessed 12 Mar 2019.

Balázsi, Ildikó, László Ostorics, Balázs Szalay, Ildikó Szepesi, and Csaba Vadász 2012. *PISA2012: Összefoglaló jelentés* [PISA 2012: Summary report]. https://www.oktatas.hu/pub_bin/dload/kozoktatas/nemzetkozi_meresek/pisa/pisa2012_osszefoglalo_jelentes.pdf. Accessed 12 Mar 2019.

Balla, István. 2016. *Ki érti ezt: Akkor most mégis jók a magyar gyerekek matekból* [Who understands it: So are Hungarian children finally good at maths]? https://hvg.hu/kultura/20161130_timss_pisa_iskola_oktatas_diakok_felmeres. Accessed 12 Mar 2019.

Ballér, Endre. 1973. Tanulói attitűdök vizsgálata [Study on students' attitude]. *Pedagógiai Szemle* 23 (7–8): 644–657.

Balogh, László. 2012. *Komplex tehetségfejlesztő programok* [Complex talent developmental programs]. Debrecen: Didakt Könyvkiadó.

Bánfi, Ilona. 1999. A háttéradatok elemzése [Analysis of background data]. In *Monitor 97: A tanulók tudásának változása: Mérés, értékelés, vizsga 6*, ed. Vári Péter, 265–321. Budapest: Országos Közoktatási Intézet.

Barbarics, Márta. 2018. Alternative assessment in teaching mathematics. https://prezi.com/kgcufmmemflm/marta-barbarics-alternative-assessment-in-teaching-mathemat/. Accessed 12 Mar 2019.

Barrington Leigh, R., and Andy Liu, eds. 2011. *Hungarian problem book*. Vol. IV. Washington: Mathematical Association of America.

Báthori, Zoltán. 2000. A maratohi reform: 2. rész [The Marathon race long reform: 2nd part]. http://epa.oszk.hu/00000/00011/00043/pdf/iskolakultura_EPA00011_2000_11_003-026.pdf. Accessed 12 Mar 2019.

Bragyisz, Vlagyimir Modesztoviccs. 1951. *A középiskolai matematikatanítás módszertana* [The methodology of secondary school mathematics education]. Budapest: Közoktatásügyi Kiadóvállalat.

Connelly, Julianna. 2010a. Hungary and the United States: A comparison of gifted education. *Hungarian Cultural Studies* 3: 18–26.

———— 2010b. *A tradition of excellence transitions to the 21st century: Hungarian mathematics education, 1988–2008*. Ph.D. thesis. New York: Columbia University.

Connelly Stockton, Julianna. 2010. Education of mathematically talented students in Hungary. *Journal of Mathematics Education at Teachers College* 1: 1–6.

Csahóczi, Erzsébet, Katalin Csatár, Csongorné Kovács, Éva Morvai, Györgyné Széplaki, and Éva Szeredi. 2013. *Kézikönyv a Matematika 5. tanításához* [Teachers' handbook for the mathematics 5 textbook]. Budapest: Apáczai Kiadó.

Csapó, Benő. 1991. *Math achievement in cultural context: The case of Hungary*. Paper presented in the Symposium on culture and mathematics learning. Eleventh biennial meetings of the International Society for the Study of Behavioral Development, Minneapolis, USA [ERIC Ref. No.: ED 367 539].

————. 2000. A tantárgyakkal kapcsolatos attitűdök összefüggései [Interrelations between attitudes and school subjects]. *Magyar Pedagógia* 100 (3): 343–366.

Csépe, Valéria. 2018. *A Nemzeti alaptanterv tervezete* [A draft version of the National Core Curriculum]. https://www.oktatas2030.hu/wp-content/uploads/2018/08/a-nemzeti-alaptanterv-tervezete_2018.08.31.pdf. Accessed 20 Apr 2019 [Propositions to NCC 2018].

Csicsigin, V. G. 1951. *A számtantanítás módszertana* [Methodology of teaching arithmetic]. Budapest: Közoktatásügyi Kiadóvállalat.

Csíkos, Csaba. 2012. Melyik a kedvenc tantárgyad: Tantárgyi attitűdök vizsgálata a nyíltvégű írásbeli kikérdezés módszerével [Which is your favorite subject: A study on students' attitudes on subjects by an open-ended written questioning method]. *Iskolakultúra* 1: 3–13.

Csüllög, Krisztina, Éva Molnár, and Judit Lannert. 2014. A tanulók matematikai teljesítményét befolyásoló motívumok és stratégiák vizsgálata a 2003-as és 2012-es PISA-mérésekben [Study of motives and strategies influencing the mathematical productivity of students in the 2003 Csüllög and 2012 PISA tests]. In *Hatások és különbségek: Másodelemzések a hazai és nemzetközi tanulói képességmérések eredményei alapján*, ed. Ostorics László, 167–211. Budapest: Oktatási Hivatal.

Dárdai, Ágnes, Katalin Fried, and Gabriella Köves. 2015. *The analyzing and evaluating of new textbooks, teaching materials and other teaching tools according to the National Core Curriculum and frame curricula within the research of the development of teaching tools*. Research report. Manuscript.

Demeter, Katalin. 1990. Egy fejlesztő kísérlet összehasonlító adatok tükrében [A developmental experiment in the light of comparative data]. *Fejlesztő Pedagógia* 1: 29–33.

Deterding, Sebastian, Dan Dixon, Rilla Khaled and Lennart E. Nacke. 2011. Gamification toward a definition. In *C. Gamification workshop proceedings*, Vancouver, BC, 7–12 May 2011, 1–4.
EGMO (2019). *European girls mathematical olympiad.* https://www.egmo.org/. Accessed 12 March 2019.
Erdős Pál Matematikai Tehetséggondozó. 2019. *A program célja* [The goals of the program]. https://erdosiskola.mik.uni-pannon.hu/index.php/altalanos-informaciok/a-program-celja.html. Accessed 12 Mar 2019.
Eszterág, Ildikó. 2010. Tantervi változások 1989 és 2010 között [Changes in the curriculum between 1989 and 2010]. *Új Pedagógiai Szemle* 60: 83–91. http://folyoiratok.ofi.hu/sites/default/files/article_attachments/upsz_2010_5_08.pdf. Accessed 5 Oct 2019.
Eurydice. 2013. https://eacea.ec.europa.eu/national-policies/eurydice/content/conditions-service-teachers-working-early-childhood-and-school-education-34_en. Accessed 20 Apr 2019.
Fábián, Mária, Józsefné Lajos, Tamásné Olasz, and Tibor Vidákovich. 2008. *Matematika kompetenciaterület* [Mathematical competencies]. http://www.kooperativ.hu/matematika/1_koncepcio/Matematikai%20kompetencia%20fejleszt%C3%A9se.pdf, Accessed 29 Apr 2019.
Forrai, Tiborné. 1972. *A matematikaoktatási eljárás ismertetése. Egyéni matematikatanulás osztályközösségben: Feladatgyűjtemény az általános iskola 5. osztálya számára* [Mathematical procedures, Individual studies, A problem book for grades 5]. Budapest: Tankönyvkiadó.
Frank, Tibor. 2011. Teaching and learning science in Hungary, 1867–1945: Schools, personalities, influences. *Science & Education* 21 (3): 355–380.
———. 2012. Acts of creation: The Eötvös family and the rise of science education in Hungary. In *The nationalization of scientific knowledge in the Habsburg Empire, 1848–1918*, ed. Mitchel G. Ash and Jan Surman, 113–137. New York: Palgrave Macmillan.
Fried, Ervin, Ivánné Lánczi, and János Surányi. 1968. *Ki miben tudós: A Magyar Televízió 1964. és 1966. évi matematikai versenyeinek feladatai* [What is your strongest subject: A collection of the problems of the Hungarian Television's mathematics competition between 1964 and 1966]. Budapest: Tankönyvkiadó.
Fried, Katalin and Csaba Szabó. 2018. *Practices for identifying, supporting and developing mathematical giftedness in school children: The scene of Hungary*. EMS Newsletter. http://euro-math-soc.eu/sites/default/files/giftedness-long-HUN.pdf. Accessed 12 Mar 2019.
Gordon Győri, János. 2018. 50 years later: Some preliminary data from research on the very first Hungarian special math class for gifted students. *The International Group for Mathematical Creativity and Giftedness Newsletter* 14: 5–9.
Gordon Győri, János, and Péter Juhász. 2017. An extra-curricular gifted support programme in Hungary for exceptional students in mathematics. In *Teaching gifted learners in STEM subjects: developing talent in science, technology, engineering and mathematics*, ed. Keith S. Taber, Manabu Suida, and Lynne McClure, 89–106. London: Routledge.
Gosztonyi, Katalin. 2015. *Tradition and reform in mathematics education during the New Math period: A comparative study of the case of Hungary and France*. Ph.D. thesis. Manuscript.
———. 2016. Mathematical culture and mathematics education in Hungary in the XXth century. In *Mathematical cultures: The London meetings 2012–2014*, ed. Brendan Larvor, 71–89. Basel: Birkhäuser.
Government Decree 1993. *Act LXXVI of 1993 on public education.*
——— 1995. 130/1995. (X. 26.) *On National Core Curriculum.*
——— 2012. 110/2012. (VI. 4.). On the Announcement, Introduction and Application of the National Core Curriculum.
——— 2010. *2010 LXXI. Act amendment on basis of Act LXXIX of 1993 on public education.*
Government Rule. 2016. *Government rol 134/2016 (VI. 10) on the bodies involved in the maintenance of public education service and the Klebelsberg Center.* http://kk.gov.hu/download/1/b7/32000/Miniszteri%20Felh%C3%ADv%C3%A1s.pdf. Accessed 29 Apr 2019.
Gyarmathy, Éva. 2013. The gifted and gifted education in Hungary. *Journal for the Education of the Gifted* 36 (1): 19–43.
Hajdu, Sándor. 1989. A Monitor'86 vizsgálat ismertetése [Report on the Monitor'86 inspection]. *Pedagógiai Szemle* 39 (12): 1142–1152.

Hajdu, Sándor, and Lászlóné Novák. 1985. Az általános iskola felső tagozatában folyó matematikatanításról [On mathematics education in upper primary classes]. *A matematika tanítása* 32 (3): 65–67.
Hajnal, Imre. 1984. *A matematika tanítása a magyar giuináziumokban az ENTWURFTÓL 1979-ig*. [Teachnig mathematics in Hungarian secondary schools from ENTWURF until 1979]. Doctoral thesis. http://doktori.bibl.u-szeged.hu/3754/1/1984_hajnal_imre.pdf. Accessed 12 Mar 2019.
Halmos, Mária, and Tamás Varga. 1978. Change in mathematics education since the late 1950s: Ideas and realisation: Hungary. *Educational Studies in Mathematics* 9 (2): 225–244.
Hersh, Reuben, and Vera John-Steiner. 1993. A visit to Hungarian mathematics. *The Mathematical Intelligencer* 15 (2): 13–26.
Horthy, Nicolas. 2000. *Memoirs*. Safety Harbor: Simon Publications.
HVG. 2016. *Itt a PISA-eredmény: 3 év alatt drámaian esett a magyar diákok tudása* [Here are the new results of PISA: The knowledge of Hungarian students dropped drastically in 3 years]. https://hvg.hu/itthon/20161206_friss_pisa_teszt. Accessed 12 Mar 2019.
International Mathematical Olympiad Foundation. 2019. https://imof.co/about-imo/history/. Accessed 12 March 2019.
Karp, Alexander. 2009. Teaching the mathematically gifted: An attempt at a historical analysis. In *Creativity on mathematics and the education of gifted students*, ed. Rosa Leikin, Abraham Berman, and Boris Koichu, 11–30. Rotterdam: Sense Publishers.
———. 2014. The history of mathematics education: Developing a research methodology. In *Handbook on the history of mathematics education*, ed. Alexander Karp and Gert Schubring, 9–24. New York: Springer.
Kim, Jinho. 2016. Korean special secondary schools. In *Special secondary schools for the mathematically talented: An international panorama*, ed. Bruce R. Vogeli, 159–169. NJ: World Scientific.
Klein, Sándor. 1980. *A komplex matematikatanítási módszer pszichológiai hatásvizsgálata* [The psychological impact assessment of the complex teaching method]. Budapest: Akadémia Kiadó.
Kurschak, Jozsef. 1963a. *Hungarian problem book: Vol. I. Based on the Eötvös Competitions, 1894–1905*. Washington: Mathematical Association of America. Transl. by Elvira Rapaport.
———. 1963b. *Hungarian problem book: Vol. II. Based on the Eötvös competitions 1906–1928*. Washington: Mathematical Association of America.
Laczkovich, Miklós. 2016. *A matematikaoktatás legégetőbb problémái egy felmérés válaszainak tükrében* [The most pressing problems of mathematics education in the light of the responses of a survey]. Manuscript.
Lannert, Judit. 2009. *Analysis of the system of research, development and innovation in education in Hungary*. Budapest: TÁRKI-TUDOK. https://tarki-tudok.hu/files/rd_final_paper.pdf. Accessed 5 Oct 2019.
———. 2014. *Gondolkodás nélkül: Miért nem megy a magyarnak a matek* [Without thinking: Why Hungarians cannot do maths]. https://magyarnarancs.hu/publicisztika/miert-nem-megy-a-magyarnak-a-matek-90531. Accessed 12 March 2019.
Laricsev, Pavel A. 1952. *Algebrai feladatok gyűjteménye* [A collection of problems in algebra]. Budapest: Kossuth Kiadó.
Liskó, Ilona. 1991. *Hogy áll a NAT* [The state of NCC]. http://beszelo.c3.hu/cikkek/hogy-all-a-nat. Accessed 12 Mar 2019.
Liu, Andy. 2001. *Hungarian problem book*. Vol. III. Washington: Mathematical Association of America.
Maróthi, György. 1743. *Arithmetica vagy számvetésnek mestersége* [Arithmetics or the art of calculations]. Debretzen.
MaTech. 2019. *MaTech matematika verseny* [MaTech mathematics competition]. http://www.matechversenyek.hu/. Accessed 12 Mar 2019.
Mathematics Connects Us. 2019. *Who we are*. http://www.medvematek.hu/en/. Accessed 12 March 2019.

Mécs, Anna. 2015. Minden medve szereti a matematikát [Everybody likes mathematics]. *Természet Világa* 5: 228–229.
Medve Matek. 2019. *Medve Szabadtéri Matekverseny* [Outdoor bear maths contest]. http://medvematek.hu/. Accessed 12 Mar 2019.
Ministerium des Cultus und Unterrichts. 1849. *Entwurf der Organisation der Gymnasien und Realschulen in Oesterreich*. https://archive.org/details/entwurfderorgan00untegoog/page/n5. Accessed 5 Apr 2019.
Ministry of Human Capacities. 2017. *Matematika: Érettségi vizsgakövetelmény* [Mathematics: Requirements of the national maturation examination]. https://www.oktatas.hu/pub_bin/dload/kozoktatas/erettsegi/vizsgakovetelmenyek2017/matematika_vk.pdf. Accessed 5 Apr 2019.
Molnár, Gyöngyvér, and Benő Csapó. 2019. How to make learning visible through technology: The eDia Online Diagnostic Assessment System. In *CSEDU 2019: Proceedings of the 11th international conference on computer supported education*, ed. H. Lane, Susan Zvacek, and James Uhomoibhi, 122–131. Heraklion: SCITEPRESS.
Németh, András. 2001. A pedagógia egyetemi tudomány jellegének kialakulása és intézményesülése a pesti egyetemen [The universalizations and institutionalization of pedagogy at the university of Pest]. *Magyar Pedagógia* 101 (2): 213–238.
Nemzeti Alaptanterv [National Core Curriculum]. 2012. https://ofi.hu/sites/default/files/attachments/mk_nat_20121.pdf. Accessed 5 Oct 2019.
Nótin, Ágnes, Judit Páskuné Kiss, and Győrő Kurucz. 2012. A matematikai szorongás személyen belüli tényezőinek vizsgálata középiskolás tanulóknál [The study of individual factors of mathematical fears among secondary school students]. *Magyar Pedagógia* 112 (4): 221–241.
Oktatási Hivatal. 2016a. *Tájékoztató a középfokú beiskolázás egységes írásbeli felvételi vizsgáinak – magyar nyelvi és matematika – feladatlapjairól* [Information on the tests of the unified entering examinations to secondary schools in the subjects Hungarian grammar and mathematics]. https://www.oktatas.hu/kozneveles/kozepfoku_felveteli_eljaras/tajekoztato_intezmenyeknek/2019_2020beiskolazas/tajekoztato_matek_magyar. Accessed 5 Oct 2019.
———. 2016b. *TIMSS 2016*. https://www.oktatas.hu/kozneveles/meresek/timss. Accessed 12 Mar 2019.
———. 2017. [On Pisa]. https://www.oktatas.hu/kozneveles/meresek/pisa. Accessed 12 March 2019.
Oktatáskutató és Fejlesztő Intézet. 2016. *Kerettanterv* [Frame curricula]. http://kerettanterv.ofi.hu/01_melleklet_1-4/1.2.3_matemat_1-4_u.docx. Accessed 12 Mar 2019.
Pálfalvi, Józsefné. 1997. A NAT és Varga Tamás komplex matematikája [NCC and the complex mathematics of Tamás Varga]. *Matematikatanár-képzés, Matematikatanár-továbbképzés* 4: 3–22.
———. 2018. *Szóbeli közlés* [Verbal communication].
Parliamentary Resolution. 2008. Number 126/2008. (XII. 4.) OGY on the adoption of the National Talent Programme, the financing principles of the National Talent Programme and the principles of the establishment and operation of the National Talent Co-ordination Board.
Páskuné Kiss, Judit. 2014. *Tanórán kívüli iskolai és iskolán kívüli programok a tehetséggondozásban* [The role of extracurricular and outside-of-school activities in talent development]. Budapest: Matehetsz.
Pornói, Rita. 2011. A tehetségmentés szerepe a Horthy-rendszer kultúrpolitikájában [The role of gifted support programs in the cultural politics of the Horthy-era]. *Iskolakultúra* 6–7: 123–133.
Prókai, Eszter. 2013. *A magyar kamaszoknak nem megy a matek* [Hungarian teenagers cannot do maths]. http://www.origo.hu/itthon/20131203-a-2012-es-pisa-felmeres-eredmenyei.html. Accessed 12 Mar 2019.
Pukánszky, Béla and András Németh. 1996. *Neveléstörténet* [History of education]. http://mek.oszk.hu/01800/01893/html/. Accessed 12 Mar 2019.
Radnóti, Katalin. 2004. Milyen oktatási és értékelési módszereket alkalmaznak a pedagógusok a mai magyar iskolában [What kind of educational and assessment methods are used by educators in today's schools in Hungary]. In *Hidak a tantárgyak között*, ed. Kerber Zoltán, 131–167. Budapest: Országos Közoktatási Intézet.

Roy, Paromita. 2017. Gifted education in India. *Cogent Education* 4: 1–18.
Servais, Willy and Tamás Varga. 1965. *UNESCO, The teaching of mathematics at secondary level. Preliminary edition.* WS/03650192-EDS. https://unesdoc.unesco.org/ark:/48223/pf0000007959?posInSet=21&queryId=97241ee4-75ab-47d2-b747-d65af692fc38. Accessed 21 Apr 2019.
Sinka, Edit, József Kaposi, and Attila Varga. 2014. Diversity of curriculum implementation tools in Hungary. In *From political decisions to change on the classroom: Successful implementation of educational policy*, ed. Therese Nyhamn and N. Hopdenbeck, 184–208. Oslo: Norway.
Somfai, Zsuzsa. 2009. *A matematikatanítás helyzete a középiskolában: A 2003-as obszervációs felmérés tapasztalatai.* [The state of the art of mathematics education in secondary schools: The results of the observational survey]. www.oki.hu/oldal.php?tipus=cikk&kod=kozepfoku-somfai-matematikatanitas. Accessed 12 Mar 2019.
Sun, W., and Xiaoxi Tian. 2016. Gifted and talented education in China: Youth classes and science experiment classes. In *Special secondary schools for the mathematically talented: An international panorama*, ed. Bruce R. Vogeli, 185–194. NJ: World Scientific.
Suppa, Ercole. 2007. *Eötvös-Kürschák competitions: Mathematical and Physical Society.* https://matek.fazekas.hu/images/versenyek/orszagos/kurschak/ek_competitions_18942003.pdf. Accessed 12 Mar 2019.
Szász, Réka. 2006. Mathematics teachers and differentiation: Results of a survey concerning Hungarian secondary schools. *Annales Mathematicae et Informaticae* 33: 189–205.
Szénay, Márta. 2009a. *Tantárgyak, tanórák és a tanulói érdeklődés* [Subjects, classes and students' interest]. http://ofi.hu/tantargyak-tanorak-es-tanuloi-erdeklodes. Accessed 12 Mar 2019.
———. 2009b. *A diákok munkaideje* [Students' schedule]. http://ofi.hu/tudastar/tanulok-munkaterhei/diakok-munkaideje. Accessed 12 Mar 2019.
Szentgyorgyi, Zsuzsa. 1999. A short history of computing in Hungary. *IEEE Annals of the History of Computing* 21 (3): 1–13.
Teklovics, Boglárka, and János Gordon Győri. 2016. Tehetségek nélkülözésben: Egy alacsony szocioökonómiai státusú gyerekeket támogató tehetségprogram Indiában [Talents in need: A talent program for children with low socioeconomic status in India]. In *A tehetséggondozás világa: 15 ország jó gyakorlata a tehetséggondozásban*, ed. Gordon Győri János, 45–54. Budapest: Család-, Ifjúság- és Népesedéspolitikai Intézet.
TIMSS. 2016. *Trends in international mathematics and science study.* https://www.oktatas.hu/kozneveles/meresek/timss. Accessed 12 Mar 2019.
Török, Judit. 2006. The Mathematics Education Traditions of Europe (METE) project. *Teaching Mathematics and Computer Science* 4 (2): 353–364. http://tmcs.math.unideb.hu/load_doc.php?p=98&t=doc. Accessed 14 Mar 2019.
Totyik, Tamás. 2016. *A magyar közoktatás a tények és adatok tükrében* [Hungarian public education in the light of facts and data]. LeMonde Magyar kiadás. https://www.magyardiplo.hu/index.php/2096-kozoktatas-a-tenyek-es-adatok-tukreben. Accessed 12 Mar 2019.
Trencsényi, László. 2012. *A Magyar Pedagógiai Társaság (MPT) Elnökségének állásfoglalása a 2012 februárjában megjelentetett "Nemzeti alaptanterv – 2012 –" nyilvános vitaanyagáról* [Resolution of the Board of the Hungarian Pedagogical Society on NCC 2012]. http://pedagogiai-tarsasag.hu/nat-2012/. Accessed 12 Mar 2019.
———. 2015. *20 éves a NAT* [NCC is 20 years old]. http://nevelestudomany.elte.hu/index.php/2015/11/20-eves-a-nat/. Accessed 12 Mar 2019.
Tutiblog. 2016. *Félelem és reszketés a Gólya presszóban* [Fear and shivering in the 'Gólya' coffee]. https://tutiblog.com/felelem-es-reszketes-golya-presszoban/. Accessed 12 Mar 2019.
Ujváry, Gábor. 2017. *Az egyetemi ifjúság útkereséses az 1930-as évek végén, az 1940-es évek első felében* [Searching for a better path among university age youths at the end of the 1930-ies and the first half of the 1940-ies].
Varga, Tamás. 1970. *Komplex matematikatanítás* [Complex mathematics education]. Budapest: Országos Pedagógiai Intézet.
Vásárhelyi, Éva. 2002. *A magyar matematikai nevelés a nemzetkezi összehasonlítások tükrében: Háttértanulmány Somfai Zsuzsa részére* [Hungarian mathematical education in the light of national comparisons: A background study for Zsuzsa Somfai]. Manuscript.

Vecseiné Munkácsy, Katalin. 2011. *Tehetséggondozás hátrányos helyzetű tanulók körében: Doktori (PhD) értekezés* [Talent management among students with disadvantages: Ph.D. dissertation]. https://dea.lib.unideb.hu/dea/bitstream/handle/2437/152664/Munkacsy_phd_titkositott.pdf;jsessionid=A803E3F9D36BC893984090C96CD0FE92?sequence=8. Accessed 12 Mar 2019.

Walker, Erica. 2007. Why aren't more minorities taking advanced math. *Educational Leadership* 65 (3): 48–53.

Wieschenberg Arvai, Agnes. 1984. *Identification and development of the mathematically talented: The Hungarian experience. Ph.D. dissertation*. New York: Teachers College, Columbia University.

Chapter 4
Changes in Polish School Mathematics Education in the Years 1989–2019

Marcin Karpiński, Ewa Swoboda, and Małgorzata Zambrowska

Abstract In Poland, changes to mathematics education took place in a very dynamic way, and the direction of the changes was often determined by changes within the government. This concerned all areas related to school mathematics. The desire to quickly transform the educational system implied that the legal bases for changes were often underdeveloped and the proposed solutions unsatisfactory. In the outlined period, an effort was made to rebuild lesson models to include a greater openness to the creativity of both the student and the teacher. A series of new textbooks was created, and new journals addressed to teachers appeared. Changes in the way mathematics teachers were educated were broad. The effects of the undertaken efforts were only partially satisfactory, which is confirmed by research conducted by the Educational Research Institute. After 30 years of transformation, Poland has still not worked out a method of "teaching mathematics" which is acceptable to most people.

Keywords Assessment · Curriculum changes · Educational research · Systems of education · Teacher training · Textbooks

---

M. Karpiński
The School of Education Polish-American Freedom Foundation and University of Warsaw, Warsaw, Poland

E. Swoboda (✉)
The State Higher School of Technology and Economics in Jarosław, Jarosław, Poland
e-mail: eswoboda@ur.edu.pl

M. Zambrowska
The Maria Grzegorzewska University, Warsaw, Poland

1 Introduction

In order to imagine the future and fate of Polish mathematics education, it is not enough to focus on the changes that took place over the last 30 years. They were initiated by political movements in Poland resulting from specific background circumstances—which we describe in Sect. 2. However, the course of recent changes has also been influenced by movements and decisions that were made even earlier. That is why we describe the changes in the Polish approach to teaching mathematics in two basic blocks. In the first block (Sect. 3), we describe the approach until 1989. In later sections (Sects. 4 and 5), we deal with the educational situation from 1989 to 2019. In each of these periods, the basic issues concerning mathematics education were viewed differently. Therefore, in each of these parts, we deal with further educational reforms and their relationship to curricula and textbooks. We analyze the education of teachers separately with subsections that are devoted to the creation of the didactics of mathematics as a scientific discipline for supporting the teaching of mathematics. In Sect. 5, we additionally deal with new phenomena that are characteristic of school mathematics education—participation in student competency tests and various methods of popularizing mathematics. We conclude our description with an outline of further perspectives, challenges, and expectations that are still facing mathematics education in Poland.

2 On the Political Transformation in Poland in the 1990s

Before 1980, the situation in Poland had been partially the result of relations between the East and West of Europe. For a long time following the Second World War, Polish foreign policy was strictly dependent on decisions made in Moscow. For many years, our country lost the possibility of normal development by being stuck in an artificially imposed economic, social, and political system. In later years, Poland tried to establish contacts with the West and use Western technologies. Thanks to this, the country has slightly modernized, but at the same time the state has fallen into considerable debt.

The largest concentration of political changes was associated with social movements resulting in a wave of strikes in the 1980s, which culminated in the creation and registration of the Trade Union Solidarity "Solidarność" (November 10, 1980). The reaction of the ruling government was to declare martial law in Poland on December 13, 1981, across the whole country. This act led not only to numerous cases of repression but also deepened the existing economic crisis. Even though martial law was abolished in 1983, the economic situation of the country was deteriorating. In 1985, the national income level was lower by 20% as compared to 1979. During the period of 1981–1985, inflation of official prices amounted to 15%. Yet in those years, some elements of a free market economy were attempted by the government, while at the same time it maintained full control over the centrally planned economy –which was contradictory.

The year 1989 marks another important date in the history of Poland—the so-called round-table talks, in which representatives of both the opposition and the ruling party signed an agreement concerning the most basic issues for the country. On June 4, 1989, elections to the parliament and senate were held, and in August, Tadeusz Mazowiecki became the new prime minister. The parliament set up a new government on September 13, 1989, and Poland entered a period of building independence from the Soviet Union's influence.

3 Changes Related to School Mathematics Before 1989

3.1 The Education Situation During the Years 1945–1960

From 1944, when the first areas of Poland were liberated from the German occupation, work on rebuilding the educational system began. It was an extremely dynamic period, full of anxiety. On the one hand, new authorities were constituted; on the other, a group of teachers educated before the war still existed, for whom the new conditions were often unacceptable.

1950 was the year in which an 11-year general school plan was developed and made public. The plan organized schools into two levels. Grades one through seven were supposed to establish a basic level of education, and grades eight through eleven composed a high school education. The introductory remarks to the planned program highlighted mathematics as a field that provided the theoretical foundations necessary for all technical sciences. It stressed the great importance of mathematics and its teaching for the development of the country, the implementation of economic plans, and the shaping of a worldview consistent with the ideas of socialism:

> The outstanding qualities of mathematics in the field of cognitive education in general, for the development of the ability to formulate and justify conclusions and strict scientific thought formulation, bring its teaching alongside the mother tongue to the forefront of the teaching plan as an outstanding factor for shaping the mind, disciplined reasoning, and scientific worldview. (Ministerstwo Oświaty [Ministry of Education] 1950, p. 3)[1]

The topics of study for grades one through four (primary education) included arithmetic using natural numbers in a range that gradually expanded up to 10,000. The program for grade five contained lessons in four mathematical operations for fractions (ordinary and decimal). In grade six, students were taught proportionality and percentage calculations and were given an introduction to using algebraic symbols (letters). In the last grade of their basic education, the students became acquainted with solving equations and solving systems for first-degree equations. They also learned the concept of functions and graphs. Geometry was treated very

---

[1] All translations from Polish are by the authors of the chapter: Karpiński, Swoboda, and Zambrowska.

marginally in grades three to five; its scope was "conditioned by the needs of the course based on arithmetic and geography." Learning about the basic information regarding flat and spatial figures occurred in grades six and seven.

In the notes on the implementation of the curriculum, it was stated that this content should be mastered by all students. The notes specified that all of the material provided for a particular grade should be mastered and the order of topics should not be changed. It was necessary to repeat the material often, so that it would be grounded in practice. On the other hand (in order to facilitate learning), the content was broken down into isolated issues, without concern for creating a systematic approach and showing relationships between various operations and concepts. An important part of learning was solving word problems, the content of which was consistent with the prevailing ideology of the ruling party.

The changes in the subsequent years in the understanding of the role and shape of school mathematics were a reflection of the general changes taking place in Europe. Reforms related to school mathematics took place in waves.

3.2 The First Wave of Reforms: 1960–1970

These reforms were the result of the so-called New Mathematics trend spreading throughout Europe under the influence of French Bourbakists. It was focused on the refinement and formalization of school mathematics education. The initiative to reform the mathematics curriculum in Poland came from a group related to the Polish Mathematical Society. The changes themselves were initiated by the work of didactic circles gathered around Anna Zofia Krygowska. However, even Krygowska approached the implementation of new ideas into practice very cautiously. She noted that teachers' preparation for the proposed changes was insufficient.

An important change was the extension of the period covered by basic school—from 7 years to 8 years. The government undertook preparations for the introduction of the "10-year-schooling" plan, which was supposed to be a response to the postulation of "mathematics for all."

3.3 The Second Wave of Reforms: 1970–1980

During the second wave of reforms of this period, the main slogan was the implementation of so-called mathematics for all. The plan was to extend the length of obligatory schooling by introducing a 10-year school system after which there would be another 2 years of high school. These changes, planned and partly introduced in 1975, in retrospect should be assessed as the most radical change in the area of compulsory mathematics education in Poland. Preparations for implementing this idea were based, among other things, on the development of teacher training, mainly at early education levels (Ciesielska 2012, pp. 33–54). Also, the media was involved in this reform activity—in the mid-1970s, TV lectures were broadcast on public television, and a series of lectures about new topics that were to appear in the curricula were prepared for teachers. The contents of these lectures were then published in special editions of the journal *Education and Upbringing* [Oświata i Wychowanie]. At the same time, work on a new curriculum continued. In 1975, the teaching of "New Mathematics" in primary education, based on theoretical approaches and set theory, was introduced in Polish 10-year schools. When the authorities attested that the teaching of New Math in the first three grades was being successfully executed (and that teachers had been properly prepared and appropriate new textbooks had been developed), they started to implement reform in grades four through eight without a clear vision for the future of 10-year schools and without certainty that changes in the structure of the education system could actually be introduced.

New proposals for didactic solutions presented in curricula aroused emotional responses. Many scholars warned about a huge overload of content, mainly in primary school. When the reform was started by the government, which disregarded the strong opposition in academic circles, protest notes were issued by the Polish Mathematical Society and the Committee of Mathematical Sciences of the Polish Academy of Sciences. Even in the press, the lack of vision for mathematics in the "10-year-schooling" plan was attacked for its fast and poor implementation of an overly ambitious program. There was a state of chaos.

As a result, the introduction of a 10-year-schooling curriculum was abandoned, but the main reason for this decision was the economic state of the country. One of the consequences of stopping the reform was the suspending of work on the prepared textbooks for grades seven through ten. However, because textbooks for grades four, five, and six were ready, they were released for use. In this way, for the first time in postwar history, in the following years, three sets of textbooks for grade four were simultaneously available, three for grade five (Fig. 4.1) and two for six. They were very different in their methodological approaches—from the traditional approaches in some to proposals that would continue a modern approach in others.

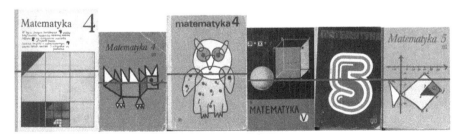

Fig. 4.1 Set of textbooks used in schools for grades four and five (Turnau et al. 1979; Łabanowska 1976; Chrzan-Feluch and Zawadowski 1978; Abramowicz and Okołowicz 1978; Szymański and Zygadło 1979; Zawadowski 1985)

3.4 Building a Scientific Background for Mathematical Education

At the same time, thanks to the efforts and international position of Professor Krygowska, a group of researchers who were scientifically involved in the didactics of mathematics was formed. The group operated under the name "Krakow's school of didactics of mathematics." Its members worked as instructors in the State Higher Pedagogical School in Krakow, which was established in 1946. The open scientific seminar, led by Professor Krygowska, gained the rank of a nationwide seminar. Polish didactics of mathematics owes Krygowska a lot. Krygowska described the philosophy of the teaching studies carried out at Higher Pedagogical Schools in the following way:

> The development of the didactics of the major subjects as autonomous, although interdisciplinary, fields constituted an important element of the concept of a higher education school for educating teachers (…). The special function of didactics as the major discipline, which, inter alia, integrated the whole of the studies, could only be fulfilled if it developed as a field of research at the scientific level. Hence, the very concept of educating teachers of mathematics in Krakow resulted in activities in the field of the didactics of mathematics with the following objectives: firstly—to create the theoretical and methodological foundations of this new scientific discipline, growing in equal degree from mathematics, psychology, pedagogy, philosophy, and methodology of science; secondly—creating the legal and organizational conditions for the education of scientific staff in this new discipline; thirdly—based on theoretical and experimental research in the field of didactics of mathematics, constantly improving the teaching of mathematics in Polish schools at various levels. (Nowecki 1984, pp. 13–14)

In 1977, her book *Outline of Didactics of Mathematics* ["Zarys dydaktyki matematyki"] was the first important scientific publication of this new field. Among other things, the author included her theory for mathematics teaching and learning, which was based on a careful analysis of various aspects of mathematics and on psychological, chiefly Piagetian, theories. This book over the years has been the main textbook for mathematics students preparing to be teachers of students aged 10–18.

3.5 Reforms 1980–1990

After 1980, there was a clear tendency to minimize the role of mathematics as a school subject (by reducing the number of hours in respective classes and by developing an abridged core curriculum instead of full syllabi). This led to great chaos; one change followed another change, and subsequent versions of "thinner" textbooks could not keep up with the changes. In 1988, the results of these actions were analyzed in a report commissioned by the Committee of Mathematical Sciences of the Polish Academy of Sciences and the Polish Mathematical Society. This report, prepared by the Committee of Experts on National Education and the Committee of

Mathematical Sciences of the Polish Academy of Sciences, described a disturbing state of mathematics education in many areas (Mathematical Messages XXIX 1990, pp. 113–121).

Changes made at the high school level on the wave of so-called New Mathematics consisted of more than just changes to the language used for formal logic and set theory in textbooks and lessons. The curriculum created an obligation to teach all high school students, including those in the humanities, such topics as probability, analytic geometry, and differential calculus. The result of these ambitious intentions was a large discrepancy between what was written in the school curricula and textbooks and the skills that the students were really able to acquire. The compulsory secondary school exit exam thus became seen as an insurmountable hurdle by many students intending to only study the humanities. Similar opinions were expressed by their parents and teachers. The math requirement in order to graduate from secondary school was then eliminated by the Minister of Education in 1982 at the end of the martial law period (Regulation of the Minister of Education and Education of the 22nd of April 1982, *Journal of Laws*, 1985 of the Ministry of Science and Higher Education No. 5, item 40). The regulation came into force in 1985 when the students who started their studies in 1982 finished their school education. From that time onward, only students who took an extended program in mathematics and physics classes had an obligation to pass mathematics on the leaving certificate. Other students could choose mathematics as an exam subject, but they did not have to. The regulation was issued at the beginning of martial law, and, of course, it did not trigger public discussion because there was no place in Poland for public discussions on any subject at that time.

This decision significantly affected the perception of math in secondary schools—its teaching began to wane in Polish schools. Another ministerial decision that abolished entrance examinations for higher education institutions completed this marginalization.

Changes to the curricula were not made based on the analysis of results from the previous solutions, because political issues blocked that analysis for a long time. It was only in 1984 that a wide discussion about the shape of school mathematics began. Everyone participated in the discussions from parents, to politicians, to academic mathematicians. An example is the ongoing debate in the periodical "Polityka"—a series of speeches under the prominent title "Checkered" desperation.[2]

In the article, "What school for this program"—an interview with four experts (Wacław Zawadowski, Roman Duda, Krystyna Wuczyńska, and Włodzimierz Waliszewski)—there was a very strong opinion expressed by Włodzimierz Waliszewski:

> Waliszewski: (…) In the teacher's opinion, a return to the traditional curriculum would be a harm done to the children. Any state of affairs that is not accepted by the teacher is

---

[2] In Polish vocabulary, the term "despair" is often emphasized by an additional adjective, e.g., "black despair," which means deep hopelessness or "despair in dots," frivolous despair, for example, a child who was offended by a friend. In this case, it is a reference to the checked notebook used in mathematics lessons.

automatically harmful to students. The teachers can't teach well, because they are convinced that the program is bad, the textbooks are bad, and that the administration that directs education is bad.

Journalist: So we have to believe that the textbooks are good, the program is good, and the administration is the best under the sun?

Waliszewski: (...) but do not move away from reforms as energetically as they were introduced. (*Polityka 28* (1429), 28 September 1984, pp. 1 and 6)

The debate over the proposed changes to the approach to teaching mathematics also took place in the pages of the journal *Mathematical Messages* [Wiadomości Matematyczne], published by the Polish Mathematics Society (PTM) (Mathematical Messages 1987, pp. 190–191). Both research mathematicians and researchers representing the new scientific discipline in Poland, the didactics of mathematics, were involved. The first group was represented, among others, by the members of the Committee of Mathematical Sciences of the Polish Academy of Sciences who officially opposed the educational situation at the time regarding school mathematics programs and textbooks. Sometimes these voices were very radical, calling for a return to the approach in the early twentieth century. Here is an example of a statement by one of the mathematicians, a university professor. The author agrees that school mathematics should be modernized; for instance, he concedes that "there is no point in learning logarithmic tables or memorizing certain trigonometric formulas." However, he is of the opinion that:

> ...the progress of mathematics is primarily on the higher levels. The anachronism of old syllabi is not as frightening as the progressives suggest (...). The canon of elementary school mathematics remains unchanged. A school must teach how to count. Also "applying psychology to the development of mathematical concepts" cannot replace the great and extremely educational satisfaction that a child gains from the ability to count effectively and the resulting opportunities to participate in adult life. (Burnat 1987, p. 327)

In the case of school curricula, Z. Krygowska also finally took the floor. She stated that the Committee should:

> ...be cautious in formulating judgments regarding specific matters in the teaching of mathematics. It is necessary to rely on objectively confirmed facts and make fully justified demands. This is very difficult in Poland, because there is no wider research into the state of school mathematics.

She agreed that the string of reforms should be stopped for the time being but— first of all—that it was necessary to carry out research not only on new curricula but also on the functioning of the current ones in order to be able to objectively assess their errors:

> The point is not to make new mistakes while developing new concepts, for example by throwing out of the program elements of discrete mathematics (combinatorics, elements of the probability theory) only because these elements were introduced as new in the current syllabus.... However, regardless of everything that brings new trends, school mathematics should be mathematics. It should teach correct descriptions of terms, correct justification of courts, generalization and specification skills, heuristic approaches to problem solving, etc. It is worth recalling today, when extreme pragmatism and minimalism are threatening us that minimalism is a reaction to the present maximalism that characterizes above all the primary school program. (Krygowska 1987, p. 333)

Table 4.1 The change of numbers of hours devoted to the teaching of mathematics between 1950 and 1985

Class year	I	II	III	IV	V	VI	VII	VIII	IX	X	XI
1950	6	6	5	6	6	6	5	5	4	4	4
1963	6	6	6	6	6	6	5	5			
1978	5	5	5	5	5	5	4	3	3	3	
1985	5	5	5	4	4	4	3	3			

A new curriculum and new textbooks were introduced in 1986. This was connected to the need to replace the existing textbooks for secondary schools and mainly concerned changes to the teaching of geometry. While it was possible, at least partially, to adopt the existing textbooks for algebra and analysis, such was not the case for textbooks for geometry. The geometry curriculum had undergone the greatest changes, and the teachers were offered manuals with sets of problems, instead of textbooks (Frycie 1989).

During the discussed period, the number of hours devoted to the teaching of mathematics in the basic school per week looked as follows (Table 4.1):

3.6 Strengthening the Importance of the Didactics of Mathematics as a Scientific Discipline

It is worth noting, however, that in 1981, the first issue of the annual *Dydaktyka Matematyki* [Didactics of Mathematics] was published, which from then became one of the journals published under the aegis of the Polish Mathematical Society. Zofia Krygowska was the first editor in chief. The main goals of the magazine were to create a forum for information on scientific, didactical research conducted in the world. The magazine wanted to make reports presented at international conferences available (translating articles and conference speeches into Polish) and to create a place where Polish researchers could publish their articles. Also, during this period, it became possible to carry out procedures for obtaining advanced academic degrees in the didactics of mathematics—PhDs and even postdoctoral degrees were given. Although the position of mathematicians regarding these degrees was unfavorable, they became a reality. Also, thanks to the efforts of the Krakow community, from 1981 onwards, cyclical meetings under the name of the School of Mathematics Didactics took place. They formed their own conference-style meetings, usually associated with a parallel mathematical conference and aimed to serve as a platform for the interaction of mathematicians and groups involved in the didactics of mathematics.

It seems, therefore, that the very problem of mathematics education in Poland (at all educational stages, also including the professional education of teachers) was one of the most important. As such, its importance was perceived both by government authorities and the scientific community. This is a good starting point to discuss the fruitful dealings with these issues in the subsequent new political reality.

4 The Beginnings of the Functioning of Education in the New Political Reality: 1990–2000

4.1 Starting Point in Education for Creating a New Reality

At the turn of the 1980s into the 1990s, a thorough analysis of the state of mathematics education in Poland was necessary and required. By order of the Committee of Experts on National Education and the Committee of Mathematical Sciences of the Polish Academy of Sciences, the team, under the direction of Professor Zbigniew Semadeni from the University of Warsaw, prepared a report. The first results were announced in May 1988; however, the final text appeared 2 years later. Its theses sounded very disturbing.

Here are the main theses of the report:

- The teaching of mathematics in Poland is in a highly unsatisfactory state.
- The current situation is to a large extent the result of poor educational decisions made in the second half of the 1970s. School syllabi are overloaded, and there is a disproportion between the actual learning outcomes and the requirements of subsequent levels of education.
- All future teaching reforms should be carried out with the participation and support of teachers. There is an urgent need to reduce the scope of mathematics education, mainly in relation to primary education. Earlier attempts in this area were fictive. Some topics were discarded without taking into consideration the fact that they would be needed and treated as a basis in higher-level education.
- The existing practice of waiting for decisions to be made about the content of the curriculum and its implementation dates and only then starting the work on textbooks is unsatisfactory.
- The practice shows that teachers often require more from students than is expected by the curriculum. This should be changed. In addition, the teacher should have more freedom in adapting the material and teaching methods to the wants and needs of students.

Another set of theses from the report concerned teacher education. The opinion was expressed that, due to several reasons, mathematics education was inadequate. Among the reasons mentioned was the poor preparation of candidates (indeed, teacher status in society was very low, and the best high school graduates rarely entered the teaching profession). The curriculum for mathematics teacher education at universities or pedagogical colleges leading to a master's degree came as a sort of compromise between the educational needs of teacher preparation and the traditions of educating scientists and proved to be unrealistic. Additionally, the dominant approach to education was found to be too theoretical, and activities offered to prospective teachers were often not connected with the challenges that awaited them at schools. Additionally, it was emphasized that the issue of the mathematics education of primary and preschool teachers was a big problem.

The new political situation had triggered the need to open up to the achievements and approaches offered in Western Europe. On the other hand, economic conditions had created hitherto unknown phenomena.

4.2 Act on the Education System in 1991, with Changes

Political changes in Poland entailed the need for changes in the educational system. On September 12, 1990, the Higher Education Act was passed, which became the first comprehensive document in the postcommunist era providing basic legal regulations (*Journal of Laws,* 1990, No. 65 item 385). Together with many European countries, Poland had started a cycle of changes to the organization of educational studies, called the Bologna process.

These changes had a significant impact on the forms and ways of teacher training. On September 7, 1991, the government of Poland passed an act affecting the education system (*Journal of Laws,* 1991, No. 95 item 425). The Act of 1991 included provisions on the possibility of setting up and running schools (at all levels) also by non-state institutions and individuals. Earlier, in July 1991, the "Concept of the General Education Program in Polish Schools" was prepared by the Ministry of National Education. This document was one of several that would help introduce educational changes related to a systemic change in Poland. The document was widely discussed—its importance for building new, democratic, Polish schools was understood. From September 1, 1992, a document entitled "Minimum Curriculum for Elementary and Secondary Schools" was in force in Polish schools (Decree No. 23 of the Minister of National Education of August 18, 1992). The adoption of this document allowed the teachers to choose a study program from several programs admitted for use in schools. One of the conditions for permission to be a part of the program was compliance with the minimum program requirements.

At the same time, it was assumed that there would be a change in the style of work done at schools, a departure from the transmissive style of teaching and, instead, a use of individualized teaching based on the idea of cognitive constructivism. The "cascading education of active teachers" strategy was adopted to enable changes in the way that mathematics classes should be taught in schools. The cascade consisted of preparation, often with foreign support and experts from the first group of teacher-specialists. They were introduced to the new program and to work techniques, equipped with new didactic packages, containing textbooks, exercises, collections of classroom scenarios, foliograms, and video cassettes. Teachers who were being educated undertook apprenticeship with the supervision of experts, and only after they were successful would they start work. Each one of them conducted classes of a least 20 seats for another group (Łysek 2003, p. 34). In this way, new ideas were to spread among teachers. From the years 1997 to 1999, thousands of active teachers participated in the cascading trainings, called "New School." It was a program developed entirely by Polish experts and coordinated by the Central Teacher Training Center, and it introduced educators to the most important elements of the program changes.

4.3 A New Approach to Teacher Education at Universities

Unfortunately, the act was created at a rapid pace and as a result had had many negative consequences. It was difficult to predict its negative results in the new reality. One such negative result of the organizational changes was the almost complete discontinuation of the discussion on teacher education at universities.

The Central Council of Higher Education also started to create minimum curriculums for specific fields of study. The work devoted to teacher education as a part of higher education was dominated by documents describing compulsory academic content and determining the number of obligatory classes. The discussion over their content encompassed the entire academic milieu and triggered the first disputes regarding the qualitative and quantitative relationship between pedagogical content knowledge and subject matter knowledge.

Regardless of the fact that the universities had autonomy that enabled them to decide on the division of hours for particular classes, the anchoring of teaching studies in the mathematic departments often resulted in wide proposals for classes in formal mathematics and a minimizing of the hours allocated for didactic subjects. Additionally, within the didactic block, the instructors were specialists in general pedagogy rather than mathematics educators who made decisions about the content to be included in the core curriculum. Mathematics didactics still did not have the status of a scientific discipline, and its representatives did not have an advisory voice in determining the scope of content to be implemented throughout its study. The first program minimums for mathematics education were introduced in 1998 (Resolution No. 200/98). These program minimums were required not only for public but also private universities. Overseeing the universities' implementation of the minimum curricula was entrusted to interuniversity committees.

In terms of pedagogical and didactic preparation, old paradigms were still in force, not keeping up with expectations, and they were *not* formulated into the teaching objectives of the "core curriculum" prepared for schools. It was not taken into account that graduates' skills should be shaped in such a way that as teachers they would be able to move away from the paradigm of "transferring knowledge" in favor of shaping their ability to work flexibly and to follow the thinking of students. Within the didactics of mathematics, the "classical" terms of general pedagogy (such as general goals of teaching, types of lessons, and assessment methods) still prevailed; there was almost no relationship between the topics in contemporary didactic research and what the future teacher was supposed to know.

The number of nonpublic higher education institutions started to grow at an alarming rate. At the end of the 1990s, there were over 130 public universities in Poland and over 300 nonpublic universities. At the nonpublic universities, only a few had a right to offer teaching qualifications, and only a few offered over 20 doctoral degrees. Often the faculty staff of nonpublic universities worked there part time and were typically recruited from the state universities. Nearly two-thirds of independent academic employees supplemented their basic (low) remuneration in this way. Nonpublic universities were keen to undertake teacher education, mainly for elementary school education, foreign language teaching, or computer science.

The spontaneous growth of the number of universities also resulted in a spontaneous increase in the number of students. This, in turn, drastically affected the level of education at universities, including the state ones. A "struggle for students" began to ensure the functioning of departments for studying and maintaining places to work for academic faculty.

4.4 Teachers Colleges: A New Form of Teacher Education

One of the very important elements that influenced changes in Polish education was a wide opening of Poland to foreign contacts. Although visas were still required for Polish citizens to travel to Western Europe, there were initiatives from the West to support changes. These changes resulted mainly in involving teachers in the processes related to education and training. The possibility of implementing new ideas was ensured by participation in the European Union's TEMPUS assistance programs (Trans-European Mobility Program for University Studies), of which Poland had been a beneficiary since 1990. Some of the programs in which teachers and university faculty dealt with teachers' education participated. The programs that could visit the western countries were *LEROPOL* 1991–1992, *TEMPUS Redesign* 1992–1994 and 1996–1997, *TEMPUS Primary Science* 1994–1996, and *Phare SMART* 1997–1999 (all of them offered the possibility of contacts with Great Britain and the Netherlands).

The contact of employees at the University of Warsaw (a group focused around Professor Wacław Zawadowski) with Jan Potworowski, working at the West London Institute of Higher Education, proved to be significant and far reaching. The first trip of a group of 12 people, who in various ways had a relation to teacher education, was financed by the TEMPUS "Recursion" program. The events and activities of the visits were designed to familiarize participants with other forms of teacher education and with a different philosophy of conducting such education. They were supposed to be aimed at developing creativity during lessons.

In 1990, the Ministry of National Education established a 3-year teachers college, a new organizational form of professional teacher training (this concerned especially early education, foreign language teaching, mathematics, and the Polish language). It was supposed to be an answer to market needs (supplementing the lack of teachers mainly in foreign languages but also computer science, as well as educating kindergarten and primary school teachers) of colleges as a form of higher vocational studies, introducing the education of teachers in two rather than only one subject, and the adaptation of a teacher improvement system to facilitate changes taking place in education. The director of one of such colleges, Anna Strzelecka, writes:

> In autumn 1989, the Teacher Training Department of the Ministry of National Education (DKN MEN) undertook work to reform teacher education, which led to the creation of foreign language teacher training colleges (1990) and teacher training colleges for educating primary school staff (1991). They were to become a recipe for solving shortages in employment in education and an alternative model of teacher education in Poland. The legal aspect of the functioning of the colleges was defined in the Regulation of the Minister of National Education of March 5, 1992, on teacher training institutions (1992). According to this document, the colleges are:
>
> - 3-year teacher training centers (public or nonpublic), operating under the academic supervision of universities and under the pedagogical supervision of the Ministry of National Education,
> - schools, subordinate to acts on education, whose graduates receive two diplomas: a college diploma and a bachelor's from a patron university. (Strzelecka-Ristow 2015, pp. 211–229)

The structure of such institutions was complicated. They had to be financed by local governments. The institution was subject to specific ministerial acts, which were not always in line with the requirements set by universities, providing scientific guidance and supervision of the colleges. This is shown in the diagram below (Fig. 4.2).

The first 2 years of the functioning of the colleges revealed that their subordination to the requirements posed by various elements of this system created a number of difficulties that were of a different nature for different groups and which led to different—generally negative—consequences.

The existence of teachers colleges was not well received by every group that was formerly involved in teacher training (Flis 2011, pp. 47–52). Professors within the educational faculties at universities had reservations. Doubts about the education in

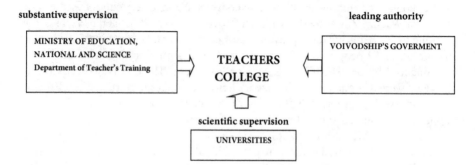

Fig. 4.2 The structure of supervision to which the colleges were subjected

teacher training colleges concerned various aspects. The most frequently formulated doubts included departure from the theory of providing teachers with a full academic education and especially the separation of education from cultural contexts and the imitation of Western standards.

In 2005, the process of closing teachers colleges began (Law on Higher Education from the 27th of July 2005, item 26, 1b). Regardless of the fact that in 2008/2009 the number of existing colleges was as high as 108, there were premises for objections to making a decision not to use this form of education. One of them was the lack of coherence in legal regulations concerning the place of teachers colleges in the higher education system. Another issue was that a period of population decline had started. The wide offerings of universities (including nonpublic) made young people reluctant to take the opportunity to study at colleges.

4.5 New Magazines Addressed to Teachers of Mathematics

Regardless of the different evaluations of changes to teacher education, this period should be assessed by how the position of the innovative teacher had strengthened.

The contacts of Jan Potworowski from the West London Institute of Higher Education with Wacław Zawadowski (Warsaw University) and his group had quite significant consequences. One of the effects of these trips was the idea of creating an association of mathematics teachers in Poland and publishing their own periodical.

In the UK, during a trip within the TEMPUS program, participants met with members of two teacher associations, the Association of Teachers of Mathematics (ATM) and the Mathematical Association (MA). As the result, the idea emerged to create a similar association in Poland. After its return to Poland, the group started preparations for publishing a new magazine for teachers. The first issue of *NiM* (*Nauczyciele i Matematyka—Teachers and Mathematics*) came out in June 1991. The text was prepared on an Archimedes computer; the Archimedes PCs were purchased using money saved from the trip, and the whole issue was printed on A4 sheets, reproduced on a photocopier, and the pages were joined with staples. The Polish Mathematics Teachers Association was formally established in November 1991 in Bielsko-Biała. Wacław Zawadowski became its first president.

Regardless of various problems, the colleges were developing. Teachers used trips with the TEMPUS program to gain new experiences for themselves. As a result, they disseminated teaching publications. The *Teacher's Studies Bulletin* (*Biuletyn Studiów Nauczycielskich*), created in 1992 in Szczytno, served this purpose ([BKN] 1992–2005). This periodical was also "homemade," assembled on an Archimedes computer and copied on a photocopier (Fig. 4.3).

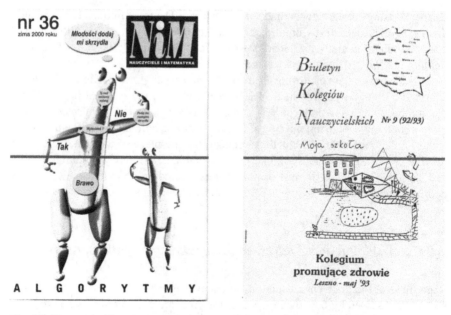

Fig. 4.3 New magazines addressed to teachers: *Teachers and Mathematics* and *Teacher's Studies Bulletin*

4.6 New Textbooks for Elementary School

In the years 1991–1993 in Poland, conditions for preparing various curricula for school subjects (as original works for teachers, schools, and other institutions) were created. One of the conditions for formal approval of such curricula by the Ministry of National Education and permission to use them throughout the country was compatibility with the then-functioning so-called minimum curriculum, which defined compulsory teaching content. Previously, teachers used curricula developed by large teams of specialists from individual disciplines organized at the School Curriculum Institute. Curricula had to be approved for school use by the Minister of Education, and their principle was to have only one curriculum used for a given subject. Seeing an opportunity, new, previously unknown educational publishing houses appeared on the market. They began to compete with the existing publishing house, WSiP (Wydawnictwo Szkolne i Pedagogiczne—School and Pedagogic Publishing House), which was once a monopoly.

Some of these new educational firms were created by teachers and enthusiasts of education. Others were created by people who decided that the development of educational materials was a good area for business activity. The quality of these new materials was varied. Some publishers proposed innovative and thoughtful concepts for teaching the subject; others limited themselves to producing materials not necessarily of high quality but willingly bought by teachers and parents.

Small publishing companies, founded by enthusiasts, were often limited only to the field which was closest to its owners. An example in the field of mathematics is the *Gdańskie Wydawnictwo Oświatowe* (GWO), created by a married couple, both mathematicians, which, over time, gained a dominant position in the mathematics textbooks market for primary and junior high schools (currently it manages about 2/3 of the market in this segment).

Before 1989, in the *socialist market economy*, not only was there a lack of many everyday products, but there was also a shortage of basic materials needed for industry. In particular, the paper needed for printing books was perceived as a scarce material. Paper was distributed centrally. Governmental institutions decided on how much paper could be allocated to each title and which titles could be printed. This applied to textbooks and other educational materials. One of the government regulations issued in the seventies prohibited the printing of single-use educational materials, which meant that it was not possible to produce, for example, some types of exercise books. Teachers and students had only textbooks and, possibly, collections of exercises. Special workbooks were the source of success for some small publishing houses beginning in 1989. At a time when the state monopoly WSiP based its activity on the sales of the same, unchanged textbooks for years, these small publishing houses made a sensation with materials that facilitated the work of teachers and, therefore, were often well received by them. A good example of this is GWO again, which published its *Notebook for Mathematics for Grade 4*, and in its first year, the book's circulation grew larger than the total number of students in the fourth grade! Of course, not only the effect of the freshness of the product but also the high substantive level of the proposed exercises made this success possible.

Until 1999, with the introduction of changes related to school reform, the most-used educational materials on the secondary level (grades four through eight) were from the same five series:

1. *Mathematics Around Us*— WSiP

    In the first editions, the series represented the traditional approach to teaching mathematics, well known to teachers and well accepted, similar to what was proposed in textbooks before 1989. The books showed a strong preference for arranging and practicing basic mathematics skills with a dominant role for instrumental mathematics.

2. *Mathematics 2001*— WSiP

    Manuals and exercise books from this series were prepared by authors involved in building the Association of Teachers of Mathematics. Thus, they presented completely new ways of approaching consecutive issues in school mathematics from both the substantive and methodological aspects. The authors offered teachers many new and ingenious tasks and exercises—all their own, completely original, and also inspired by teaching methods used in other countries. However, the textbooks were very demanding of teachers. In order to work well with them, a teacher had to adapt some topics to the capabilities of the students and think about the order and the development of some issues. Nevertheless, each year, such an approach became more and more popular. In particular, those teachers who wanted to develop their professional skills and search for new teaching methods chose *Mathematics 2001* proposals.[3]

3. *Blue Mathematics* — KLEKS Publishing

This series of textbooks was prepared by a group of mathematicians referencing the work of Zofia Krygowska. Among the authors were also her former students and colleagues. A noticeable feature of the textbooks was a strong emphasis on a consistent mathematical construction of theoretical material and the implementation of key principles of didactics of mathematics. The sets of exercises and tasks proposed to students were quite traditional.

4. *And You Will Become Pythagoras*—ADAM Publishing

This was a series of collections of problems that was very popular at that time. The collections were extremely extensive, and for each lesson topic, the teachers could find many exercises that they could solve with students. In the eyes of teachers, the great number of suitable math problems was a big advantage of these books. It gave them confidence that they would always find an example appropriate for their students. The vast majority of exercises are tasks that check and practice methods of using learned typical procedures.

5. *"Gdansk Notebooks," Exercise notebooks for mathematics* — Gdańskie Wydawnictwo Oświatowe

These notebooks gained enormous popularity at the time when the form of the workbook itself was a novelty in Poland. However, it was not only because of the form that almost every student in Poland used these books. The authors presented a nontraditional approach to teaching mathematics and used the form of a workbook to offer a complete mathematics course for primary schools. The layout of the material was structured as well as subsequent lessons. The careful ordering of the exercises took into account the gradual development of a student's skills with a strong emphasis on modeling the most important concepts and mathematical properties. The level of difficulty of each of the following tasks was selected with great care, so that teachers could easily choose the ones that were in the correct zone for the immediate development of their students. The tasks themselves posed mathematical problems in an attractive way, with frequent reference to situations in the world around them. All of this meant that teachers felt unusually confident when their students used the *Gdansk Notebooks* as they were called at this time (Fig. 4.4).

4.7 Functioning of a Community Dealing with the Didactics of Mathematics as a Scientific Discipline

In 1988, Professor Z. Krygowska died, which, among other things, resulted in research mathematicians becoming more reluctant to treat didactic works on mathematics as appropriate scholarly pieces. Participation in international conferences

---

[3] http://www.oskwarek.pl/index.php?option=com_content&task=view&id=118&Itemid=128)

Fig. 4.4 Textbooks used in schools from 1992 to 1999: *Mathematics Around Us* by Lewicka and Rosłon (1999); *Mathematics 2001* by Zawadowski et al. (1995); *Blue Mathematics* by Wachnicki and Treliński (1995); *And You Will Become Pythagoras* by Łęska amd Łęski (1991); *Gdansk Notebooks* by Dobrowolska and Zarzycki (1999)

remained too expensive. The system of scientific grants offered by the Ministry of Science and Higher Education still did not promote activities related to the didactics of mathematics, and university authorities were reluctant to subsidize foreign trips. Visas were needed to visit many countries. The most available contacts were those with the countries of the former Eastern Bloc. Among all of the possibilities, relations with the Czech Republic seemed to be the most interesting (mainly with Charles University in Prague); the Czech Republic was more open to cooperation with the West, and Czech researchers could find support by using government cofinancing for didactic research.

5 Amendments to the Act on the Education System, 1998

5.1 Reform of the Educational System

Frequent changeover of the Ministers of Education did not allow the initiated changes to be implemented into practice at schools. From the years 1989 to 1999, the position was held by seven different people. Each minister had his own ideas which he usually did not manage to implement.

During the years 1997–2001, the government, headed by Jerzy Buzek, prepared and conducted four major reforms: educational, retirement, administrative, and health. Among other things, the idea of centrally implemented education reform emerged. Internal arrangements of the final version of the May 15, 1997, core curriculum on "key competences," "education," and "tasks of schools" meant that educational reform was very radical. In May 1998, the Ministry of National Education presented its Reform of the Education System (including the results of a 6-month consultation on the initial draft version). It presented a description of a new structure for the education system (*Journal of Laws*, 1998, No. 117 item 759). According to the authorities, the educational reform was aimed at popularizing secondary education and raising its level. It was the beginning of another large reform to education (Rura & Klichowski 2013). Changes to the education system were, among others, related to the following:

- Reduction of the period of primary school education to 3 years, introduction of 3-year lower secondary schools, and shortening education in general secondary schools by 1 year
- Introduction of a new examination system: the Central Examination Committee was to prepare the standards for examination requirements
- Reform of curricula
- Changes in the school financing system.

This new system was in force in Poland from 1999 (*Journal of Laws*, 1999, No. 14 item 129). Junior high schools and specialized (profiled) high schools returned to the Polish education system. The school system became responsible for children from the age of 6 who were to attend the "0" class.

The introduction of new curricula was a consequence of organizational changes in schools. The minimum curriculum from 1992 was replaced with the new core curriculum (*Journal of Laws*, 1999, No. 14). In addition to the standards for examination requirements, the novelty of the core curriculum owed to the introduction of "educational paths" to deliberately show students the same topic in various subjects. Changes involved the following:

- Class "Zero," called "Zerówki," was organized in kindergartens but also other schools had the right to set up such a branch. For the children, it was supposed to be a period of stimulating their development and preparing them for further learning.
- The basic school was to last 6 years. This course of education was to be divided into two stages. The first 3-year stage included grades one through three. Learning in these classes was to take place in blocks, as of "integrated teaching," and the teacher was supposed to select the form of classes and decide what subject matter the course would include. The content of the education was regulated by the ministry-issued document, "core curriculum." The next stage was composed of grades four through six. There, classes were to take place in a classroom setting (45-min lessons) and were to be taught by teachers responsible for teaching separate subjects. Education at basic schools was to end with an external examination. According to the provisions of the Act, the 6-year primary school was to end with a test whose passage entitled the student to further education in middle school.
- The next stage of compulsory education was gymnasium (junior high school). Students attending junior high school were young people aged 13–16. Education at this stage was of a general nature. Young people graduating from junior high school had to sit for an external (state) examination.
- Students could continue their education by choosing one of the different types of high schools (general or specialized (profiled)—such as a music, art, or, for example, chemistry high school) or a technical school. Learning at this school was to last 3 years (high school) or 4 years (technical school) and prepare the

students for the matriculation examination: in Polish, a foreign language, and mathematics. In this way, mathematics had once again become a compulsory subject in high school examinations. It was assumed that in 2002 the new "Matura," or secondary school certificate, would combine the old secondary school exit examination and the entrance examination for universities.

The Ministry of Education (MEN) allowed for diversity in the school curricula. It could be created both by teachers themselves and by educational publishers. Thus, it created the danger of a great diversity in the levels of teaching. Therefore, maintaining a necessary common standard of achievement was ensured by a system of state examinations that students were to write at the end of each stage of education. The examinations were to be prepared and conducted by new institutions: central and regional examination commissions.

Initially, the introduction of a new educational system resulted in the focusing on organizational issues (the need to provide separate buildings for each type of school, the appropriate number of teachers, and so on). Neither teachers nor parents had any experience with how to function in the new structure. The Ministry of National Education was convinced that 13-year-olds required a specific approach, that is why—after consultation with developmental psychologists—it was determined that after 6 years of primary school education, a different learning approach should be offered to young people. Again, the media contributed to agitation over the new education model. *Gazeta Wyborcza*, in a daily news cycle, titled "Gymnasium: The Dispute Whether to Enter Them. 1999," discussed with parents the benefits of this concept in the form of questions and answers. Here are some of the questions and answers:

7. *Will my child reach the gymnasium level?*

The Ministry promises "yes." But opponents do not believe this is true. Here they see the biggest failure of the reform. The number of Junior high schools will be—according to the Ministry of Education—between five and seven thousand, two to three times less than the current number of primary schools. From this it follows that a 13-year-old's commute to school will be extended. When it exceeds four km, the community has to organize transportation or money for bus tickets. The reform does not mention the latter option—Ministry (MEN) promises the "orange gimbus" for this. These "orange gimbuses" are supposed to be buses throughout the country painted orange that have special road rights in traffic and would be used to transport high school students (as well as primary school students). Skeptics do not believe that it will be possible to organize an efficient delivery system; however, 80% of communities are interested in the transport. Will all communities find the money to buy a bus and support its maintenance? Will the buses be able to reach everywhere—on bad roads and in bad weather? Finally, they warn that children will spend a lot of time away from home (because they will have to travel far for school).

9. *Will there be new teachers in junior high school?*

Nobody knows that yet. By April 15, the municipalities will appoint middle school directors, and only then will they complete the school's pedagogical council. Since the majority of junior high schools will be located in buildings previously belonging to primary schools, one can expect that they will find mainly teachers from "fraternal" primary schools. Especially given that a junior high school housed in the primary school will not be able to fill the schedule of a full-time teacher, there will be too few junior high school classes (whereas in independent gymnasiums, which have classes from "a" to "m," there are enough hours).

But there is hope. Teachers employed in middle schools will need to have the same qualifications for teaching high school—i.e., higher education (master's or bachelor's degree). A teacher' college degree or a high school diploma itself—allowing a teacher today to teach elementary school—will not be enough. Directors of all schools—not only gymnasiums—will get a little more freedom in the selection of qualified staff on this occasion. Today, employing an engineer as a mathematics teacher or a doctor as a biology teacher requires the consent of the school superintendent. From September onwards, the director himself will be able to decide (*Gazeta Wyborcza*, 23.02.1999).

The newspaper also published interviews with Ministry officials, the point of which was to convince teachers that they would manage in the new reality.

5.2 New Textbooks, in Line with the New Core Curriculum

But the situation was very uncertain and perhaps anxious. After 1999, the changes to the documents defining the school curriculum concerned the objectives of education and teachers' duties. Among other things, teachers had to "adapt the transfer of appropriate knowledge, skills and attitudes to the natural activity of students" and "inspire students to express their own thoughts and experiences" (*Journal of Laws*, 1999, No. 14 item 129, p. 585). School mathematics classes had to teach how to explore the world and how to act in it. The boundaries between what activities were recommended as a part of mathematics education and what should not be done were blurred. There was full acceptance of various independent student strategies. School mathematics was to be diverse and flexible to create favorable development for every child. From the perspective of years, these proposals have been evaluated in the following way:

> The core curriculum of May 97 was based on an idealistic hope that everybody would understand each other, and teachers would collaborate and jointly assume responsibility for the implementation of the school's task. Previous school subjects, usually fulfilling the role of an elementary introduction to individual academic disciplines, were to be replaced with an education aimed at equipping students with key skills (competences). It was about organizing, evaluating and planning one's own learning, about communicating effectively in different situations, about effective cooperation in a team, about solving problems in a creative way. (Legutko 1999)[4]

---

[4] http://nauczyciel.wsipnet.pl/serwisy/reforma/arch/ref071.htm

However, it was not clear how the new school organization could influence the implementation of the core curriculum (Regulation of the Minister of National Education and Sport of April 18, 2002). Because the Ministry opened the possibility of teaching according to a teacher's own curriculum (created in relation to the imposed "core curricula"), there was a flood of curricula created by teachers and publishing programs and soon also textbooks tailored to these proprietary programs:

> Already by March 1999, the Minister of Education approved 135 programs, and teams of consultants submitted amendments and additions to 191 more proposals.... At the end of December 1999, a list of new textbooks for primary and junior high school, approved by the minister of national education for school use, contained 282 items. (Kraś 2007–2008, pp. 303–318)

Another aspect of the creation of a free market for educational materials was the conviction of at least some publishers that a textbook was a commodity. New publications began to use marketing, which was previously nonexistent: representatives made direct contact with school directors, and promotional meetings were held. Many publishers started to offer interesting didactic tools, task sets, and teacher materials in parallel. The quality of textbooks varied.

The most disturbing and troubling were the textbooks for early school education, generally written by general educators. The number of proposals alone would cause difficulties for assessing the suitability of textbooks for the individual needs of teachers and their classes. However, the most important problem was the attempt to align these textbooks with the conception of "integrated teaching" which was popular at this time and suggested integrating mathematics with other subjects. This concept was completely unknown to teachers. In the proposed textbooks and notebooks for students, mathematics often died in the flood of other content. It also happened that mathematics was included with other topics in a completely artificial and sometimes bizarre way. This approach was highly criticized by researchers connected to mathematics education. Professor Edyta Gruszczyk-Kolczyńska repeatedly appealed for a change from this approach:

> For many years, in primary classes, children have been educated in the convention of integrated education.... The order and rhythm of the content of mathematical education are regulated by the seasons and the calendar of social events in which children participate.... For years, books have been developed and, recently, children's notebooks as well for the teaching of integrated education. These publications are interwoven with pages containing content in the field of Polish language, natural science, and mathematics education, etc.... The problem is that the order of the education of mathematical content in the subsequent months according to the seasons and social events disturbs the substantive order of shaping messages and mathematical skills. And the shattered layout of the content of education is not conducive to the continuity of instructive mathematical education for children. (Gruszczyk-Kolczyńska 2019, p. 54)

The phenomenon of the uncontrolled creation of various versions of textbooks for one level of education concerned not only mathematics. New textbooks were created for each school subject. The difficulty of containing the deluge of textbooks aroused concern from scientists associated in relevant fields of knowledge; this mainly concerned textbooks for secondary schools. Reactivated in 1989, The Polish

Academy of Arts and Sciences (Polska Akademia Umiejętności), located in Krakow, established an interdisciplinary Commission for the Assessment of School Textbooks at the end of the year 2000. The commission dealt with the substantive evaluation of textbooks approved by the Ministry for school use. Beginning in 2001, the commission published opinions in the annual *Educational Opinions of the Polish Academy of Arts and Sciences—Works of the PAU Committee for the Assessment of School Textbooks* (Ciesielska & Szczepański 2017, pp. 101–122). Annually, about 50 textbooks were included in this assessment.

Each year, the title of "Best Textbook" was awarded. The title of best textbook was obtained by the GWO-issued handbook three different times: in 2002 (series for middle school), in 2006 (series for high secondary school), and in 2016 (new version of the series for high secondary school).

5.3 Teacher's Preparation for Work in the New System

During the years 2005–2018, all issues related to teacher's higher education were regulated by the Law on Higher Education from July 27, 2005. Two years later, legal regulations concerning educational standards were approved (Regulation of the Minister of Science and Higher Education of September 13, 2007, on education standards for specific fields and levels of education). The two-stage form of study assumed that after completing the first degree, the graduate would have the right to work in kindergarten and to teach in basic school, while a graduate of the second-cycle completing master's studies would be able to work in all types of schools (Regulation of the Minister of Science and Higher Education of January 17, 2012, on standards of education preparing for the teaching profession). However, teacher training at universities had become problematic. In the opinion of the senate and the university authorities, the teaching profession had ceased to be popular and could not have been enough of a magnet to attract students. In addition, the method of parametric evaluation of scientific institutions (and universities were recognized as such), which was developed and applied at this time, not always took teaching in consideration. Universities had to demonstrate the implementation of research, the results of which were to be published in high-ranked publications listed by the Ministry of Science; these were periodicals with a high impact factor index. Education was still not treated as a scientific discipline in Poland. Hence, the academic achievements of faculty in this field were generally not recognized. The faculty of mathematics teacher education decreased in size, and even the people interested in the didactics of mathematics did not see the possibility of scientific development; therefore, they gave up hope of promotion (Hejnicka-Bezwińska 2011). The few people involved in didactics acted without support (neither grants nor financial support for taking part in international conferences was available to them). They often lost the motivation to self-improve. For example, in the area of the didactics of mathematics from 2010 to 2018, only three people obtained the title of doctor habilitatus, including one in Košice (Slovakia), a second in Olomouc (Czech

Republic), and a further one in Poland, but at the pedagogical university (Maria Grzegorzewska University in Warsaw—APS). All these people had problems getting their scientific achievements recognized by mathematical departments. The habilitation, obtained in Germany, was not respected in Poland at all because of a lack of official agreement between the Polish and German governments. This situation led to a slow shrinking of university offerings addressed to people interested in becoming mathematics teachers. Contemporary teaching qualifications for teaching mathematics can be acquired by enrolling in the following Universities:

- Pedagogical University in Krakow (the only university in Poland univocally focused on educating teachers)
- Jagiellonian University in Krakow (Faculty of Mathematics and Computer Science, second-cycle studies + a separate block of pedagogical subjects)
- University of Rzeszow (Faculty of Mathematics and Natural Sciences)
- University of Wroclaw (Institute of Mathematics)
- Adam Mickiewicz Poznan University (Faculty of Mathematics and Computer Science)

At some of these universities, not every program on teacher education is guaranteed to have the number of interested students that would be sufficient to launch that teaching specialization. The teacher education studies are very traditional. It is still not taken into account that building a foundation on psychological and pedagogical subjects, without considering specific mathematical education needs, is pointless. In the meantime, we believe a good substantive and didactic-methodical preparation of the teacher must be the starting point; first, a program should build a base on the form of didactics of mathematics and then improve, extend, and supplement it with pedagogy and psychology. Pursuing content that refers to research areas in the didactics of mathematics ("communicating in mathematics classes," "problem-solving," "gifted students," "algebraic thinking," "geometric thinking," "proof and proving," and so on) as a part of a university education typically is not easy—it depends only on the personal competences and interests of the person conducting classes, as no one suggests such topics in the curriculum.

Many universities have decided to offer teacher education in passing, offering various forms of postgraduate studies. Typically, they offer three semesters of extramural studies (classes conducted on Saturdays and Sundays). In this mode, the right to enter the teaching profession could be obtained by the graduates of such programs in which a certain number of hours of mathematical studies and a short course in mathematics didactics were provided. General pedagogical subjects are a separate unit, as there is an erroneous belief that the students themselves can transfer general statements to specific problems related to teaching a specific school subject—mathematics. The motivations for offering courses were often of an economic nature. As a result, many institutions offering courses or postgraduate studies appeared without having qualitative verification.

The postulate of "continuing teacher education" is still overlooked although never denied in theory. Obviously, during 5-year studies, it is hardly possible to provide knowledge and skills for the entire professional life of graduates. But the system of professional development is not in good shape (Pawlak 2011, pp. 99–104).

Against this unpleasant background, an initiative worth mentioning is the establishment of an institution in Warsaw called the School of Education. It is a joint initiative of the University of Warsaw and the Polish-American Freedom Foundation. The curriculum was developed in collaboration with Teachers College, Columbia University. Studies at SE are not typical postgraduate studies. Classes last 10 months and are conducted on a daily basis. From Monday to Thursday, students take apprenticeships at schools, and in the afternoons and on Fridays, they have academic classes conducted by the SE staff. After 1 year of study, they can gain teaching qualifications.

5.4 Compulsory Final Examinations in Various Types of Schools

The only state examination that students had to pass in the pre-1989 system was the secondary school exit examination (Matura) for students graduating from high school or technical school. At the end of this period, 50% of students were matriculated (Marciniak 2018). Only two examinations were mandatory: one on Polish language and one on an optional subject selected by the student from a choice of mathematics, biology, history, foreign language, etc.

The way of preparing questions and checking Matura examinations changed from time to time. At the end of this period, the questions were prepared in voivodeships (in each case there was a different set of tasks), and the examinations were checked by teachers from the same school as the examinee. This system of organization could not provide comparison on a national scale and objective information on the skill level of secondary school graduates. In order to take up studies at a university, one had to pass a secondary school exit examination in the relevant subjects, but the exam's result did not influence admission to university. University admission was based on the results of a separate exam organized by each institution. The examination requirements of some universities were significantly higher than the requirements of the core curricula or school curricula.

Mathematics ceased to be a compulsory subject in the secondary school exit examination in 1985. Since then, the number of those who chose this subject on the examination has decreased significantly. In 2009, less than 20% of high school graduates chose mathematics. The effects of this state of affairs began to be felt by technical universities, because there were no candidates for engineering studies. After a few years, it turned out that those who planned to study pedagogy to prepare for teaching grades one through three mostly did not pass mathematics in the secondary school exit examination. This meant that their involvement in the science of mathematics was already small in high school. In principle, such teachers avoided mathematics and were often afraid of it. They then went to school to learn to teach mathematics, among other subjects, in the early grades of primary school. It was one of the important factors in the decreasing level of mathematical skills of Polish students. Teachers who entered the profession at that time would remain in it for several dozen years. Since 1998, other examinations have started to appear in Poland in addition to the school exit exams. The new examination system was introduced along with the reform of education in 1998, which established gymnasiums. The new system implemented three compulsory examinations. The first of them took place after the 6th year of education and the next ones every 3 years: after the 9th year of study, at the end of junior high school, and after the 12th year, the end of high school (or after 13 years of education for technical secondary school students). Initially, mathematics was not a separate subject in the examinations after the 6th and 9th years of education. Mathematics questions appeared on the same examination sheet as other subjects. For example, the examination taken after the 9th year of schooling consisted of two parts, one of them devoted to humanities and the other to mathematics and natural sciences. In the second part, among the tasks in biology, chemistry, geography, and physics, there were also problems on mathematics.

All these examinations were prepared, carried out, and evaluated under the supervision of a specially appointed institution—the Central Examination Board. Licensed examiners checked their work. This system was introduced gradually with the appearance in subsequent years of the new core curriculum. The first sixth grade and junior high school student examinations took place in 2002 and the first new type of exam in 2005 (mathematics was not a compulsory subject in this exam). The sixth grade examination was compulsory, but its results had no effect on students' futures. The students' achievements in the lower secondary school examination were taken into account for admission to secondary schools, but this was not the only factor. Students' cumulative grades also determined 50% of admissions success to these schools. The results of the new matriculation examination were respected by universities and replaced the entrance examination stage of the recruitment process.

Regardless of the changes described above, efforts were made to restore mathematics as a compulsory examination subject in the secondary school exit examination. The reversal of the nonobligatory status of the mathematics exam was difficult to carry out. The first attempts were made at the end of the 1990s during the first postcommunist educational reforms. The return of mathematics to the matriculation exam was planned for 2002, as part of a wider project called the "New Matura." A year earlier, a new government was formed as a result of parliamentary elections, and the government withdrew from the project. Only in 2010 did mathematics again become a compulsory subject in the matriculation exam. The team that worked for several years on preparation for this change was headed by Professor Zbigniew Marciniak. Support for the undertaking was provided by the academic community, including rectors of the most important universities. University staff, especially at technical universities, were also leading supporters of this change. The introduction of the compulsory math exam was preceded by an expensive government promotional campaign on television and the Internet. In short videos, famous sportsmen, artists, architects, and businessmen talked about how mathematics was useful in their professional life.

After several years, along with the program reform of 2012, the system of examinations was also adjusted. First of all, mathematics became a separate subject in all three external examinations. Each graduate now must pass mathematics at least at an elementary level and may also additionally choose to take an examination at an advanced level. In 2018, 27% of high school graduates took mathematics at this higher level.

Another important element of the changes from 2012 was the combination of examination standards and the core curriculum into one document—a core curriculum was written in such a way that the requirements described in it became examination requirements, too.

5.5 Participation in International Student Assessment

PISA (Program for International Student Assessment) is the largest international study of student skills in the world. The survey has been carried out every 3 years since 2000 in all OECD countries and in partner countries. Poland has participated in the study since the beginning. In each variant of the study, one of the following areas would lead: mathematical skills, reading and interpretation, or scientific reasoning.

It is interesting that in 2000, the first surveys of 15-year-olds in Poland included students who were not covered by the reform of education. The subsequent variants of the study were held every 3 years. Such a "start" of the international survey in Poland allows the readers to objectively see the impact of the reforms undertaken. The success of the reformed schools can be seen in the fact that the difference in student achievements between schools has been significantly reduced as the level of teaching has leveled out. In the PISA 2000 study, Polish students obtained 470 points, which placed them on the 25th position among 41 countries which took part in the survey. Twelve years later, in 2012, the average mathematical result of Polish students was 518 points. This put them at 13th place among 65 countries and economies participating in the survey and fourth among European Union countries. Polish students have achieved a level of mathematical skills identical to Canadian students and statistically indistinguishable from Finnish students.

The results of the study from 2012 show that Polish students found themselves among the best in the European Union. Fifteen-year-olds with the highest and the lowest math skills improved their results. They achieved a marked improvement in tasks requiring complex skills: reasoning, argumentation, and creating their own strategy to find solutions.

Another international study in which Poland participates is the TIMSS (Trends in International Mathematics and Science Study). The aim of the study is to understand the level of knowledge and skills of students in the fields of mathematics and natural sciences. Poland participated in the study twice: in 2011, when the study included students from the third grade, and in 2015, when the pupils of the fourth grade were included in the survey. The results of the study, like the results of other national surveys, show that the teaching of mathematics in the first period of school education needs change. Polish third graders in 2011 achieved a result of 481 points, which was well below the average for the countries and territories taking part in the survey (500 points). Their score ranking was 34th out of 50 countries. 13% of Polish students obtained a result below the lowest level of 1 (400 points), and only 2% ranked in the highest 5 levels. In 2015, the results were much better. Polish students scored 18th out of 49 countries (535 points). Only 4% of the students were below level 1, and up to 10% were above level 5.

However, the results of these two variants of the TIMSS study can't be directly compared. The surveyed population in Poland in 2011 (third year students) had completely different experiences learning mathematics than the fourth year students in 2015. The latter ones not only studied mathematics a year longer, but they also had separate math lessons with a mathematics teacher. In Poland, until grade three, children have only one teacher (with a pedagogical education) who teaches all subjects. The subjects themselves are integrated—there is no rigid division between math, mother tongue, and other subjects. Lessons in the fourth grade are taught by a teacher with a university education.

5.6 Educational Research

Since people involved in research in the area of the didactics of mathematics still could not count on institutional and financial support for realizing their aims, the Institute for Educational Research should be regarded as the only institution conducting such research. This institution was established on December 9, 1952, under the name Institute of Pedagogy. Since 1990, it has been called the Educational Research Institute (IBE). It is an institution that is subordinate to the Ministry of National Education. It is supposed to, among other things, support educational policy and practice.

In 2010, the Educational Research Institute began implementation of the project "Research into the Quality and Effectiveness of Education and the Institutionalization of Research Facilities" (Enthusiasts of Education). This project was supported by the resources of the European Union under the Human Capital Operational Program. The main objective of the project was to strengthen the education system in the field of educational research and to increase the use of research results in educational policy and practice, as well as in educational management (Karpiński & Zambrowska 2015).

By 2015, as part of the project, over 100 research and analytical projects were carried out in five areas:

- Core curriculum and development of subject didactics
- Psychological and pedagogical foundations of school achievements

- Sociological and legal aspects of educational policy
- Economics of education: public and private outlays as well as educational markets
- Education and the labor market

In 2013–2014, IBE also carried out research, cofinanced by the European Union through the European Social Fund, Human Capital Operational Program, Priority III "High Quality of Education Systems," and Action 3.2 "Development of External Examination Systems."

The research conducted by the institute from 2010 to 2016 is the first such large empirical educational research in Poland. A dozen of the studies concerned mathematics. The activities carried out by IBE regarding mathematics can be divided into three types:

- Diagnoses of the knowledge and skills of students
- Research on teaching mathematics
- Teacher support

The study of students' mathematical knowledge and skills had to provide teachers with information on the strengths and weaknesses of their students. Often, the form and the type of tasks used in the diagnosis referred to the questions from a post-primary school test or a gymnasium exam. Sometimes, tips on how to interpret the obtained results were included. Results of classes participating in the study were presented to teachers against the background of the results of all students, in addition to the individual and class results of their students. The teachers received tips for further work with students. Participation in the diagnosis was voluntary, and the study was free. The school that started the diagnosis received the diagnostic sheet prepared by IBE and carried it out on a particular day. School teachers checked students' answers in accordance with the provided assessment scheme and provided information on solutions using a computer program. Schemes for assessing open tasks focused not only on whether the student gave a good or bad answer but also on what errors he/she made. This allowed for the preparation of individual recommendations for the further work of the teacher with the students. The study included students of various classes, and 50–70% of schools in Poland participated in them. The tests were conducted among third, fifth, sixth grade, and junior high school students (Grudniewska et al. 2013). The assumption was that the results of the classes and students were provided to schools participating in the research only. Often, however, local governments pressed schools to share the results with them and used the results to assess the work of schools, although the diagnostic tests were not intended for that purpose. This aroused worries among teachers who felt that their work was being evaluated and compared in this way.

General information about the results of the diagnostic tests appeared in the press, usually under alarming titles, such as "One-fifth of Third Grade Students Cannot Add and Subtract," "Problems with the Mathematical Thinking of Fifth Graders," and "Sixth Grade Test 2015 IBE Diagnosis Shows Pupils' Problems with Mathematics."

During the years 2011–2016, IBE carried out 11 diagnostic research studies:

- School of Independent Thinking (2011)
- Diagnosis of Junior High School Competences (2011, 2012)
- Nationwide Survey of Third Grade Skills (2013, 2014)
- Diagnosis of the Mathematical Skills of Fifth Graders (2014)
- Diagnosis of Sixth-Year Skills (2014)
- Third Graders' Competences (2015, 2016)
- Competences of Fifth Graders (2015, 2016)

Research on the teaching of mathematics was supposed to provide information on what teaching mathematics at different levels of education in Poland looked like. The results of research in this area were supposed to identify strengths and weaknesses in the teaching practices used in primary and middle schools. The results were to help the preparation of recommendations, the implementation of which would help improve the quality of mathematics teaching (Czajkowska et al. 2015). During the years 2012–2014, three additional studies were prepared and carried out at IBE:

- Research on teaching mathematics in primary school
- Research on teaching mathematics in middle school
- Research on the needs of teachers of early childhood education and mathematics in the field of professional development

The research tools that were used in the "Research in Primary and Secondary Schools" were designed to get the most accurate picture of mathematics education in primary and middle school. The most important recommendations resulting from the conducted research were:

- Changing the way of educating future mathematics teachers
- Ensuring a high quality of textbooks and other educational materials
- Developing a system of assessing mathematical skills
- Implementation of systems for the cooperation of mathematics teachers and teachers of early school education

The goal of "Research on the Needs of Teachers of Early Childhood Education and Mathematics in the Field of Professional Development" was to measure the mathematical and didactic competences of teachers and to find the areas where teachers who teach mathematics in primary and middle schools needed the greatest support. The most important conclusions of the study were:

- About 20% of teachers do not have the basic mathematic knowledge necessary to teach mathematics at the educational level at which the teacher works.
- About 20% of teachers do not have sufficient didactic knowledge, e.g., they cannot interpret mistakes made by students, they impose their way of solving a problem on a student, and they give too much importance to the formal record of a solved task.

- Mathematics laboratories in schools are not well equipped with traditional and multimedia aids to support mathematics teaching.

The results of the research described above also appeared in the newspapers and were also accompanied by alarming headings: "They Teach Our Children: They Divide by Zero and Do Not Know Percentages," "Report: Every Fifth Mathematics Teacher Should Not Teach," and "Fatal Results of Mathematics Teachers: They Are Not Able to Solve Tasks for Elementary Students."

In addition to diagnoses and tests, the IBE also provided support for teachers. The most important initiatives in supporting teachers for the teaching of mathematics included the following:

- "Children Think: How to Effectively Teach Children to Think Mathematically"
- "Tools in Action"

The goal of "Children Think: How to Effectively Teach Children to Think Mathematically," also known as the "Bydgoszcz Mathematical Bubble," was to launch a change to the teaching of mathematics in grades one through three of elementary school. The desired change was to introduce a peer tutoring method in teams of teachers as a tool to strengthen children's mathematical thinking and stimulate independent student activity. As part of the implementation of this measure within participating schools and between schools, methods for teacher cooperation were developed, and the development of methodological training for teachers was supported. An important element of the activities was the cooperation of IBE and local government authorities—which ensured the sustainability of this undertaking.

As part of the implementation of "Tools in Action," special sets of lesson tasks were prepared. Over 200 teachers participating in this activity were expected to use the tasks in class, discuss them with the class, and propose modifications or create similar tasks. The work of teachers was supported by mentors. The tasks prepared were concerned with those areas which—as was implied by other studies conducted by IBE—required support. Among them were the following:

- Tasks which required working in groups
- Tasks which had several correct solutions
- Tasks which required careful analysis of content
- Tasks which could be solved in very different ways

All tasks were placed in the deliberately created "Database of Good Practices."

5.7 *Mathematical Competitions as an Increasingly Common Form of Popularizing Mathematics*

Poland's accession to the European Union has opened up new areas of action to interested people. One of these areas is activities to support education. Mathematics quickly began to be an important area for the implementation of projects that support both students and teachers. Work with gifted students is a task on which a great deal of emphasis has been placed. This emphasis has been reflected in the wide promotion of various mathematical competitions for students in various age groups. Here are some of those that have been popularized or created in Poland:

The International Mathematical Kangaroo Competition—a competition that was invented by an Australian mathematics teacher very quickly became one of the best-known mathematics competitions. Its form—closed questions, marked by students on a special answer card—allows thousands of students to participate in the competition at the same time. The competition appeared in Europe in 1991 and in Poland in 1992. It is the most popular mathematical competition in Poland. In 2018, at all levels, from the first grade of primary schools to all other students, over 370,000 participated in it.

Nationwide and local projects supporting mathematical education:

- The Mathematics Olympiad is the oldest and most prestigious mathematical competition in Poland. Its initiator was the Polish Mathematical Society, and the first competition took place in the 1949/1950 school year. Students with high mathematical abilities take part in the Olympiads. Tasks that are to be solved often take on the form of open problems or statements that need to be proven.
- The project, Gifted from Pomerania. Pomorskie Province—a Good Education Course: Supporting students with special predispositions in the areas of mathematics, physics, and computer science, was carried out in 2010–2013. Its aim was to provide support to particularly gifted students. As part of the project, a model for a support system for gifted students was developed. Elements of the project were:
  - Involvement of universities in cooperation
  - Organization of extracurricular activities for students
  - Science camps
  - Scholarships and competitions for pupils
  - Workshops and training for teachers and parents

The project was cofinanced by the European Social Fund and the state budget under Priority IX of the Human Capital Operational Program 2007–2013.

- One of the activities of the Foundation of Wroclaw Mathematicians—an organization that was established in 2007 but continues the activities of the Scholarship Foundation of Wroclaw Mathematicians that operated in 1992–2007—is Mathematical Marathons. Participants solve various types of tests with mathematical problems "until the end." The competition is played in five age categories. Typically, the competition in each category lasts 12–18 h.
- The I Play, You Play, We Play (Gram, grasz, gramy) project has been implemented by the Let's Develop Society (Stowarzyszenie ROZWIŃ SIĘ Edukacja, Kultura, Sport) since 2007. The main goal of the project is the development of mathematical skills of children, youth, and adults. As part of the project, participants in the workshops play intentionally selected, generally available board games. While playing, they develop many mathematical skills: logical thinking,

anticipation of consequences, creation of a problem-solving strategy, counting in memory, spatial imagination, concentration of attention, and perceptiveness. The project implementation is cofinanced by funds from the capital city of Warsaw.
- The aim of the project The Secrets of Mathematics—A Comprehensive Program Supporting Students in the Second Stage of Education in Zawiercie is the development of key mathematical and social competences in pupils. As part of the project, students can participate in developing mathematics classes and benefit from psychological and pedagogical help. Teachers in the project raise their skills within the established cooperation network. The project was implemented in 2018–2019 and cofinanced by the European Union under the Regional Operational Program of the Śląskie Voivodeship for the years 2014–2020.
- *Banachiada* is a team mathematical competition for junior high school students, referring—both in the name and form of organization—to the forms and methods of work of Stefan Banach. It is organized in Jarosław and addressed to the students of Podkarpackie voivodeship. Students spend 5 h working on mathematical problems and use tabletops covered with paper as notebooks. At this time, teachers (mentors) participate in any form of professional development (listening to a lecture for example) and later invent tasks—problems that they write down in the "Scottish Book."

In addition, there are many local competitions for students of the first educational level, e.g., Little Abacus (Bydgoszcz),PajDej (Jarosław), Little League of Tasks, and similar ones.

6 Instead of Conclusion

It seemed that the reform of 1998–1999 would stabilize the educational structure. Unfortunately, it did not. While the functioning of gymnasiums as one of the places for education started to settle in with the social consciousness, the matter that still aroused much controversy was the decision to commence schooling for children starting at age 6.

The Act on the Education System assumed that from September 2014, all 6-year-olds would have to go to the first grade of primary school. Going to the first grade was to be preceded by the obligation to participate in classes for 5-year-old children. The battle over change in the educational system lasted until 2009, when a gradual reduction of the school age began—from 7 to 6 years old. Initially, for the period of those 5 years, the decision on the age of children to commence schooling (at age 6 or age 7) was to be made by parents.

In the years 2012–2013, the Minister of Education, then part of the government headed by Donald Tusk, made several controversial decisions that were evaluated by society as political.

In the 2012/2013 school year, 6-year-olds were required to start education in primary schools, while 5-year-olds were to be subject to compulsory preschool preparation in the so-called zero grades (formerly for 6-year-olds). Many Polish people did not accept such a solution, believing that parents should continue to decide when a child is ready to begin compulsory education. In addition, there was a belief that neither school nor teachers were ready for such a solution. There were no properly prepared and discussed core curricula or textbooks aimed at being used by 6-year-old children. Civic committees and associations demanding the postponement and even complete abolition of the law began to be formed. One such association was the Association of Children's Rights, which collected over one million signatures from parents under the petition "Save the Kids and Older Children Too" demanding a referendum. However, the government approved the matter of compulsory schooling for 6-year-olds.

Another controversial decision concerned the introduction of a free textbook for first grade, prepared by the Ministry of National Education. From September 2014, children in Polish schools began to study using free textbooks. In the following year, subsequent classes of primary schools as well as the first grade of junior high school joined in. In 2017, free textbooks were planned to be the basic source of knowledge for all students in primary and junior high school classes.

This decision was also rejected by society as having a political basis. In addition, the time to prepare the textbook was frighteningly short, and it was difficult to assemble a team that wanted to undertake this task. The Ministry defended itself by announcing that the versions being prepared would be made public on websites so that everyone could submit their comments and corrections. Such a solution was considered to be questionable. In addition, this decision was met with great opposition from the publishers, who already had their own textbooks and had a lot of didactic materials supporting the work of teachers. Some of them announced that they would conduct the free distribution of their own textbooks to schools, regardless of the Ministry's campaign.

The textbook for the first grade of primary school was written utilizing the integrated teaching approach, and the pages designed for mathematical problems were chaotic. Subordination of mathematical issues to other social themes, which was supposed to demonstrate a modern approach to school issues, resulted in a very low evaluation of this publication by both experts and a large number of teachers, who just put the free textbooks away. Publishing houses began to boost their financial condition by issuing workbooks, adapting them to the government's free textbooks.

Below (Fig. 4.5) is one page from the book, dedicated to elaboration of the number 5. The questions under the picture are:

1. What color is the thumb painted?
2. Paint your index fingers with any paint and press them on a piece of paper. Using a magnifying glass, compare your fingerprints in pairs. What do you notice?
3. Paint a picture with your fingers. Use your fingers instead of a brush.
4. What sense do you associate your hand with?

The next ruling party decided to completely abandon the decisions of earlier ministries. They announced this during the election campaign, and the results of the polls showed the need for change.

One of the new administration's decisions affected the statutory age for beginning compulsory education. According to data from a survey, carried out by

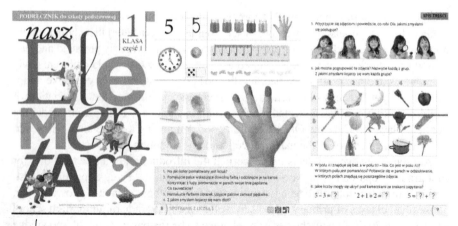

Fig. 4.5 The free textbook prepared by the Ministry of Education and its two pages devoted to the number 5

Millward Brown SA[5] on August 24, 2015, on a nationwide representative sample of 1003 adults and published in the newspaper Rzeczpospolita, 75% of Poles were in favor of abolishing the general statutory schooling obligation for 6-year-olds and restoring the start of the universal statutory school duty from the age of 7. In the 2016/2017 school year, based on the Act of December 14, 2016, the expected change came into effect: abolition of compulsory education for 6-year-olds. According to the amendment, children were required to start education at the age of 7, after an annual obligatory preschool preparation for 6-year-olds.

However, this change brought some chaos (Wojnar et al. 1999). Apart from ensuing disputes (not everyone accepted this solution), the Ministry had to find a place in the system for children who—as 6-year-olds—had already begun their school careers. The Ministry allowed parents of 6-year-old children who had been sent to first grade that school year to reenroll them in the same grade the next year, and such a child was not subject to annual classification.

Another thing was the issue of free textbooks and, more broadly, textbooks in general. The free textbooks for schools were dropped. Educational publishers could prepare proposals, three of which could be recommended by the Ministry. Those recommended (and all others) had to take into account the content of the core curriculum (which was prepared in a great hurry again), by separate, non-cooperating teams responsible for the various educational stages.

In the social consciousness, however, it was not the case of the 6-year-olds or the free textbooks which is perceived as the biggest change of the last few years. The Ministry of Education gave up what seemed to be the most important part of the

---

[5] http://www.millwardbrown.com/. SMG/KRC (untranslatable proper name) Poland was founded in 1990 as one of the first independent research institutes in Poland. Since 2000, SMG/KRC Poland has been a member of the global Kantar Millward Brown network and has adopted the name Kantar Millward Brown. SA—abbreviation of "Spółka Akcyjna" (Joint-Stock Company).

1999 reform; it made another change to the school system (Regulation of the Minister of National Education of February 14, 2017, Regulation of the Minister of National Education of January 30, 2018). The educational stages were to be as follows:

1. An 8-year primary school
2. A 4-year high school, a 5-year technical school, or a school preparing students for a profession
3. One of the consequences of this change was the elimination of the test for sixth graders. The test is now to be taken by students finishing the eighth grade, although there is still little knowledge as to what form it should take and how it should affect the further fate of the pupil.

According to the Ministry, "the prepared reform of the education system meets the expectations of the majority of Poles who want a modern school, and at the same time, one which is strongly rooted in our tradition."[6] The Ministry also believes that "the changes are well thought out and were planned for many years."[7] However, such statements are not always able to convince society. On many social forums on the Internet, both the opponents of this change and the people who support and accept the new shape of the system are still arguing. Members of various political groups, among others, spoke to the press. They asked for the reasons behind the reform, with questions like: "Why are we all doing all of this? Because we are not talking about the core curriculum, about progress, about jumping forward. We live in sentiments. What do you foresee as the results?" Another politician, from the ruling party, when asked for information about the consequences of the reform and whether it would result in the dismissal of teachers, said, "A statutory solution will be adopted so that junior high school teachers can easily go to work for an 8-year elementary school or secondary school."[8] Discussions generally concern social and organizational issues. The fact is that the organizational change will not be easy for schools. Due to the reform of education in 2019, students from the past 2 years will enter the first year of high school at the same time equaling a total of 700,000 students. According to the Ministry of National Education, students will enter into separate departments (some will go to a 3-year high school, and others will go to a 4-year high school). The situation of students repeating classes will be very complicated—it may turn out that there is no class in which a student has something to learn. Contrary to assurances that no teacher will lose his or her job, the reorganization of education has already caused many problems with employment.

---

[6] http://reformaedukacji.men.gov.pl/—Special website prepared by Ministry of Education with information about reform in education.

[7] http://reformaedukacji.men.gov.pl/

[8] https://wiadomosci.wp.pl/poslowie-pytali-minister-anne-zalewska-o-powody-i-skutki-wprowadzenia-reformy-edukacji-6039756550902401a [Deputies-asked-Minister-Anna-Zalewska-about-reasons-and-effects-of-educational-reform-implementation]

All this means that substantive issues related to teaching methodology, teacher training, and even research work in the area of mathematics education will be pushed to the background.

In Poland, we still do not have a good situation for reliable and calm work in mathematical education.

References

Abramowicz, Tomasz, and Marek Okołowicz. 1978. *Matematyka V* [Mathematics V]. Warsaw: WSiP.
Burnat, Marek. 1987. Listy do Redakcji: Głos malkontenta, czyli jeszcze o programach matematyki [Letters to the Editor: The voice of malcontent regarding mathematics programs]. *Wiadomości Matematyczne* XXVII: 326–330.
Chrzan-Feluch, Barbara, Wacław Zawadowski. 1978. *Matematyka 5* [Mathematics V]. Warsaw: WSiP.
Ciesielska, Danuta. 2012. Koncepcje kształcenia nauczycieli matematyki w Polsce po II wojnie światowej [Conceptions in the educating of mathematics teachers in Poland after the Second World War]. In *Opinie edukacyjne Polskiej Akademii Umiejętności, Prace Komisji PAU do spraw Oceny Podręczników Szkolnych*, ed. Grzegorz Chomicki, vol. vol. X, 33–54. Cracow: PAU.
Ciesielska, Danuta, and Jerzy Szczepański. 2017. Szkolny Podręcznik do matematyki z perspektywy czterdziestu lat (1975–2015) [School Handbook for mathematics from a forty-year long perspective (1975–2015)]. In *Opinie edukacyjne Polskiej Akademii Umiejętności, Prace Komisji PAU do spraw Oceny Podręczników Szkolnych*, ed. Grzegorz Chomicki, vol. vol. XV, 101–122. Cracow: PAU.
Czajkowska, Monika, Marzenna Grochowalska, Margaret Orzechowska. 2015. Potrzeby nauczycieli edukacji wczesnoszkolnej i nauczycieli matematyki w zakresie rozwoju zawodowego [The needs of teachers of early childhood education and mathematics teachers in the field of professional development]. Warsaw: IBE.
Flis, Marek. 2011. Szkoła wyższa w toku zmiany a wewnętrzny system jakości kształcenia [University in the course of changes and an internal system of education quality]. In *Szkoła wyższa w toku zmian*, ed. Janina Kostkiewicz and Mirosław Szymański. Cracow: Impuls.
Frycie, Stanisław. 1989. *Przemiany w treściach kształcenia ogólnego* [Transformations in the content of general education]. Warsaw: WSiP.
Gazeta Wyborcza. 1999, 23 Feb. Wiadomości. *Gimnazja, Spór czy je wprowadzać. 1999.* [Gymnasium: The dispute whether to enter them. 1999].
Grudniewska, Alina, Marcin Karpiński, Małgorzata Zambrowska. 2013. *Raport z badania Nauczania matematyki w gimnazjum* [Report of the survey findings on Teaching Mathematics in Middle School]. Warsaw: IBE.
Gruszczyk-Kolczyńska, Edyta. 2019. O fatalnym stanie edukacji matematycznej dzieci i o konieczności podejmowania szybkich działań naprawczych [On the disastrous state of mathematics education for children and the need for undertaking quick remedial actions]. In *Wybrane obszary praktyki w edukacji wczesnoszkolnej*, ed. Grażyna Kwaśniewska and Ewa Swoboda, 51–75. Jarosław: PWSTE.
Hejnicka-Bezwińska, Teresa. 2011. Próba odczytania strategii procesu zmian w szkolnictwie wyższym (przeprowadzonych w ostatnim dwudziestoleciu) [An attempt at explaining the strategy of change to the process of higher education (carried out over the last twenty years)]. In *Szkoła wyższa w toku zmian*, ed. Janina Kostkiewicz and Mirosław Szymański. Impuls: Krakow.

Karpiński, Marcin, Małgorzata Zambrowska. 2015. *Raport z badania nauczania matematyki w szkole podstawowej* [Report of the survey findings on Teaching Mathematics in Grammar School]. Warsaw: IBE.

Kraś, Jacek. 2007–2008. *Reforma systemu oświaty w III RP. Założenia i realizacja* [Reform of the education system in the Third Republic of Poland: Assumptions and implementation]. *Resovia Sacra. Studia Teologiczno-Filozoficzne Diecezji Rzeszowskiej* 14–15: 303–318.

Krygowska, Zofia. 1977. *Zarys dydaktyki matematyki* [Outline of the didactics of mathematics]. Warsaw: WSiP.

———. 1981. *Koncepcje powszechnego matematycznego kształcenia w reformach programów szkolnych z lat 1960—1980* [Concepts of public mathematics education in the reforms of school programs from 1960–1980]. Cracow: Wydawnictwo Naukowe WSP.

———. 1987. Listy do Redakcji: W sprawie programów szkolnych [Letters to the Editor: On school programs]. *Wiadomości Matematyczne* XXVII: 331–333.

Krygowska, Zofia, Bolesław Rok. 1984. *W czystym powietrzu matematyki, z prof. Z. Krygowską rozmawia Bolesław Rok, z cyklu Rozpacz w kratkę*, [In the clean air of mathematics: prof. Z. Krygowska is interviewed by Bolesław Rok, from the series "Checkered" desperation]. *Polityka*, 28 no. 25.

Łabanowska, Hanna. 1976. *Matematyka. Podręcznik dla klasy czwartej* [Mathematics. Manual for fourth grade education]. Warsaw: WSiP.

Legutko, Marek. 1999. *Krótka historia trzech fal reform polskiej oświaty* [A brief history of three waves of reforms in Polish education]. Warsaw: WSiP

Łysek. Jan. 2003. Proces zmian kształcenia i doskonalenia nauczycieli w Polsce [The process of change in teachers education and improvement in Poland]. *Nauczyciel i Szkoła* 1–2 (18–19): 31–49.

Marciniak, Zbigniew. 2018. *How Poland moved ahead, "21st century mathematics."* Geneva, Switzerland: conference on May 25, 2018.

Ministerstwo Oświaty [Ministry of Education]. 1950. *Program nauki w 11-letniej szkole ogólnokształcącej (projekt) Matematyka* [Curriculum for an 11-year general school (project) on Mathematics]. Warsaw: PZWS.

Nowecki, Bogdan. 1984. *Krakowska Szkoła Dydaktyki Matematyki* [Krakow School of the Didactics of Mathematics]. Krakow: WN WSP.

Pawlak, Ryszard. 2011. Nowa koncepcja kształcenia studentów w ramach przygotowania do wykonywania zawodu nauczyciela [A new concept for educating students in preparation for teaching]. In *Rocznik Pedagogiczny 34, Polska Akademia Nauk, Komitet Nauk Pedagogicznych*, ed. Maria Dudzikowa. Radom: Instytut Technologii Eksploatacji – BIP.

Rura, Grażyna, and Michał Klichowski. 2013. Założenia programowo-organizacyjne reformy oświaty z 1999 roku w zakresie edukacji elementarnej [Assumptions of the program and organizational reform to education in 1999 in the field of elementary education]. In *Dziecko w szkolnej rzeczywistości. Założony a rzeczywisty obraz edukacji elementarnej*, ed. Halina Sowińska. Poznan: Wydawnictwo Uniwersytetu im. A. Mickiewicza w Poznaniu.

Strzelecka-Ristow, Anna. 2015. *Kolegia nauczycielskie jako alternatywna forma kształcenia nauczycieli – idee założycielskie, oczekiwania, niepokoje* [Teachers' colleges as an alternative form of teacher education – founding ideas, expectations, anxieties]. *Forum Oświatowe* 27 (2): 211–229.

Szymański, Karol, Jakub Zygadło. 1979. *Matematyka 5* [Mathematics 5]. Warsaw: WSiP.

Turnau, Stefan, Marianna Ciosek, Maria Legutko. 1979. *Matematyka 4* [Mathematics 4]. Warsaw: WSiP.

Wiadomości Matematyczne [Mathematical Messages] XXIX. 1990. *Tezy raportu o edukacji matematycznej w Polsce* [Theses on the report on mathematics education in Poland], 113–121.

Wiadomości Matematyczne [Mathematical Messages] XXVII.1. 1987. *Stanowisko Komitetu Nauk Matematycznych PAN w sprawie szkolnych programów i podręczników matematyki przyjęte na posiedzeniu plenarnym w dniu 28 lutego 1986 r.* [The position of the Committee of Mathematical Sciences of the Polish Academy of Sciences regarding school programs and mathematics textbooks, adopted at the plenary session of February 28, 1986], 190–191.

Wojnar, Irena, Andrzej Bogaj, and Jerzy Kubina. 1999. Strategie reform oświatowych w Polsce na tle porównawczym [Strategies for educational reforms in Poland against a comparative background]. Komitet Prognoz "Polska 2000 Plus", Warsaw.

Zawadowski, Wacław. 1985. *Matematyka 5* [Mathematics 5]. Warsaw: WSiP.

Cited Legal Acts

Dziennik Ustaw 1985, Ministerstwa Nauki i Szkolnictwa Wyższego nr 5, poz. 40 [Journal of Laws 1985, of the Ministry of Science and Higher Education No. 5, item 40].

Dziennik Ustaw 1991, nr 95 poz. 425 [Journal of Laws 1991, No. 95 item 425].

Rozporządzenie Ministra Edukacji i Edukacji z 22 kwietnia 1982 r. [Regulation of the Minister of Education and Education of the 22nd of April 1982].

Rozporządzenie Ministra Edukacji Narodowej z 5 marca 1992 r. [Regulation of the Minister of National Education of March 5, 1992].

Rozporządzenie Ministra Edukacji Narodowej z dnia 15 lutego 1999 r. w sprawie podstawy programowej kształcenia ogólnego. Dz.U. 1999 nr 14 poz. 129 [Regulation of the Minister of National Education of February 15, 1999 on the core curriculum of general education. Journal of Laws 1999 No. 14 item 129]

Rozporządzenie Ministra Edukacji Narodowej i Sportu z dn. 18 kwietnia 2002 r. w sprawie określenia standardów nauczania dla poszczególnych kierunków studiów i poziomów kształcenia (Dz. U. Nr 116, poz. 1004) [Regulation of the Minister of National Education and Sport of April 18, 2002 on determining teaching standards for specific fields of study and levels of education].

Rozporządzenie Ministra Nauki i Szkolnictwa Wyższego z dn. 13 września 2007 r. w sprawie standardów kształcenia dla poszczególnych kierunków oraz poziomów kształcenia [Regulation of the Minister of Science and Higher Education of September 13, 2007 on education standards for specific fields and levels of education]

Rozporządzenie Ministra Nauki i Szkolnictwa Wyższego z dnia 17 stycznia 2012 r. w sprawie standardów kształcenia przygotowującego do wykonywania zawodu nauczyciela [Regulation of the Minister of Science and Higher Education of 17 January 2012 on standards of education preparing for the teaching profession].

Rozporządzenie Ministra Edukacji Narodowej z dnia 14 ltuego 2017 r. w sprawie podstawy programowej wychowania przedszkolnego oraz podstawy programowej kształcenia ogólnego dla szkoły podstawowej, w tym dla uczniów z niepełnosprawnością intelektualną w stopniu umiarkowanym lub znacznym, kształcenia ogólnego dla branżowej szkoły I stopnia, kształcenia ogólnego dla szkoły specjalnej przysposabiającej do pracy oraz kształcenia ogólnego dla szkoły policealnej (Dz.U. z 2017 r. poz. 356). [Regulation of the Minister of National Education of February 14, 2017].

Rozporządzenie Ministra Edukacji Narodowej z dnia 30 stycznia 2018 r. w sprawie podstawy programowej kształcenia ogólnego dla liceum ogólnokształcącego, technikum oraz branżowej szkoły II stopnia (Dz.U. z 2018 r. poz. 467) [Regulation of the Minister of National Education of January 30, 2018].

Uchwała nr 200/98 RGSW z dnia 16.04.1998 r. w sprawie określenia minimów programowych dla studiów magisterskich na kierunku matematyka [Resolution No. 200/98 of The Main Council of Higher Education (Rada Główna Szkolnictwa Wyższego: RGSW), April 16, 1998 on determining the curriculum minima for master's studies in mathematics].

Ustawa z dnia 12 września 1990 o szkolnictwie Wyższym [Journal of Laws 1990, No. 65, item 385 from 12th of September 1990 on Higher Education].

Ustawa z dnia 25 lipca 1998 r. O zmianie ustawy o systemie oświaty, Dz. U. 1998. Nr 117, poz. 759 [Act of July 25, 1998 amending the act on the education system, Journal of Laws 1998. No. 117, item 759].

Ustawa z dnia 27 lipca 2005 r. Prawo o szkolnictwie wyższym (Dz. U. Nr 164, poz. 1365) [Law on Higher Education from the 27th of July 2005]

Ustawa z dnia 14 grudnia 2016 r. – Prawo oświatowe (Dz.U. z 2018 r. poz. 996) [Act of the 14th of December 2016, Journal of Law, 2018, item 996].

Zarządzenie nr 23 Ministra Edukacji Narodowej z dnia 18 sierpnia 1992 r. [Decree No. 23 of the Minister of National Education of August 18, 1992].

Chapter 5
Russian Mathematics Education After 1991

Alexander Karp

Abstract The aim of the present chapter is to trace the recent history of Russian mathematics education, considered as part of the social history of Russia. Although by contrast with what has taken place in politics, in mathematics education there has never even been talk about "perestroika" or restructuring, let alone any radical changes, the changes that have taken place over the last 30 years have in fact been significant and are clearly connected, even if often not directly, with social-economic changes. This chapter will focus on the main aspects of mathematics education: the organization of the learning process, textbooks, exams, the preparation and professional development of teachers, and so forth.

Keywords Russia · Reform · Standards · Textbooks · Uniform State Exam · Teacher education · Gifted education

The Soviet Union officially collapsed in December 1991. Of the countries that came into being on its former territory, the largest is Russia or the Russian Federation, which will be the subject of the present chapter. In the subsequent almost 30 years of its existence, it has seen all manners of political as well as cultural changes. It does not follow, of course, that these changes were always accompanied by changes in the teaching of mathematics; the changes that took place in this sphere—and such changes undoubtedly did take place, even if they might not have seemed as dramatic as the political ones—were produced by a combination of factors, including both domestic developments in mathematics education and processes common to the whole world; what is clear, however, is that the development of mathematics education did not occur in isolation from what was taking place outside the doors of mathematics classrooms. The aim of the present chapter is to trace the recent history of Russian mathematics education, considered as part of the social history of Russia as a whole.

A. Karp (✉)
Teachers College, Columbia University, New York, NY, USA
e-mail: apk16@columbia.edu

© Springer Nature Switzerland AG 2020
A. Karp (ed.), *Eastern European Mathematics Education in the Decades of Change*, International Studies in the History of Mathematics and its Teaching,
https://doi.org/10.1007/978-3-030-38744-0_5

Studying recent history is usually not very easy and not only because certain documents are inaccessible—we still have an enormous quantity of various publications at our disposal—but because these publications often exude a political fervor, which may infect their authors and cause them to distort the facts. Up until a certain time, the author of the present chapter himself took part in what was happening—as a teacher, as an author of textbooks, and even as an administrator—and fully acknowledges the fact that he must not elevate either his experience or his observations into absolutes, without, however, excluding them from consideration. To a certain extent, the goal of this chapter is precisely to collect various observations and opinions, whose collocation with official information might help to paint a true picture.

It should also be noted that, as far as the author of this chapter knows, there have hitherto been practically no attempts at a survey study of what has occurred in mathematics education in Russia over the past three decades. The closest approximation to such a study is M.I. Bashmakov's (2010) chapter, to which we will repeatedly refer in what follows.

It is necessary to voice yet another caveat: Russia is the largest country in the world in terms of area, and living conditions in it—and therefore also the conditions of teaching mathematics—are by no means everywhere uniform. Furthermore, it is evident that what is taking place in mathematics education in different parts of the country is not treated with the same level of detail in the press and in other types of publications. In writing about Russia below, the author will attempt not to confine himself to Moscow and St. Petersburg, but, undoubtedly, more will be said about them than about any other parts of the country.

Below, we will address various directions in the changes that have occurred, but clearly we must begin by discussing what mathematics education was like in Russia during the last years of the USSR.

1 Mathematics Education in the Last Years of the USSR

By the late 1970s and early 1980s, Soviet mathematics education, although formally preserving its main former features, in fact had become significantly different from what it had developed into during the 1930s–1940s (Karp 2010). Undoubtedly, the system was still absolutely centralized—curricula and textbooks were approved and endorsed in Moscow, at the Ministry of Education of the RSFSR (in keeping with uniform Soviet Union requirements). The streamlined vertical system of methodological direction—consisting of a specialist in mathematics education at the ministry, regional methodologists, and district methodologists—was supported by and was a part of a general vertical system of control, from the Ministry, to the regional Departments of Education, and further to district departments. Mathematical methodologists, working with departmental inspectors, were tasked with, as it was said, direction and control over methodological work, making sure that curricula were followed in the manner prescribed in Moscow and that the requisite results were consequently achieved.

In practice, however, in the 1980s, the system worked even less smoothly than it had originally when it was first established. In many respects, human resources had already been exhausted. A shortage of mathematics teachers had already begun, and in general, it had become necessary to soften the former severity (although, to be sure, the system could by no means have been characterized as "soft"). For all the virtues of the developed methodology, it could not provide the desired level of results; but if during the 1940s, along with fighting underachievement, it was deemed necessary to fight against "window dressing" and complacency, so that 15–20% of students in a class failing was not something unheard of (Karp 2010), now times had changed, and it became undesirable to have even 1–2% of students in a class fail. The constitution of the country, from 1977 on, guaranteed the provision of universal mandatory secondary education. A consequence of this was the development of what at the time was called "percentomania," in other words, the awarding of grades above what a student deserved according to existing norms, so that the class and the school, and thus also the district and the city, might achieve higher results. The measurement of results (above all, examinations) could turn into a yearslong virtually open racket. The author of this chapter remembers, for example, how beginning in the second half of the 1970s and for approximately 10 years the problems on final examinations in mathematics for mass-scale schools would become known in Leningrad (St. Petersburg) to both teachers and students several days before the examination. Again, drawing on my own experiences, I can recall hearing a talk at the Collegium of the Ministry (State Committee) by V.D. Shadrikov, one of the heads of Soviet education at the time, who spoke about the fact that no one actually knew what results had been achieved—the official figure was that virtually 99% of students in Russia learned elementary calculus. But how many learned it in reality? Even if it was only 50%, that was wonderful, but the information simply did not exist.[1]

By the mid-1980s, the system had abandoned one of its main principles—the single textbook. Formerly, the whole country had been taught using the same textbook; now, as a result of a struggle among various influential groups following the abandonment of the so-called Kolmogorov reform (which to a certain extent paralleled New Math in the United States), it became necessary to conduct an open competition for textbooks and subsequently to allow for the parallel existence of several textbooks on the same subject for the same grade (Abramov 2010; Bashmakov 2010). In fact, a quarter of a century earlier, it had already become clear that the acknowledgment of certain differences among students was inevitable: there appeared schools with an advanced course in mathematics (Karp 2011), to which the Soviet educational system largely owed its high international reputation.

The Kolmogorov reform was an attempt to change the content of mathematics education—an attempt that was inevitable, since to continue forever with a course that had effectively taken shape even before the revolution of 1917 was impossible; but it was a reform whose most innovative components—for example, the introduction of elementary discrete mathematics—failed (Bunimovich 2011). The unified centralized system was not well suited for updating content (recall that classic Soviet textbooks—above all, the legendary textbooks of Kiselev—had come into the schools even before 1917 in the context of an open competitive struggle, which no internal review process could replace). It cannot be said, of course, that everything done by the reformers failed: despite the negative discussions in the press, including such leading communist party periodicals as *Kommunist*, and the denunciation of certain approaches employed by Kolmogorov, for example, set theoretical symbolism, much in the course survived; but nevertheless, the general view was that it was necessary to return to the tried and the old—by creating, as it were, an updated Kiselev (Karp and Werner 2011).

Attempts were also made to reform the manner in which classes were conducted: thus, V.F. Shatalov, a mathematics teacher from Donetsk who published several books (Shatalov 1979, 1980, 1987) that advocated the method of *supportive abstracts*—which allegedly gave incredible results—became popular in the late 1970s. Although Shatalov's arguments were harshly criticized (Dadayan et al. 1988) and the wave of enthusiasm for his methods gradually subsided, their very popularity was indicative and attested to a sense of incongruity in what was happening in mathematics education (which had traditionally and largely justifiably been considered a strong or even the strongest part of Soviet education) in the hope of finding some kind of miraculous approach that would remedy everything, without, however, really changing anything. We should also note that a sense of crisis grew with the initiation and development of Gorbachev's "Perestroika," which called for a reconsideration of the traditional Soviet system as a whole.

2 Reforms in Education

2.1 *The General Situation*

Reforms were viewed as changes not specifically of mathematics education but of the whole educational system. Plans for such reforms had already been drawn up relatively long before 1991. As early as 1988, the State Committee on Education had formed a temporary scientific research collective named "The Basic School" (and later called simply "The School"), which was directed by Eduard Dneprov—who later became Russia's minister of education—and which became a center for the preparation of reforms. Pertinent ideas were expressed relatively quickly, but their realization in the USSR kept getting postponed. But they were put to use when Russia became an independent state. Dneprov himself (Dneprov 1998) articulated the causes of the crisis quite clearly:

> The critical condition of the schools, which had become apparent already by the early 1980s, reflected an analogous condition of society and stemmed from the same basic cause—a crisis in the totalitarian regime, the exhaustion of resources for its development[1] (p. 36).

---

[1] All translations from Russian are by the author.

In his opinion, the reform had to be based on the following principles (pp. 46–52):

- Democratization of education (including destatization of the schools and decentralization of their administration)
- Pluralism of education (including its multistructurality, variability, and alternativeness)
- The people and national character of education
- Openness of education (in particular, emancipation from dogmas)
- Regionalization of education
- Humanization of education (the school must turn toward the child)
- Humanityzation of education (the school must pay more attention to humanities)
- Differentiation of education
- The developmental and practical character of education
- Continuity of education.

This and similar phraseology became the basis of legislative measures passed at the beginning of the Yeltsin period, including the Education Act of 1992 and the famous Decree No. 1 on Top Priority Measures for the Development of Education in the RSFSR (Russian Soviet Federative Socialist Republic), which was enacted in July 1991 and which by the very number in its title emphasized the fact that the government's problems of top priority were no longer in the sphere of the arms industry but in the area of education and spiritual development.

What, then, did all of this mean for mathematics education? Here, we must immediately acknowledge that school life was by no means regulated exclusively by the legislation that governed education but also simply by life itself—the unfolding economic crisis (not to say catastrophe) inevitably had an influence on the schools. Teachers, whose position had recently still been quite stable, even if they did not belong to the wealthiest sections of society, suddenly found themselves in a situation in which their salaries fell to fractions of what they had been previously and moreover were paid very irregularly. This could not but have an impact on teaching and teachers' professional ethics. A blow was dealt to parents as well, whose preoccupation with work also began to change (and therefore also their attitude toward schools and their opportunity to help children in their studies). Not the least significant role was also played by psychological factors—students and their parents saw that the former goal of becoming an engineer (for which one needed to be a good student in mathematics) was now quite questionable—engineers were losing their work and livelihood en masse. Completely different professions became prestigious. It must also not be forgotten that mass emigration began—many scientists left the country, whose influence on schools, in one way or another, had been considerable. Dneprov (1998, p. 106) especially notes that education reformers, by contrast with other reformers of the Gorbachev and subsequent periods, had a clear plan and program, and while others, having gotten the airplane off the ground, as it were, did not know where it would land, education reformers supposedly knew everything. The problem with this view, however, is that education is not an isolated airplane—consequently, there neither was nor could there have been any "pure" experiment to determine what would have happened if education had been changed according to Dneprov's plan, while everything else remained stable and sound. The overall effect from all the transformations (and not just directly in education) was not sufficiently considered, and it is naive to reduce matters to the incompetence of specific leaders—after all, Dneprov himself particularly praises Moscow mayor. Luzhkov, who, by contrast with other local politicians, found money for education (Dneprov 1998, pp. 152–155), ignoring Moscow's special position in the country.

But let us return to specifically educational issues. Even before the passing of the Education Act, which Dneprov regards as the starting date of the reform (Dneprov 1999, p. 13), there was a feeling that the pressure of prohibitions had weakened, and consequently all kinds of local experiments that affected the teaching of mathematics were beginning to take place, and people also simply sensed that there was less oversight and control.[2] The administrative component of the national reform that was probably most tangible to mathematics teachers—already in June 1993, under Minister Ye.V. Tkachenko, who had replaced Dneprov—was the appearance of a new basic teaching plan or, more precisely, different versions of a basic teaching plan (see, for example, Committee 1994). The corresponding Decree No. 237 from the Ministry of Education legalized the changes that had already started taking place and pushed them further. Embracing the principles of pluralism, the ministry ceased specifying a precise number of hours to be allocated to each subject in each classroom across the whole country. Certain areas of education were defined—social sciences, natural sciences, Russian language, mathematics, and others—and for each of them, the minimal possible number of hours per week was indicated. For example, for mathematics in the two upper grades, this number was 3 h. Along with mandatory "ministry" hours, there were also regional hours, allocated according to the decisions of the regional governments (for example, the government of St. Petersburg could decide that all of the city's students were required to study the history of St. Petersburg, while the government of some autonomous republic might make the republic's history a requirement, thus fulfilling the principle of attention to national character), and finally, hours allocated by each school itself.

In this way, schools all at once acquired the right radically to change the existing structure—it became possible to teach mathematics in the upper grades for 3 h, 5 h, 8 h, or even more. It became possible radically to reduce the teaching of physics, which was evidently connected with the teaching of mathematics, replacing it either with some kind of integrated course or with some subject that could be related to the natural sciences.

In principle, the newly acquired freedom could only have been a cause of rejoicing; but not sufficient thought had been given to how and why each school would make its specific choices, particularly given the numerous constraints that each

---

[2] See, for example, the article by Eidel'man (2007) with the characteristic title "The Year of Realized Utopias: Schools, Teachers, and Education Reformers in Russia in 1990," in which the author, copiously quoting other teachers, tells about the changes that occurred at that time (even though, despite her obvious admiration for what took place then, after quoting certain discussions from the period, she sometimes adds: "today one can detect a certain touch of madness in these words").

school was under, beginning with the fact that everything that a school had at its disposal (for example, its teaching staff) had been organized with another system in mind. Thus, if a school was short of a physics teacher, for example, the principal could decide that there was no need for this teacher. The choice of a teaching curriculum might stem from simple administrative convenience, rather than from any higher considerations. There was no mechanism for accommodating existing needs and opinions nor were there any mechanisms for gathering information, without which even the formation of such opinions became problematic (to be sure, what such mechanisms should be like is a difficult question and unlikely fully solved anywhere).

Schools actively competed for "good" children, inventing ways to attract them, or at least not to lose them, keeping them from transferring to other schools—because such transfers also became easier and freer. Consequently, schools often started making use of a "brand" that had earned a good reputation—an advanced course in mathematics. New "humanities" schools also appeared, which promised at the outset not to torment children with mathematics. Some schools offered something altogether unexpected, hoping to attract students by this means—the author of this chapter once visited a school that offered the study of the Coptic language (already in elementary school), which in the opinion of the school's directors would guarantee genuine depth of preparation in the humanities. To repeat, such "specialized" schools and classes began appearing even somewhat before the law was passed—people felt that this was already possible.

There was less time to pay attention to children who were "not good," especially since there was less enforcement of such attention, just as there was less enforcement in general, while the time that a teacher had available to spend hours with "weak" students—and probably the desire that a teacher had to do so—was often reduced to nothing: it was simpler to give a student the coveted passing grade of "3" (out of a possible 5). On the whole, one might say that both "good" and "bad" teachers acquired greater freedom to act as they pleased.

The following information—to a certain degree and with certain caveats—conveys an idea of the state of affairs that existed at the time.

2.2 Diagnostic Work in St. Petersburg in 1992

In October 1992, so-called diagnostic work was conducted in St. Petersburg in all tenth grades in two districts, in 49 schools altogether (here and below we will rely on the publication Committee 1994). The diagnostic work was based on materials of basic schools (that is, 9-year schools). It must be emphasized that the results of the work pertain to St. Petersburg, and there is no reason to infer that the results would have been identical everywhere else in the country; moreover, even the selection of two districts out of the twenty that existed at the time—a large and somewhat peripheral one (Kirovsky) and a small and central one (Dzerzhinsky, which was subsequently eliminated as a separate administrative unit)—can be subjected to criticism; for example, Dzerzhinsky District happened to have one of the best, if not the best, schools with an advanced course in mathematics, which drew children from the whole city and whose presence obviously considerably influenced the average figures. Even so, however, the data below help to understand the emerging picture. And although it is naturally impossible to rule out a certain amount of cheating, copying, and the like, nonetheless, we can assume that it was limited; a second diagnostic work was planned (but never carried out), so that if a teacher's own work was going to be judged, then it would be judged only on the basis of the changes that had occurred since the first diagnostic work, which merely determined the existing level of the classes, which had been formed only a short time beforehand; thus, teachers had no incentive to inflate the outcomes.

The diagnostic work consisted of two parts and was to be completed in 3 h. The first (main) part contained five groups of problems (on identical transformations, solving inequalities, solving systems of equations, studying graphs and functions, and word problems, respectively). Each group contained a problem A, worth a maximum of three points; a problem B, worth a maximum of six points; and a problem C, worth a maximum of nine points. Only one problem from each group was counted. The students themselves decided which problems to solve. Thus, a student could get a maximum of 45 points. For example, the fourth group contained the following problems:

(A) Construct a graph of the following function $y = x^2 - 6x + 5$ and find the coordinates of the points whose y-coordinate equals 5.
(B) Find the least value of the following function $y = x^4 + 5x^2 + 2$.
(C) Given the function $f(x) = \begin{cases} x-1, & \text{if } x \leq 0 \\ ax^2 + 2x + a, & \text{if } 0 < x < 1 \\ x+2, & \text{if } x \geq 1 \end{cases}$

Find all values of parameter a such that this function is an increasing function.

One could say that problems A corresponded to the standard requirements of general education schools; problems B corresponded to the standard requirements of schools with an advanced course in mathematics; and problems C corresponded to heightened requirements of such schools.

The second part consisted of ten problems (the first of them had three parts, that is, there were 12 questions in all), which were far simpler than the problems in the first part from a technical point of view but which required a certain capacity for reasoning—for example, in one of them, students were asked to give an example of an equation that has exactly four roots on the segment $[-3, 5]$. For each question (again, there were 12 in all), a student could get a maximum of three points.

This diagnostic work was not graded, but one could say that 15 points on the first part were equal to the highest grade of "5" at a general education school, while 9 points were equal to the lowest passing grade of "3" at such a school.

The outcomes showed an extreme stratification among the students. Out of 2175 people who took part in the diagnostic work, 952 (that is, 43%) got fewer than nine points on the first part, that is, were unable to complete the work in a way that met the minimal standard requirements of general education schools. Moreover, there existed entire classes in which no one got nine points. Meanwhile, the students' grades for the ninth grade, on the basis of which they were admitted to the tenth grade, were usually quite good (certainly not lower than "3"). On the other hand, 572 students (that is, 26.3% of those taking part in the diagnostic work) got 15 points or more, that is, would have gotten a grade of "5" by general education school standards. Seventy-five people or 3.4% of those taking part in the diagnostic work got 30 points and more (note that 65 of them attended the same school).

Two tables show the figures for schools of different types. Table 5.1 presents the results for the first part of the test. Table 5.2 – for the second (with a maximum score of 36). The results are quite similar.

As can be seen, very many students at that time left for schools that were specialized in one way or another (and to repeat, it can be assumed that in other districts of the city, and even more so in smaller cities, the corresponding figures would have been somewhat lower). Furthermore, even in humanities-oriented schools, the outcomes in mathematics turned out to be somewhat better than in general education schools. Note that even the technically simple problems in the second part often presented serious difficulties—instances in which students completed computational and algorithmic problems relatively well but demonstrated a complete incapacity for even the simplest reasoning were by no means a rarity.

Let us also give the figures separately for boys and girls (Table 5.3).

As can be easily seen, among those unable to complete the assignments, even by the standards of general education schools, girls were more numerous than boys.

Table 5.1 Results for schools of different types (first part)

School type	Number of students	Average score	Average grade (if it had been given)
General education	838	6.80	2.27
Humanities	634	9.54	3.21
Physics-mathematics	475	18.45	5
Total	2175	10.70	3.57

Table 5.2 Results for schools of different types (second part)

School type	Average score
General education	11.44
Humanities	13.73
Physics-mathematics	24.56
Total	15.62

Table 5.3 Results for boys and girls separately (first part)

School type	Boys	Girls
Got fewer than 9 points	375	577
Got 15 points or more	307	265
Got 30 and more points	56	19

This author does not know of any statistical figures of this type for previous years—in the USSR, people took little interest in such questions, automatically declaring and assuming equality between the sexes. However, informal teachers' opinion had regarded "slackers" as boys rather than girls—based on the reasoning that girls deviated to a lesser degree than boys from the requirements of the school. It may be argued that the changes stemmed from a change in public opinion: mathematics stopped being considered a subject necessary to everyone and in particular a subject necessary to girls (we might add that, based on our observations, the number of girls entering schools specializing in mathematics fell sharply). In general, as was noted in the study cited here, the interest of students and their parents in obtaining a high-quality education in mathematics was a crucial factor in their "educatedness." The old Soviet slogan—"If you don't know how to, we will teach you. If you don't want to, we will force you"—was disappearing into the past. Somewhat simplifying matters, we might say that those who "didn't know how" were not always taught, while those who "didn't want" were not only not forced, but often not even helped to begin to want.

2.3   From 1993 to 2000

Eduard Dneprov, who was the head of the ministry of reformers, believed that the reforms were fully realized only during his tenure:

> With the passage of the Education Act, the first, groundbreaking phase of the educational reforms ended. The reforms entered a new phase—the phase of their technological realization. But this phase was effectively aborted. Thanks to the efforts of "velvet restorers," the educational reforms were transformed into **pseudoreforms** with clear tendencies toward backsliding (Dneprov 1999, p.13).

Nonetheless, however one evaluates what occurred, it is clear that changes during this period continued (again, without immediately assessing what led away from the Soviet model and what led back to it and without in any way assuming that everything in the Soviet model was bad). In addition, the changes that Dneprov took pride in did not reach the schools immediately (as has already been said, even the new basic plans arrived after Dneprov retired). Consequently, even agreeing with Dneprov about the growth of backsliding tendencies, we can still view the period from 1991 to 2000 (the Yeltsin years) as a whole and study the processes that took place—by no means all at once—during this period.

Despite the tendency to refer to these years as "wild," which became entrenched in the propaganda of the subsequent period, it cannot be said that everything that happened in mathematics education then was unsound: just the opposite, the unfolding processes, as is clear from what has already been said, were very contradictory. New schools opened, new ideas were expressed, new textbooks and problem books were written, and teachers learned and experimented. At the same time, it is impossible not to note that, for example, the financing of education out of the state budget shrank by 48% between 1991 and 1999[3] (Dneprov 1999, p. 5), which inevitably had an impact on education. And even strictly educational measures were by no means always considered beneficial by everyone.

Below, we will examine certain important aspects of the mathematics education of that time, without deciding in advance whether they were reformist or counter-reformist in character.

3 The Content of Education and Standards

Mathematics was obviously not among those subjects in which educators needed to struggle especially hard against former Soviet ideology, reconstructing its whole content. Nonetheless, the question of the content of education became relevant in this field also. The Soviet model indicated the knowledge, abilities, and skills that had to be achieved by teaching. With the beginning of the period of reforms, such lists became unwelcome—they started being referred to as the "notorious ZUNs" (acronym for "*znaniya, umeniya, navyki*"—"knowledge, abilities, skills")—and it was explained that educators had to move away from meaningless drills, rote memorization of facts that no one needed, and the like. In the monograph by Stefanova et al. (2009), this position was formulated as follows:

> Reform of the education system in contemporary Russia is a way to overcome the crisis in the sphere of education, which used to be characterized by a high level of politicization of the learning process, an orientation toward conserving the ideals of the existing sociopolitical system, the determination of all aspects of the educational process by a prevailing subject-centered conception of education. Within the framework of this conception, the assimilation of subject (mathematical) content was seen as the main goal of education. As for the content of mathematics education, it was structured as a kind of model of scientific knowledge suited for both general education schools and higher educational institutions. (p. 181)

And further:

> At the contemporary stage of the development of society, the main goal of general education consists in the formation of a many-sidedly creative personality, capable of realizing its personal potential under dynamic socioeconomic condition both in its own interests and in the interests of society. (p. 182)

The idea of developmental teaching had been assimilated by Russian education at least since the time of Vygotsky (and in reality, much earlier), but it would be beneficial to be less general and to give it some kind of concrete form. All the same, the finding that almost half of the tenth graders in St. Petersburg were unable to meet standard requirements in a subject could not be considered a proof of the fact that they had benefited from a many-sided development.

---

[3] How this decrease should be understood given the radical changes in the purchasing power of the ruble is open to discussion.

The way out was found in the use of an expression that was new to Russian education: "standard." The Federal Standard Law was passed already in 1992. It called for developing standards that would regulate education. The aforementioned Dneprov (1999) notes that, "in itself, the idea of mandatory standards [was] undoubtedly positive" (p. 19), since they were supposed to:

> describe a sphere of basic education in the interests of the child and of society, to stipulate the bounds of the greatest acceptable academic workload, to formulate mandatory requirements for school graduates. In addition, the standards were supposed to serve as a bulwark against pressure from the "methodological lobbies," which ceaselessly strove infinitely to expand the content and limits of their academic subjects. (p. 19)

The outcome, however, was not what had been envisioned, in Dneprov's opinion, and the developers of the standards were to blame. In his opinion, the standards were "implemented in accordance with the old 'ZUN' philosophy" (p. 21) and became a tool for the conservation of the old, the overloading of the students, and other evils.

The first drafts of standards began to appear as early as 1993. The draft prepared by the Institute of the General Education School at the Academy of Education (Institut 1993) did not significantly differ from the requirements that had already been formulated by the Ministry back in Soviet times (Ot Glavnogo upravleniya 1991). As in the older document, the requirements were presented at two levels. The first level was called the *level of possibilities*: "it describes the outcomes toward which students who study the general education course may aspire" (Institut 1993, p. 10). "The second [level] is the *level of mandatory preparation*. It describes the unquestionable minimum that all students must achieve and defines the lowest acceptable threshold for the outcomes of a mathematics education" (p. 10).

The level of mandatory preparation was formulated not only verbally but also through models of typical problems. For example, the section on equations at the high school level indicated the following:

> *The level of mandatory preparation is defined by the following requirements:* solving the simplest exponential, logarithmic, and trigonometric equations; using the interval method to solve simple rational inequalities. (p. 19)

And further, a small set of problems was given, the ability to solve which was considered mandatory. For example, the following problems on using the interval method were included:

Solve using the interval method:

a) $\dfrac{(x-1)(2x+3)}{4-x} \geq 0,$ b) $\dfrac{x^2-4}{x+5} < 0.$ (p.19)

Working teachers found special significance in the concluding sections, which discussed how the implementation of the standards was to be monitored and verified and, in general, how the teaching process was to be structured in light of the standards. The idea of differential learning and open requirements was variously advocated. It was emphasized that testing to ascertain that a desired level had been achieved had to take place at boundary stages, for example, upon completion of the basic school (ninth grade) and high school (eleventh grade), and that educators must not limit themselves to testing for the achievement of the mandatory level but that they also need to investigate how was achieved a higher level.

In the ensuing discussion, various critical views were expressed. It was noted that, for teachers, the normative document ought to consist not of standards but of programs and curricula (Timoschuk and Nozdracheva 1994); that it would be worthwhile to write multilevel problem books, which would themselves define the standards (Dubov 1994); that the teaching of probability theory needed to be substantially strengthened (Gnedenko and Gnedenko 1994); and much else. Professor Gladky (1994) of Moscow wrote about the legal illegitimacy of the draft—standards had to be approved on a competitive basis. Moreover, he went on, although the draft paid lip service to democratic phraseology, in fact it proposed the introduction of a uniform methodological system for the whole country—something unseen either under the tsars or under Soviet rule. This system itself he regarded as absurd:

> It needs to be pointed out that the principle of "free choice of the level of assimilation," proposed by the authors, is fundamentally flawed. This can be seen especially clearly when it is applied to the study of mathematics. Imagine that a child with good, but not exceptional abilities, at the age of ten carelessly chooses the lowest level, and the "democratic" teacher, paying heed to the instructions of our authors, does not attempt to influence the child's choice. If the child remains at the same level for two or three years, for the rest of his life he will never have any choice with regard to mathematics, except perhaps if he is lucky enough to find himself under especially favorable circumstances of some kind. This person will be spiritually robbed and many paths in life will remain permanently closed to him. This kind of "humanism" and "democratism" is in reality nothing other than a profound indifference to the child and the child's fate. (p. 7)

Along with common federal standards, regional standards were also published. Thus, for example, the Moscow regional mathematics standards education were promulgated, which allowed Viktor Firsov, the head of the group that prepared them and one of the theoreticians behind their composition, to express his views one more time. For example, he deliberately pointed out the difference between standards and curricula. As he wrote: "a teaching curriculum expresses a concrete, methodologically conceived strategy and tactics for teaching" (Firsov 1998, p. 4). Consequently, there might be many different curricula. Standards, on the other hand, are introduced for the specific purpose of "establishing a manageable multiplicity of curricula." He went on: "Standards must become a normative foundation enabling a transition from schools that are still excessively uniform to schools of a new type" (p. 5).

Further, he answered those who came out against reducing the quantity of the content of education. Opponents of such a reduction were divided by Firsov into two groups, the first of which allegedly equated a large quantity of requirements with a large amount of assimilated knowledge—such opponents were invited to present genuine, rather than falsified, facts about the knowledge of Russia's schoolchildren and were additionally informed "about the quantity of content that is considered sufficient for mandatory education by educators in most developed countries" (p. 6). (Note that Firsov thus took the unanimity of these educators for granted.) The second group, according to Firsov, was made up of those who were concerned only with "the elite," with those who in their opinion "determine the prospects for the country's development" (here, Firsov caustically pointed out that the supporters of these views fail to specify which country they have in mind, thus hinting that these "elite" schoolchildren can later move to other countries, likely the very ones whose educators he had cited just a little while earlier). This group was invited to recognize that reducing the scope of standards did not rule out preserving the scope and size of specific curricula.

In general, the article allowed for various alternative ways of following and not following standards; for example, Firsov noted that it was possible to allow for following standards with delays if the parents assumed responsibility in the event of any difficulties (for example, in the event of a transfer to a different school). Moreover, he allowed for not following standards at all but then without government financing and without a state certificate (no one else, however, was ready to grant such rights).

The very term "standard" at that time was understood quite broadly, nor was it entirely clear what new regulatory measures might appear. In St. Petersburg, for example, it was deemed sufficient to publish examination materials under the title "Standards" (Standarty 1993) and leave it at that (these "Standards," which will be discussed below, did in fact establish what was required of the students to a certain extent).

It may be said that this period of the development of standards ended with a competition, which was won predictably by the draft prepared by the Russian Academy of Education, the materials of which were published in 1998 (Lednev et al. 1998). This document, however, had little impact on schools at the time. Schools continued to function as they had done before, using curricula prepared and published by the ministry or, to put it perhaps even more precisely, using textbooks that in one way or another corresponded to these curricula.

Kuznetsova's collection (Kuznetsova 1998) contains some curricula and some other documents, including a "Mandatory Minimum Content in Mathematics for Basic Schools," written, as the text explains, "on the basis of existing curricula that follow the temporary standards and the methodological letters of the Ministry Directorate of General Secondary Education…which were introduced into teaching practice between 1992 and 1997" (p. 60). This document lists various topics and concepts that students are required to learn (it is not indicated how fully or how deeply) and is quite traditional. The curricula include curricula for classes with an advanced course in mathematics and, conversely, curricula for classes with insufficient preparation (sometimes called "correctional"). But, to repeat, teachers were guided first and foremost by textbooks, which were supposed to correspond to these curricula.

4 Textbooks and Certain Basic Tendencies in the Development of Mathematics Education

The main textbooks at that time were those prepared in the mid-1980s, which had gone through a nationwide competition (Abramov 2010). It should be noted that some of them continue to play a dominant role to this day (for example, the geometry textbooks edited by L.S. Atanasyan), while others are currently being used in somewhat revised versions prepared by somewhat altered lists of authors. The changes that took place, including the increase in the number of publishers that put out literature for schools, led to the appearance of new textbooks. Without attempting to list all or even many of the books that came out during those years (see, for example, Karp and Vogeli 2011), we will point out some important trends.

One of them—which pertained to the organizational side of things—must be mentioned at once: even if monitoring over which literature was being used by teachers and their classes did exist, it was incomparable to what had been in place during the classic Soviet period. Consequently, a teacher could require parents to acquire one or another textbook (relatively cheap ones) and use this textbook in class, pretty much without asking anyone. Naturally, such a system was not officially encouraged, nor were parents always pleased about having to spend additional money, but this was quite possible and even widespread. Specific organizations specialized in approving textbooks and labeling them with various certifying designations were formed; for example, textbooks were approved at the federal level by the Ministry on the recommendation of the so-called Federal Council of Experts. Such a certifying designation made it possible to purchase textbooks using state resources, which, of course, radically influenced their distribution (such resources, however, might still be lacking, for various reasons, in various regions). But, to repeat, there were other ways of doing it as well.

Below, we will discuss three trends that seem to us the most significant. The first of them was the rapid increase in the number of so-called classes with an advanced course in mathematics.

4.1 On Classes with an Advanced Course in Mathematics

Schools with an advanced course in mathematics, which appeared in the late 1950s and early 1960s, quickly and deservedly won a very high reputation. They selected gifted students who were interested in mathematics; their curricula were designed by wonderful research mathematicians along with highly qualified teachers; these curricula were supplemented by extracurricular activities; and indeed, the teaching of even nonmathematical subjects and the general atmosphere in these schools were noticeably different from what was found in ordinary schools (Karp 2011). The authorities had a contradictory attitude toward these schools—their special atmosphere, and their selectivity in general, which was based on abilities and results, irritated the authorities; but on the other hand, in the era of the scientific-technological revolution, trained professionals were indispensable. What was evident, in any event, was that the graduates of these schools were splendidly prepared for college entrance examinations, and admission to colleges was based specifically on the results of such examinations, which were conducted by each college independently, which made the preparation for such examinations a goal of paramount importance for students and parents, giving rise to an enormous market of preparatory courses, tutors, special textbooks, and the like.

The new liberalization now allowed many schools, as has already been said, to increase the number of hours allocated to mathematics in certain upper grades and to call them "mathematical" (while possibly certain other grades with a small number of hours for mathematics were given a "humanities" label). Not everything was identical everywhere, of course, but on the whole, it can be said that in a very large number of cases, schools that did so did not set particularly difficult educational goals for themselves. If schools that had appeared under the supervision of outstanding mathematicians like Kolmogorov, Gelfand, Smirnov, and others decades earlier had aimed at educating future research mathematicians, deliberately developing difficult courses to this end (which at the very least always included calculus with proofs), the new classes now being formed usually did not aim past college entrance examinations, and these examinations focused on materials from general education schools—which, of course, did not rule out the inclusion of difficult problems.

Colleges were also interested in the new schools. The prestige of engineering professions was falling at this time—engineers were losing their jobs en masse and retraining for new occupations. Every engineering college was concerned with securing a sufficiently large number of matriculating students. Nor must it be forgotten that the matriculating students' level of preparation usually also did not make the colleges very happy. The way out in large cities was found in the creation of the "school-college" system, in which certain classes were designated as being connected with certain colleges, which in some way influenced the curricula of these classes, and most importantly offered certain benefits to the graduates of these classes when they applied to college.

While the "old" schools with an advanced course in mathematics usually allocated at least 8–9 h, per week to mathematical subjects, the "new" schools could as easily allocate 6 or 7 h, and most importantly even when mathematics was granted 8 or 9 h, often it turned out to be preferable to use "ordinary" textbooks from general education schools, designed for a smaller number of hours (say, five for all mathematical subjects). Additional time was allocated for solving problems, which teachers could, for example, draw from manuals for those applying to college.

Such classes with an advanced course in mathematics gradually started to appear not only in high school (grades 10–11) but also earlier—beginning with the eighth grade. There were isolated attempts to start even earlier, but they met with no success. Curricula began to take shape that were recommended for such classes (even if they did not possess a strictly mandatory character), see, for example, Karp and Nekrasov (2000, pp. 32–39). Special textbooks for such classes also began to appear—especially for grades 8–9, since books for students who were applying to colleges were not well suited for use at these grade levels. These were, first and foremost, problem books. We should mention, for example, such texts as Galitsky et al. 1992 or Karp 1993, whose tables of contents correspond—or correspond with certain minor caveats—to standard textbooks but which contain problems whose level and difficulty are substantially higher than those of ordinary problems.

Later, so-called supplemental chapters to textbooks began to appear. As Atanasyan et al. (1996) explain, for example:

> The manual contains additional chapters for the course in geometry for eighth grade. For each chapter in the basic textbook, there is an additional chapter in the manual. The additional chapters, as a rule, do not repeat the material presented in the basic textbook, but this material is broadly used; therefore, before beginning to study any chapter, the students must study the [corresponding] chapter from the basic textbook. (p. 3)

In reality, it was probably better for students not to study first one chapter from one book and then one chapter from the other but to examine parts of the additional chapters as they studied the basic textbook. But in any event, the manual contained additional problems and additional theorems.

We do not possess (and it is unlikely that anyone possesses) information about how many of the students in the country used these "advanced" curricula in their studies during those years. We would argue (based on the figures we know from St. Petersburg) that certainly this number was not more than 10%. Nonetheless, it is evident that the appearance of classes with an advanced course in mathematics and the possibilities they opened up raised the mathematical preparedness of a considerable proportion of teachers and likely benefited many students as well. At the same time, it is also clear that the confusion and befuddlement about what was meant in each particular case by the term "an advanced course" could only cause harm.

4.2 Textbooks for Humanities-Oriented Classes

While studying an advanced course in mathematics had a big history by the 1990s, the goal of special instruction in mathematics for future humanities students was only now formulated for the first time. The reformist principles cited above included humanityzation of education; moreover, the reformers explained that the need for this was connected:

> *First,* with the rejection of technocratic and scientific traditions, which over the last 200 years have evolved within the global system of education under the influence of a rationalistic view of the world as a kind of inanimate mechanism that may be disassembled under analysis— whether it be a human being, society, culture, etc. And *second,* with the ambition to overcome the clearly observable schism in the culture of education between a humanities and a technical component, to overcome their growing separation. (Dneprov 1998, p. 50)

Even Dneprov himself, however, wrote specifically about the humanities and corresponding school subjects, arguing that their role had to be expanded and their teaching fundamentally changed. Now it appeared, however, that it was desirable to "humanitize" mathematics as well. Expanding the role of the humanities was often equated with reducing the role of the mathematical subjects and even with changing the very style of mathematical instruction. Practically speaking, however, the problem consisted of writing a textbook that could be used for teaching mathematics in grades 10–11 where only 3 h per week were allocated for mathematics.

There arose a confusion (or, if one wishes, a discussion) about what exactly ought to be done (Sarantsev 2003). Probably the first textbook for humanities-oriented students that appeared, Butuzov et al. (1995), began with the following announcement:

> We see the purpose of our book as consisting first and foremost in conveying an idea of the most fundamental mathematical concepts, whose knowledge, in our view, must be a part of the general cultural literacy of a person of any profession. We have attempted as far as possible to tell about the application of mathematics in various spheres of human activity, to acquaint you with certain pages of history and the creators of this remarkable science. This book examines a number of questions that do not belong to traditional school programs in mathematics, but which are important for certain professions related to the humanities. Among them is elementary probability theory, the basic concepts of statistics, and certain others. (p. 4)

Clearly, it is not easy to implement such a program. Moreover, such a program itself can give rise to questions. How does one determine which mathematical concepts are parts of general cultural literacy? In what level of detail should those who know almost no mathematics be told about its applications? Can an account of the life of, say, Galois, which was indeed quite dramatic, replace knowledge of mathematics (even at a much more modest level than is required for understanding Galois theory)? How seriously should one take the goal of teaching schoolchildren those sections of mathematics which might be useful to humanities students?

In practice, however, such questions were never raised nor indeed were special textbooks typically used—in very many so-called humanities-oriented classes, ordinary textbooks for general education schools were used, whose content was impossible to cover in the allotted time. But no one especially insisted on mastering the material either.

The textbook written by the author of this chapter and A.L. Werner (Karp and Werner 2000, 2001), which came out at the end of the period that we are discussing and which reflected the experience of working with humanities-oriented classes and their graduates, which will be discussed below, was written partly in order to return to the traditional order of things—that is, to achieving certain clear outcomes in teaching. At the same time, the orientation of traditional Russian textbooks toward developing computational and other algorithmic skills (necessary, at least at that time, to future engineers) was seen as unnecessary to future professionals in the humanities. The authors' view was that future professionals in the humanities had to be acquainted with the mathematical way of thinking and in general with the activity involved in mathematics, not by being told about the way in which mathematicians think but by involving students in mathematical work (even if such work obviously could be carried out only on such very modest materials as were accessible to such students). Technically, the course was substantially lightened; many proofs disappeared; but at the same time, the authors attempted to offer students many opportunities for mathematical analysis and reasoning, for constructing examples, for comparing various situations, as well as histories of the development of various concepts, and so on. Consequently, the textbook addressed not so much applications of mathematics, as the very concept of modeling, that is, the translation of what was observable in the ordinary world around us into mathematical language.

Other textbooks also appeared (for example, Bashmakov 2004). It should be noted that in the late 1990s and early 2000s, the problem of writing textbooks for humanities-oriented schools was recognized as one of paramount importance, and the ministry held several competitions, with the support of the World Bank, to produce sets of textbooks for such schools (which were won by the aforementioned textbook by Karp and Werner and the problems books and other materials that supported it). In itself, the methodological problem posed then remains difficult and needs further study, although some outcomes were achieved. Unfortunately, during the next phase, in connection with new transformations, this work was interrupted and halted (Karp 2011).

4.3 New Textbooks for General Education Schools

Mass-scale general education schools themselves, however, also needed changing. One area in which changes were necessary has already been mentioned: the teaching of probability theory and statistics was in fact absent from Russian general education schools. During the Kolmogorov reform period (that is, in the late 1960s and 1970s), attempts were made to change this situation and to introduce elementary discrete mathematics into the schools. They did not meet with success, however, and the cause of this likely did not lie only in teachers' lack of preparedness. It is impossible to avoid the thought that the uniform determinate system of thinking, which was supported by the state, was poorly suited for fundamentally different, "probabilistic" thinking. We know of no official decisions not to cover probability theory, but no such decisions were necessary. On the contrary, what was needed was the will to begin to teach such theory, and until a certain point such a will was lacking. With the beginning of the 1990s, gradually and not without difficulties, discrete mathematics began arriving in the schools (Bunimovich 2011).

Another demand of the age was connected with the development of electronic technologies. We can point to various attempts at making use of the new possibilities, but we can detect no notable changes in the textbooks or workbooks of that time. Assignments that required the use of any kind of technology were practically nonexistent.

New textbooks, however, did continue to appear. Kuznetsova et al. (2011) specifically compare the new textbooks by Dorofeev et al. with textbooks that had appeared earlier. Below, we will likewise confine ourselves to discussing this textbook, without looking into others (about them, see Karp and Vogeli 2011), including textbooks in other subjects (geometry or algebra and elementary calculus).

Looking at the textbooks by Dorofeev et al., one can note differences with previous textbooks of various types. First, there are differences that are purely methodological and pertain to content. Such differences are many (not to mention the appearance of a systematic presentation of stochastics, which begins in the textbook for the fifth grade): for example, as Kuznetsova et al. (2011) point out, there is a change in the balance between the arithmetical and algebraic components of the beginning of the course in favor of the former. More attention is devoted to arithmetic and to working with numbers (if one wishes, one can detect here the influence of the technological revolution—we now work more directly with numbers than with algebraic symbols in real life, too). Ideas that effectively belong to analytic geometry, that is, to the connection between graphs and equations, appear quite early on. In general, various graphs appear in the textbooks earlier and in greater numbers than in previous textbooks (it may be argued that here, too, there is a certain influence of technology—although one that is not typically explicitly discussed—since children now encounter graphs more often than they did before). On the other hand, the study of certain topics is organized more gradually and sometimes introduced later than it had been previously. Here we see what is arguably the second fundamental distinctive feature of these textbooks, which is organizational in character: these textbooks, more than previous ones, take the individual characteristics of the schoolchildren into account. Note, for example, the appearance of the sections entitled "For those who are interested" (which subsequently, in one way or another, appeared in other textbooks as well). These sections offered additional "theoretical" material and additional problems at the same time. And the main sections, too, contained two groups of problems: more simple problems (A) and more difficult problems (B). The possibility for differentiation was already embedded in the textbook. The textbook also began paying more attention to independent work, and above all to self-checking—there appeared special assignments aimed at this.

And yet, there were no revolutionary changes either in these or in other textbooks of the time. They continued the existing tradition as they had done before, quite noticeably differing from American textbooks, for example, both in terms of their outward appearance (more modest, smaller in dimensions, cheaper) and in terms of content (more systematic and more proof-oriented).

5 Innovations in the System of Conducting Examinations

The system of conducting examinations was changed repeatedly under Soviet rule (Karp 2007a). By the end of Soviet rule, the system consisted of three rounds of exams: the first, upon completion of the basic school—eighth grade, in the system of numeration that existed at the time (later it began to be called ninth grade); the second, a graduation exam, upon completion of secondary school—grade 10 (later 11); and finally, those who wished to enroll in institutions of higher education also took entrance exams there. Naturally, there were no examinations in certain subjects, but in mathematics there were often even two examinations—for example, upon completion of school, a written examination in algebra and elementary calculus, and an oral examination in geometry (by the 1980s and 1990s, this last examination had often become what was called an elective examination, that is, it was taken by those who wanted to take it, while others could choose to take an examination in a different subject, say, history).

The graduation examination in algebra and elementary calculus, at this time, lasted 4 or 5 h but contained only five (later six) problems, which not only had to be correctly solved, however, but whose solutions also had to be written out in a correct and well-argued manner. Meanwhile, the grading criteria were mainly based on the notion that if five problems had been fully solved, a student was given the highest grade of 5; if four problems had been fully solved, a student was given the grade of 4, and so on. Naturally, things did not always go as smoothly as this—a problem could be solved partially or with deficiencies; there were special criteria for how many deficiencies were acceptable for each grade; it was specified which deficiencies were considered simply omissions and which were considered mistakes, etc. (Chudovsky et al. 1986). Note that the appearance of a sixth problem on the examinations constituted an important development: students acquired a choice, even if only a small one. From the 1970s on, graduation examinations were conducted at two levels—a basic level and an advanced level (for general education school and for schools with an advanced course in mathematics, respectively).

Understandably, such a system came under a great deal of criticism, which became harsher as the role of examinations in the new post-reform country grew: it was said that schools had acquired more freedom and had been liberated from trifling minute-by-minute control, yet in the end—on the exams—they had to show what they had taught their students. The following considerations were among those expressed at this time:

- Why are graduation and entrance exams conducted separately? Shouldn't they be unified, as is done in many countries? Why, in addition, are entrance exams conducted separately by each college? Are there really two separate mathematics for two quite similar engineering colleges? (We should recall the appearance of the "School-College" system, which on the whole supported at least a partial and limited unification of graduation and entrance exams.)
- Along with classes with an advanced course in mathematics and general education classes, humanities-oriented classes also began to appear: was a special exam necessary for such students?
- Since schools and students were acquiring greater rights than they had previously with regard to structuring the material being studied, and could pay somewhat more attention to certain topics and consequently less attention to others, the exams had to reflect this as well. The students must have greater choice about what to solve.
- In general, the role of each isolated skill is shrinking compared to what it had been before. Naturally, when studying trigonometry, it is important to get the students to be able to solve trigonometric equations of the required types, but upon graduation from school, and even more so later, what becomes important is

not the specific ability to solve these types of problems but a broader ability to work with mathematical texts. Therefore, it would be desirable to structure the problems in some new way.
- Discussions about partial solutions went on for years, as did discussions about the strictness with which solutions should be presented (sometimes the demands for strictness were completely exorbitant). Proposals were made to switch to multiple-choice tests or to problems with short answers that could not be argued about and would also be far easier to check.

The issues and questions posed above were discussed both in the period examined here (before 1999) and later, so we will return to them below. Here, we would merely note that answers to these questions did appear, in one way or another, and they appeared not only in a centralized manner. At a certain time, the ministry introduced examinations for humanities-oriented classes (Zvavich and Shlyapochnik 1994), but at this time, it was also possible to organize examinations at a regional level. Thus, Moscow and St. Petersburg began to organize their own graduation examinations (Karp 2003).

St. Petersburg's examinations, for example, were conducted at four levels: there were two advanced levels—in addition to the usual examination for classes with an advanced course in mathematics, a so called *elite-specialized* examination was offered, which was indeed very difficult but which from a certain time on began to be counted for all those who passed it as an entrance examination to a leading college, St. Petersburg University. The main difference between these examinations and others was their structure: students were offered not isolated problems, but blocks of interconnected questions, and they were furthermore given a relatively broad choice. Thus, the version of the examination for classes with an advanced course in mathematics offered two so-called mandatory blocks of four problems and three more blocks, of which students had to choose one (such a system could accommodate for existing differences between curricula). In the view of those who wrote these examinations, systems of problems composed into blocks taught students to look for connections between problems, by comparing and checking their solutions, learning to generalize what they had noticed, and so on (Karp 2002, 2003).

Local approaches were also employed in other places. And the ministry itself prepared new offerings, publishing a collection by Dorofeev et al. (1999), which became the basis for examinations. Every student and every school owned this book, and problems from it were systematically solved both in class and at home. On examinations, students would be asked to solve several problems from this text (the numbers were given out only before the beginning of the exam). This practice was referred to as "open exams." Exams based on this text were meant to be conducted at two levels—level A for students who had 3 h of mathematics per week and level B for those who had 4.5–5 h. (The idea of "open exams" had already appeared earlier for exams for eighth (ninth) grade; see below.)

Graduation examinations in schools of the "School-College" system were also counted as entrance examinations for schools connected with a college, which was hardly fair with respect to other schools, but at the same time could serve as a step toward the formation of some kind of more general system of uniting graduation and entrance examinations, but at later stages, these experiments were abandoned.

From today's vantage point, the practice of conducting examinations in ninth grade during this time also appears as something transitional. At the beginning of this period, examinations were based on the manual MP RSFSR (1985) (the latest edition by Chudovsky and Somova 1995), which introduced the idea of "open exams," which, it was believed, helped to reduce stress, eliminate unpredictability, prepare for the examination systematically, and so on. Later, a new collection by Zvavich et al. (1994) appeared, which contained somewhat more varied and difficult problems. But this collection was not used for long, and it was replaced by the problem book by Kuznetsova et al. (2002), while the number of problems given on the examination increased somewhat (from 6 to 10)—although their difficulty probably decreased. Behind these changes was the question (possibly not always acknowledged): what exactly should be tested—specific skills or the ability to operate with their more complex combinations? As an example of a problem from the collection by Zvavich et al. (1994), consider the following (I.599) from the first part, which was considered the easier part:

Determine whether the following inequality is true: $\cos(149^0 + x) \cos x + \sin(149^0 + x) \sin x < 0$.

This somewhat cumbersome problem nonetheless requires the execution of several steps, the most substantive of which is specifically the determination of what needs to be done, that is, the development of a plan. Undoubtedly, students need to know the formula for the cosine of the difference of two angles; then they can see at once that the question concerns whether it is true that $\cos 149^0 < 0$. And yet the very formulation of the problem is somewhat more open than, say, "simplify the expression and check whether such-and-such an inequality holds," which at once steers students down the path of following learned algorithms.

In fact, the manual by Kuznetsova et al. (2002) also contained not a few interesting new methodological approaches. But it, too, was destined for a relatively brief life.

6 Public Opinion About the Teaching of Mathematics

Before moving on to an examination of the new political period that began with the coming to power of Vladimir Putin, we should say a little more about public opinion concerning the changes that occurred (to a certain extent, we have already said something about it when discussing standards and the content of mathematics education). It is hardly possible to represent all existing points of view here, but we should still like to describe those opinions which were widespread among research mathematicians. Again, without claiming to give an account of all opinions, we will confine ourselves to opinions expressed in a book published already at the beginning of the Putin period, *Education We Can Lose* (below we will quote from its second edition, Sadovnichy 2003). This volume, which was edited by Sadovnichy, academician and rector of Moscow University, contains articles by the academicians Anosov, Arnold, Kudryavtsev, Nikolsky, and by Igor Sharygin, a prominent figure in mathematics education (it contains other materials as well, for example, an article by Professor Melnikov, a prominent figure in the Communist Party and a mathematician by training, and even an interview with the writer Alexander Solzhenitsyn). It should be noted that although the collection contains a speech given by Putin in 2001, and although the authors sometimes refer to documents that were fresh at the time, their main pathos is directed against the changes that took place during the previous period and the plans for their continuation (even if only in rhetoric).

The common trait shared by all the authors, which is also expressed in the book's title, is their confidence in the merits and even world leadership of Soviet mathematics education. "Programs outlining what should be taught when in secondary school took shape in Russia over the last two centuries. The fact that the choices made were quite good is attested to by the fact that, with respect to the fundamentality of education in the natural sciences and mathematics, the Russian school has undoubtedly occupied the first place in the world," writes L.D. Kudryavtsev (p. 125). Admittedly, I.F. Sharygin ironically comments that people will say that the theorem about the supremacy of the Russian school has not been proven, but neither does he himself offer any proofs of this nor does he allow for any doubts.

Consequently, the authors take it for granted that what was done before, on the whole, is what must continue being done. "Why abandon that which has so recommended itself?" asks the same L.D. Kudryavtsev, and he goes on:

> An analogous situation exists in schools with algebraic and trigonometric transformations of quite complex expressions: in the opinion of certain critics, too much time is spent on this in schools, while later on students rarely have to do with such matters. This objection is again beside the point, since the main purpose of solving exercises with algebraic and trigonometric transformations is to cultivate the skills necessary for carrying out goal-oriented analytic transformations. (p. 130)

More generally, the narrowly utilitarian approach—"we will teach that which will be useful in life"—is decisively rejected by practically all of the authors, who emphasize the developmental value of mathematics, including even its value for developing morality (Sharygin); furthermore, the authors emphasize the ignorance of the reformers, who do not understand and do not know the applications of mathematics, and therefore, for example, wish to eliminate logarithms from the schools as something useless (Arnold).

The need for reforms is sometimes acknowledged, but with very sizable reservations. D.V. Anosov notes that in the 1930s, approximately one fourth of all children graduated from school, while later the proportion of those graduating from school increased—"Under such circumstances, schools inevitably must change" (p. 95). But then he goes on to remark that the goals of the reforms have not been sufficiently thought through and articulated and that their implementation has not been successful. L.D. Kudryavtsev praises Kolmogorov, who reformed the content of education, for including vectors and elementary calculus in the school program, but at the same time laments the disappearance of Newton's binomial theorem and complex numbers. S.M. Nikolsky welcomes the appearance of probability theory and statistics in schools but notes that this concerns only 2% of school mathematics education.

It appears the destruction began quite a long time ago, when the share of materials involving proofs began to diminish, but the changes of recent years are especially irritating to the authors because of their explicitly formulated aim of fighting "scientification," "the scientific approach," "subject-centeredness," and the like, which they interpret as hostility to science, which is responsible for the reduction in the number of hours allocated for mathematics and its role in education in general.

Moreover, the authors by no means confine themselves exclusively to the teaching of selected students. L.D. Kudryavtsev writes:

> I regard the tendency to concentrate serious education in mathematics and the natural sciences in specialized higher grades as deeply misguided—on the one hand, this lowers the overall level of education for those who will not study in such specialized classes, and on the other hand, it makes learning more difficult for those who will study in such classes because they have not acquired sufficient necessary knowledge at an appropriate age. (p. 127)

The author has specifically mass-scale schools in mind. On a different note, the authors reject the idea that "the school program must be abridged so as to be made accessible to all." They are concerned with "normal children," as S.M. Nikolsky puts it: "These children are quite healthy and quite capable of handling the difficulties of elementary mathematics: fractions, equations, sines, logarithms. Such children are many, and progress depends on them—they are the ones I am speaking about" (p. 160). What to do with the rest, however many they might be (and S.M. Nikolsky himself notes that they not only exist but will continue to exist), is not discussed—Nikolsky merely remarks that a class composed of such children should be called not a general education class but something else.

Without entering into a debate about what seems justifiable in the cited pronouncements of these Russian research mathematicians, we should merely note that the change in the perception of what was happening was itself important: the conviction that whatever else might be wrong, the teaching of mathematics in Russia was excellent and the finest in the world, was clearly giving way to the view that things were not going well. It is noteworthy that the collection from which these pronouncements are drawn is quite anti-American in tone (naturally, there are differences among the authors in this respect): the United States turns out to be the source of the foolishness taking place in Russia, while simultaneously, we are told that American educators want to change and improve their own mathematics education (certain documents to this effect are included in the collection), and, additionally, some of the authors (I.F. Sharygin) voice the thought that, by forcing such a flawed approach onto Russia, the United States wishes to turn Russians into its slaves. All of these notions are not very easy to reconcile, but it is important to note that methodological conservatism has at least sometimes gone hand in hand with political conservatism.

7 New Times

On December 31, 1999, Vladimir Putin became the acting president of Russia, and several months later, he was elected president. New times began, which quite quickly began to be contrasted with the previous times, despite the fact that the whole career of the new leader had taken place precisely during the period that was now being denigrated. People began saying that Russia had to rise from its knees, that it was necessary to construct a "vertical of power"; there was talk of "sovereign democracy" and much else. Looking at this period as far as possible as a whole, without attempting to reconstruct the changes that took place year by year, we should repeat that there is no reason to think that each change in politics gave rise to a change in mathematics education. Nonetheless, the connection between the two is evident.

Official rhetoric about mathematics education and mathematics has been very positive. We might cite, for example, the "Conception for the Development of Mathematics Education in the Russian Federation," which was endorsed by the Russian government on December 24, 2013. It opens with a discussion of the special role of mathematics and its indispensability both for the development of society and for individual development. It is noted that the Russian system of education is the heir of the Soviet system, which, however, needs to be improved, even as its virtues need to be preserved. Problems are listed: motivational problems, connected with the underestimation of mathematics education; content-related problems, for example, that the needs of future experts in mathematics are not sufficiently taken into account; and personnel problems, including the shortage of teachers and their poor preparedness. Aims are formulated: to elevate Russian mathematics education to a leading position in the world. How this is to be achieved, however, is discussed in terms that are rather broad: it is merely stated that education must offer "each student an opportunity to achieve the level of mathematical knowledge necessary for further successful life in society," "provide each student with developmental intellectual activity at an accessible level," and so forth.

At the same time, the sense of a crisis in mathematics education has probably been even stronger than it was in previous years. In support of this claim, we will confine ourselves to listing several recent articles in Russia's main journal of mathematics education, *Matematika v shkole* ("Mathematics in the School"): "Diagnosis: Mathematically Illiterate" (Bogomolova 2014), "The Decay of Mandatory Knowledge" (Novikov 2018), "The Crisis in Secondary Mathematics Education Through a Teacher's Eyes" (Ryzhik 2013, 2014), "Signs of a Crisis in the Domestic Methodology of Mathematics Instruction" (Savvina 2017), and "The Mathematics Curriculum of 2015, or the Triumph of Unprofessionalism" (Shevkin 2015a, 2015b).

On the other hand, explanations of this crisis and recommendations about finding a way out of it vary widely. The cited article by Savvina (2017), for example, contains the following appeal:

> to make sense of the worldview that is the cause of the crisis. To understand that borrowing a conception of education as a commercial service from Protestant Western civilization with its market ideology goes against Russia's tradition and national interests. "The main function of schools is to reproduce the civilizational code, transmit traditions, fortify the country"

[Kaiumov 2014]. In connection with which, it is necessary to appreciate the danger of introducing the latest innovations (competencies, universal learning activities, and other consequences of the market ideology in education) into pedagogical discourse." (p. 7)

Instead, the author proposes focusing on Russia's own historical experience, which she had previously associated with the idea of the comprehensive development of the personality, whose origin she attributes to Byzantine influences. It may be said that the very direction and character of the "discourse" proposed by the author reflect the changes that have taken place in the country.

8 Once More About Standards

Whereas during the previous phase, standards functioned more as an educational idea, which acquired practical significance only to a very limited extent, during the new phase, they have acquired much greater force. The standards are updated—changes in them are authorized by ministerial decree—and even different generations of standards are spoken of. Without going into all the distinctions between them, we can confine ourselves to an analysis of one of the existing documents—the Federal State Educational Standard for Basic General Education (FGOS 2018). There are also standards for elementary education and high schools.

As the very first paragraph of the document states, the Standard "represents the totality of the requirements that must be fulfilled in the implementation of a basic education program of basic general education" (p. 6). The same general part lets it be known that the Standard "was developed with due regard for regional, national, and ethnic distinctions among the peoples of the Russian Federation" (p. 7). The Standard is oriented toward the formation of a graduate who "loves his region and his Homeland, knows the Russian language and his native language, respects his people, its culture and spiritual traditions" and is also ready for self-development and continuous education; "who actively and with interest learns about the world, recognizes the value of work, science, and creativity" and is also "socially active, respects law and order, and holds his actions up to moral values" (p. 8).

In order to achieve all this (and much more), the Standard puts forward requirements for demonstrating the assimilation of the program. These requirements are assembled into three groups: requirements for personality-oriented results, meta-subject results, and subject results.

The first group of requirements includes the following:

1. The formation of a Russian civic identity: patriotism, respect for the Homeland, for the past and present of the multinational people of Russia;...
2. The formation of a responsible attitude toward learning...
3. The formation of a holistic worldview corresponding to the contemporary level of science...

......

7. The development of communicative competence...

......

11. The development of an aesthetic consciousness (pp. 10–11).

The meta-subject results, which the Standard stipulates, must demonstrate the following:

1. An ability independently to determine the goals of one's education ...
2. An ability independently to plan the means for achieving the goals, including alternative means ...
3. An ability to correlate one's actions with the planned outcomes (p. 12).

And much else—including, for example, reading comprehension; the ability to create, employ, and transform signs and symbols, models, and schemas for solving academic and cognitive problems; and the development of ecological thinking.

Subject requirements are formulated separately for each subject area. The subject area of "Mathematics and Informatics" contains a comparatively large number of requirements, of different types. These include the following:

> The formation of conceptions of mathematics as a method for comprehending reality, enabling the description and study of real processes and phenomena;
> The recognition of the role of mathematics in the development of Russia and the world;
> The possibility of giving examples from Russian and world history of mathematical discoveries and their authors (p. 25).

But there are also requirements such as the following:

> Mastery of the system of functional concepts, the development of the ability to use functional-graphic representations for solving various mathematical problems, for describing and analyzing real-world relations:
> > defining the position of a point based on its coordinates and the coordinates of a point based on its position in a plane;
> > determining the values of a function based on its graph: its domain, range, zeroes, intervals of sign-constancy, increasing and decreasing intervals, maximum and minimum;
> > constructing graphs of linear and quadratic functions;
> > operating at a basic level with the concepts of sequence, arithmetic progression, geometric progression;
> > using the properties of linear and quadratic functions and their graphs to solve problems from other academic subjects (p. 26).

Note that the strictly mathematical requirements differ from what was required during the 1990s or even the 1980s for the most part only due to a certain imprecision and lack of concreteness—everything pertaining to functions listed above was also required previously, except now the Standard does not specify at what level these knowledge and skills are required. Of course, one might ask whether a basic school graduate must be able to find the domain of a function *not* based on its graph. This is not required explicitly, although the Standard does require proficiency with "techniques for solving inequalities" (p. 26). The requirements pertaining to discrete mathematics are clearly (relatively) new—as has already been said, discrete mathematics was not studied in school in the USSR. Also new, of course, is the inclusion of informatics (something close to computer science) in the same subject field as mathematics and the formulation of corresponding requirements for computer scientific and algorithmic literacy or for the formation of skills and aptitudes enabling safe and effective conduct in working with computer programs.

The Standard goes beyond the two sections described above—it contains two other sections, "Requirements for the Structure of the Basic Educational Program" and "Requirements for the Conditions of Implementation of the Basic Education Program." The latter deals mainly with financial-economic, informational, personnel-related, psychological-pedagogical, and other requirements. As for the section on structure, it not only contains formal requirements (for example, what a program must include and how the system for assessing results must be described) but also underscores certain substantive requirements, including the need to develop so-called universal learning activities, that is, activities not specific to the subject—the focus here is not on the ability to solve inequalities but, for example, on developing the students' capacity for self-development and self-improvement and on other results prescribed by the requirements in the first two groups (which are not subject-oriented).

The authors of the textbooks used in schools must follow the requirements of the Standard. The subject requirements are arguably less rigid than the requirements from the other groups. In 2014, a great deal was written in the press about the fact that Ludmila Peterson's mathematics textbook was not included in the so-called Federal List of Textbooks (that is, textbooks that are approved by the state and may be purchased using state funds), since in the expert opinion of Lyubov Ulyakhina, "The content of the textbook fails to facilitate the formation of patriotism. The protagonists of the works of Gianni Rodari, Charles Perrault, the Brothers Grimm, A.A. Milne, Astrid Lindgren, Erich Raspe, dwarves, elves, fakirs with snakes, and the three little pigs are hardly suited to cultivate a feeling of patriotism and pride in one's country and one's people." Responding to reporters' questions, the expert stated:

> I conducted an expert evaluation in accordance with the state educational standard for all textbooks in all subjects. One of the first questions that I had to answer was whether a textbook developed [students'] personal qualities. And another clause contained the word "patriotism"—and it was necessary to give an answer. "Does the textbook cultivate patriotism, love and respect for the family, the homeland, one's people, one's region?"—this was the complete formulation [of the question]. It was impossible to shun this clause, and the question had to be approached honestly. (see Znak 2014)

Whether or not we share the expert's opinion about the fact that the mention of Snow White—a representative of a foreign culture—in a fourth grade textbook is not conducive to the formation of patriotism, we must acknowledge that normatively requiring each textbook systematically to cultivate patriotism and conducting expert evaluations of such cultivation can only lead at the very least to contentious debates—a single, universally accepted hierarchy of values is hardly conceivable in such matters. A certain bias is therefore inevitable, which is unlikely conducive to improving the quality of teaching (note that even during the years of Stalin's anti-cosmopolitan campaign, excessive demands for patriotism in mathematics classes—as opposed to literature or history classes—were not welcome: for the cultivation of patriotism, extracurricular time was set apart; Karp 2007b, Karp 2010).

Without entering into a discussion of all the opinions voiced about these issues, we should only like to note that sharply negative assessments of the Standard were also expressed. Consider, for example, an article entitled "Mathematics Education

Under the Chariot of the FGOS [Federal State Educational Standard]" (Malyshev 2016). Its authors believe that, by concentrating on teaching universal learning activities, the writers of the Standard and those who follow them have undermined the teaching specifically of mathematics, offering students unsubstantive and artificial assignments in its place. But even in this article, the author rather quickly begins to talk not so much about the Standards, as about the assessments carried out in the schools, and first and foremost about the so-called EGE (Edinyi Gosudarstvennyi ekzamen – Uniform State Exam).

9 Final Assessment: The Uniform State Exam (EGE) and the State Final Assessment (GIA)

Probably the most frequently used word in discussions about education in Russia during the first and second decades of the twenty-first century has been "EGE." The idea of combining entrance and graduation examinations began to be discussed, as we have already mentioned, back in the 1990s, but only in 2001 did the ministry begin conducting experiments in a number of regions. At the same time, other local experiments in the field of examinations were discontinued: the examinations had to come from Moscow. The number of regions encompassed by the experiments grew, and from 2008 on, the EGE began to be conducted across the whole territory of Russia and no longer as an experiment. The essence of the procedure, as it took shape in 2009, is that students take examinations in a centralized manner, based on universal monitoring and measuring materials developed by a central agency for the whole country (although this agency may employ people from different regions, not only from Moscow), which are then checked in a centralized manner—by special local commissions, with possible subsequent emendations and changes by a national commission—with a view to giving an assessment (number of points) based on developed common criteria. A student who has received a certain number of points in mathematics—usually even a very low one—acquires the right to obtain a secondary school diploma (if there are no problems with other subjects, of course), and this same number of points is sent to the colleges to which the student has applied.

The practice and the very idea of the EGE has not only political and organizational but also methodological aspects. We should say at once that one of the main virtues of the EGE was proclaimed to be the reduction of corruption. One of the theoreticians behind the introduction of the EGE, Moscow Higher School of Economics rector Yaroslav Kuz'minov, said in an interview in 2009:

> The first consequence of the introduction of the EGE is a quite serious decline in corruption in admissions to colleges. In the mid-2000s, we estimated admissions corruption, i.e. bribes given for admission to colleges, in the amount of $300-$400 million. At present, this flow of money has diminished approximately fourfold... We observe no growth of corruption in the schools in recent years. And this is understandable. Schools have no direct relation to the administration of the EGE. (Kuz'minov 2009)

College teachers, whose incomes had declined substantially since the early 1990s, were accused (and probably not unjustifiably) of making the materials of their college entrance examinations available for purchase, in one form or another. Naturally, once these materials became inaccessible to these teachers, they could no longer sell them.

The second achievement of the EGE was seen to consist in the fact that it gave talented children from the provinces an opportunity to enter the leading colleges in the country, to which they had simply been unable to travel to take their examinations previously. The same Yaroslav Kuz'minov (2012) even said that objections against the EGE stemmed from the fact that Moscow and St. Petersburg residents had lost their privileged positions:

> A deep-seated cause of public antagonism toward the EGE consists not in methodology, but in its social effect. The interests of considerable groups—specifically, the populations of the country's largest cities—have been encroached upon. The residents of Moscow, St. Petersburg, Yekaterinburg, Novosibirsk, Nizhny Novgorod, in the 15 years following the fall of the USSR, had the opportunity almost monopolistically to enjoy an important social good: free higher education for their children in the best colleges. They were simply nearer to them and had average incomes that allowed them to pay for preparation courses for college entrance examinations. Residents of other regions, small cities and towns, on the other hand, found themselves outside the system of preparation for the best colleges—both in terms of their places of residence, and in terms of their incomes, which were two or three times lower than those of people residing in the megalopolises.

Of course, much was also said about the need for an honest tool for measuring the outcomes of education, which the EGE is supposed to become. The director of the Tsaritsyno Center of Education in Moscow, for example, expressed himself in the following cautious terms:

> We have never had a single objective tool for measuring the quality of children's education—neither during the Soviet, nor during the post-Soviet period. The EGE can be such a tool, provided its content is sound and the procedure for administering it absolutely clear, objective, open, effective, and transparent. (Materialy 2004, p. 274)

We should note at once that altogether different views about these questions were expressed as well. We will confine ourselves to just one quote (albeit a long one) from an article by Alexander Abramov (2009), "Three Myths That We Have Lost":

> I believe that the collapse of three myths has been documented this year. Actually, the myth of the EGE's democratic character never needed debunking. There are few student dormitories left in Moscow and St. Petersburg. Do-gooders who would finance the education of talents from the boondocks for five-six years are also nowhere to be seen…
>
> Few people believe the myth of the EGE's anti-corruption influence. College corruptionists have more than compensated for their losses on entrance exams by creating a far more profitable system of proceeds from 10-12 exam sessions. Meanwhile, the market in diplomas, tutoring, and "services" related to taking the EGE has been flourishing.
>
> The novelty of the season is the collapse of the myth about the EGE's objectivity… The fact is that in the heat of the struggle for objectivity and quality of education, the following decision was made: schoolchildren who received two failing grades—in mathematics and in the Russian language—would be given no school diplomas. Clearly, however, blackballing multitudes of angry young men in times of crisis is a risky proposition… The problem

found a three-stage solution. At the first stage—during the administration of the exams—a blind eye was turned to "shenanigans." Judging by reports on the internet, the means of acquiring decent grades are quite varied. Nor is this due only to the vast upsurge in mobile communication technologies available to exam-takers. At the second stage, an extremely low threshold for a passing grade was set by force—below ground level, as they say. At the third stage—when students re-took exams they had failed—two thirds of them overcame their illiteracy in the Russian language and mathematics in a couple of weeks. As a result, the projected exam figures were achieved: only 2–3% of graduates did not get a diploma.

Methodologically, the examination has changed several times and quite probably will yet change more than once. The earlier versions contained multiple-choice assignments (called "assignments A"). Here is an example of one such assignment (Nekrasov et al. 2007, p. 14):

Find the value of the following expression: $\sqrt[6]{3^7 \cdot 4^5} \cdot \sqrt[6]{3^5 \cdot 4}$

1. 24
2. 6
3. 36
4. $4\sqrt{3}$

These assignments were denounced as going against the Russian tradition, which precisely required students to demonstrate their reasoning, rather than to choose an answer; they were met with a wave of criticism and were gradually removed from the materials.

A second group (for a long time called "assignments B") consisted of assignments in which short answers had to be given. Here are a few examples:

- Find the maximum of the function $y = \dfrac{x^3}{3} + \dfrac{x^2}{2} - 2x - 2\dfrac{1}{3}$ (Nekrasov et al. 2007, p.18).
- Triangle ABC is inscribed in a circle with a center at O. Angle BAC is 32°. Find angle BOC. Give the answer in degrees. (Demonstration version 2019 https://www.examen.ru/add/ege/demonstracionnye-varianty-ege/).

Finally, there are also assignments (group "C") that require detailed answers (these, by contrast with the preceding ones, are checked by expert commissions). Here are two examples of such assignments (https://www.examen.ru/add/ege/demonstracionnye-varianty-ege/):

- a) Solve the equation $2\sin\left(x+\dfrac{\pi}{3}\right) + \cos 2x = \sqrt{3}\cos x + 1$. b) Indicate the roots of this equation that belong to the segment $\left[-3\pi; -\dfrac{3\pi}{2}\right]$.

- Find all positive values of a, for each of which the system $\begin{cases}(|x|-5)^2 + (y-4)^2 = 9 \\ (x+2)^2 + y^2 = a^2\end{cases}$
  has a single solution.

Here, we should note that after the EGE had been given for several years, it was recognized that it was somewhat strange to offer the same version of the examination to all students. Those who had covered the minimal program had no chance of even understanding many if not all of the assignments in which a detailed answer had to be given. The gap between "easy" and "difficult" problems was very great. The mandatory nature of the EGE destroyed "humanities-oriented" classes, since their graduates did not prepare for problems that were comparatively technically difficult. But even classes that used ordinary textbooks, which contained such problems, were not particularly helpful when only 3 or even 4 h were allocated per week for the entire course in mathematics (both algebra and geometry). On the other hand, it was necessary to include in exam problems above the minimum level in order to meet the requirements of colleges with a technological or economic orientation. In 2015, the proposal was made to conduct the examination in mathematics on two levels—a basic level and a specialized level—with only the results of the specialized exam being suitable for college applications. The specialized version approximately followed the versions that had been offered previously, while the basic version was easier—it consists (for 2018–2019) of 20 short-answer problems at the so-called basic level of difficulty. The specialized version contains eight short-answer problems at a basic level of difficulty, four short-answer problems at an advanced level of difficulty, and seven problems requiring detailed answers at a high level of difficulty. The time allowed for the exam is 240 min or during some years 235 min.

Naturally, over decades of conducting the EGE, an enormous quantity of data has been collected about its results. We might confine ourselves, for example, to statistics pertaining to Tomsk region in 2014 (Sokolov 2014). During this year, the lowest score demonstrating assimilation of the school curriculum in mathematics was 20 (in 2013, it was 24)—this referred not to the initial, "raw" score (the greatest possible raw score was 33), but to the score after a recount according to some rules (the greatest possible being 100). In other words, the lowest "passing" score was one fifth of the total. In Tomsk region, 120 graduates (2.26%) got less than this lowest score, and of them, 13 graduates did not get a single point. One graduate got the maximum score (100) in 2014, while eight had gotten the maximum in 2013, zero in 2012, and eight in 2011. The average score in 2014 in the region was 48.11 (in the Russian Federation as a whole, it was 39.6); in 2013, it was 48.04 (in the RF, 49.6) (p. 51). Also, Table 5.4 seems highly indicative (p. 53).

Also worthy of note are statistics about the distribution of high scores (80 or above) in the schools. Altogether, there were 109 students who scored at this level, of whom 46 attended the same Tomsk lyceum, 7 others the same gymnasium, 6 others two other gymnasia (3 students in each); 13 other educational institutions had two such graduates each, and 24 other educational institutions had one such graduate each (pp. 53–54). It is not difficult to compute that in dozens of Tomsk region schools, there was not a single student who scored 80 or higher. On the other hand, we also have statistics about schools with the best average score (pp. 53–54): only in 1 school it is higher than 80, in 10 schools it is between 60 and 69, and in 42 schools it is between 50 and 60. Once again, it is not difficult to compute that over half of this year's graduates in Tomsk region attended schools in which the average score was below 50.

Without reproducing the other statistics, we will confine ourselves to noting that problem C1, which required a detailed answer, was solved by 22.4%, problem C2 by

Table 5.4 Distribution of graduates over ranges of test scores

Range of scores	Number of graduates with scores within the range	Percentage of graduates with scores within the range
0–10	46	0.86%
11–20	231	4.33%
21–30	606	11.37%
31–40	1225	23.0%
41–50	805	15.1%
51–60	960	18.01%
61–70	769	14.42%
71–80	616	11.54%
81–90	67	1.26%
91–100	6	0.12%

3.4%, problem C3 by 7.7%, problem C4 by 1.6%, problem C5 by 0.8%, and problem C6 by 3.9% (p. 59). The report cited here gives no figures about the students who failed to solve a single problem from this section, but they constituted at the very least 60%.

Naturally, these figures should not be extrapolated onto all of Russia and across all years, but nonetheless in our view they convey a certain sense of the results.

Comparing the results from different regions by year, and noting sometimes significant improvements or unexpected differences in results between regions (oblasts), it is impossible not to wonder whether the examinations are always conducted fairly, whether the hope of obtaining an honest tool for assessing achievements, which was mentioned above, has been fulfilled. "Anomalously high results" on the EGE have been repeatedly mentioned in the press; there have already been reports (https://echo.msk.ru/news/2277336-echo.html) about lawsuits against officials who, in one way or another, sold exam materials; there have also been accounts of versions of examinations appearing on the Internet before the examination was given. Of note, for example, is the assertion made by St. Petersburg teacher, Dmitry Gushchin, in 2018 (for example, https://sibmama.ru/EGE-2018-2.htm), who stated that, prior to the examination, problems appeared online which were then reproduced with only slight differences on the distributed copies of the examination. For example, the materials published beforehand contained the problem: solve the inequality $2\log_3(1-2x)-\log_3(1/x-2) \le \log_3(4x^2+6x-1)$; while the examination itself contained the problem: solve the inequality: $2\log_2(1-2x)-\log_2(1/x-2) \le \log_2(4x^2+6x-1)$, in which 3 in the base of the logarithm was changed to 2. The ministry, however, denied any leak, claiming that the problems were different, and that the resemblance between them was only natural and due to a certain standardization.[4]

---

[4] At the ministry's behest, legal proceedings were begun, in the course of which Gushchin's charges were found to be groundless, since the court found nothing unlawful in the coincidences indicated by him. As the book is going into production, the case is being reviewed a second time.

Standardization does indeed exist—long before the examinations, students know (and completely legally) not only that, for example, some problem will be devoted to plane geometry, but also that this problem will involve, say, two alternative arrangements of the configuration of figures being discussed. There are entire collections devoted to solving specific problems, for example (Yashchenko and Zakharov 2014), "EGE 2014. Mathematics. Problem B8. The Geometric Meaning of the Derivative: Workbook." Mathematics instruction is thus inevitably structured by the examination: we study not mathematics in general, but specifically how to solve problem B8. Let us add that when entrance examinations to colleges existed, in one way or another fundamentally new problems appeared each year—a high-level college considered it necessary to invent them; now, by contrast, this erstwhile variety has vanished. The uniformity that has been introduced in a certain sense surpasses that of Stalin's times.

Writing about the EGE, Bashmakov (2010) rightly compares the educational system with a physical system whose state changes when it is measured. It is evident that the monopoly of the EGE could not but change the system. Note that both the supporters of the EGE and its opponents, in comparing the examination with the systems that exist outside of Russia, usually fail to notice that in these systems there is usually no monopoly with regard to conducting the examinations: in Britain, there exist various examination commissions; in France and Germany, examinations are conducted at the regional level; in the United States, there could not even be a uniform examination, since there is no uniform curriculum. But the monopoly—or, to use another expression, the unified vertical of power—was a political decision.

Criticism of the EGE has been varied. It has been noted that first-year college students admitted based on their EGE scores could not subsequently solve the same kinds of problems. Such an assertion was made, for example, by Sadovnichy, academician and rector of Moscow State University, who stated in 2009: "We have conducted two tests using materials from the EGE in two departments—the mechanics-mathematics departments and the computational mathematics department. About 40% passed the test; 60% failed" (Sadovnichy 2009). Goldina and Gil'derman (2010) tested students at a far less prestigious college (Moscow Automobile and Road Construction State Technical University) and also reached the conclusion that EGE scores reflecting the knowledge of first-year students had been greatly inflated. Not without irony, they reported that an elective course had to be offered in which first-year students were taught to complete the square and the like. There have been many such articles (see also, for example, Deminsky 2010).

The decline in requirements pertaining to the culture of reasoning has also been mentioned. Nesterenko (2009) noted that the increase in the number of problems brings with it a decrease in their difficulty and continued as follows:

> The sharp decline of requirements on the exams leads to a hollowing-out of the content of education in the schools. Teachers, contrary to their calling, are forced to drill schoolchildren to solve standard sample exercises, distributed in advance. There is no need to teach that which is not very important for obtaining a positive grade on the exam. (p. 68)

In the same article, however, he also noted that the examination as a whole is somewhat difficult—and he remarked that the difference in difficulty between problems of various groups has only limited relation to how they are scored.

It has also been said that the EGE produces crammers instead of creative human beings, that it pushes students away from an understanding of mathematics, and so on (for example, Smolin et al. 2009). Despite this criticism, however, the EGE remains in use.

Upon graduating from the basic school (nine grades), students must take the so-called State Final Assessment (Gosudarstvennaya Itogovaya Attestatsiya—GIA), which in its new form is also called the Basic State Exam (Osnovnoy Gosudarstbennyi Ekzamen—OGE). Without discussing the examination in detail, we should say that, for example, in 2019 it consists of two modules: "Algebra" and "Geometry." The examination contains 26 problems in all. The "Algebra" module contains 17 problems: 14 problems in part 1 and 3 problems in part 2. The "Geometry" module contains nine problems: six problems in part 1 and three problems in part 2. The problems include multiple-choice problems, short-answer problems, and problems that require a detailed solution. The following exercises from a sample examination that appears on the GIA official site (http://gia.edu.ru/ru/) convey an idea of the level and types of problems found on the examination:

Algebra, part 1

- The following table shows how ninth-graders are graded for running 30 m:

	Boys			Girls		
Grade	«5»	«4»	«3»	«5»	«4»	«3»
Time, seconds	4.6	4.9	5.3	5.0	5.5	5.9

- What grade will a girl receive if she runs this distance in 5.62 s?
  (1) grade "5" (2) grade "4" (3) grade "3" (4) a failing grade.
- There are buns on a plate. They are identical in appearance. Four have meat inside, eight have cabbage, and three have apples. Petya takes one of them at random. Find the probability that it will have apples inside.
- Solve the equation $x^2 + x - 12 = 0$. If the equation has more than one root, write the greater of the roots as the answer.

Part 2

- A fisherman sets out in a motorboat from a dock at 5 a.m., traveling upstream. After some time, he drops anchor, fishes for 2 h, and comes back at 10 a.m. on the same day. What distance from the dock does he travel, if the speed of the river current is 2 km/h, and the speed of the boat is 6 km/h?
- Construct the graph of the function $y = \dfrac{x^4 - 13x^2 + 36}{(x-2)(x-3)}$ and find the values c for which the line $y = c$ has exactly one point in common with the graph.

Geometry, part 1

- In an isosceles triangle ABC with base $\overline{AC}$, the exterior angle at vertex C is $123°$. Find the angle BAC. Give the answer in degrees.
- Which of the following assertions is true?

    1. Through a point external to a given straight line, a straight line can be drawn parallel to the given line.
    2. A triangle with sides 1, 2, 4 exists.
    3. Any parallelogram has two equal angles.

In your answer, write down the numbers of the assertions you have chosen without spaces, commas, or other symbols.

Part 2

- In a parallelogram ABCD, point E is the midpoint of side $\overline{AB}$. It is known that EC = ED. Prove that the given parallelogram is a rectangle.

As can be seen even from these examples, the gap between the first and the second parts is very substantial. At the same time, the overall knowledge required of the students is not very great (in the past, far more was covered in the Soviet 9-year (8-year) school, including, for example, trigonometry). A certain amount of space is allocated for what may be called real-world problems.

10 Specialization

An analysis of the examination problems reproduced above shows that some students are expected to learn only the most basic facts (the first part is for them), while others are expected receive a relatively full education in mathematics—something like what all students were claimed to have received in the past—and often even something at a much higher level (the last problems in section C or the last of the algebraic problems from the OGE reproduced above would have been considered very difficult even in Soviet times, of course). Connected with these expectations is a key idea that has become widespread in recent decades: the division of education into a basic and a specialized level.

As Lukicheva and Mushtavinskaya (2005) explain:

> Specialization in education is a means of differentiating and individualizing education by changing the structure, content, and organization of the educational process, in which students' interests, inclinations, and abilities are more fully taken into account. (p. 6)

Technically, in the opinion of the same authors, specialized education can include, for example, education in physics-mathematics, physics-chemistry, biology-geography, social sciences-humanities, industry-technology, art-aesthetics, defense-athletics, and so on, and so forth (pp. 10–11). At the same time, one can distinguish between basic courses in mathematics (in general education classes for 4 h per week, and in specialized classes for 6 h per week) and the elective courses that actually make the class a specialized one. Among the possible elective courses in mathematics, the authors list the following: "Mathematical Models in the Natural Sciences," "Methods of Mathematical Modeling in the Humanities," "Theory and Practice of Consumer Behavior," "Ecology in Numbers," "Applications of Trigonometry," "Functions. Graphs," "Remarkable Theorems and Facts of Algebra," "Optimization Problems," "Problems with Parameters," "Absolute Value," and many others (pp. 23–27). In other words, elective courses can vary greatly: they can be devoted to further study of the questions of school mathematics; they can present certain sections of mathematics non-usually studied in school; or they can be devoted to certain general questions or to the applications of mathematics in one or another specific field.

The book cited above represents only one possible view of specialized education. The difference between the teaching of basic courses in schools and classes with different specializations, for example, has also been discussed. As Bashmakov (2010) writes:

> The term specialization is typically interpreted as follows. Any high-school student must select one specialized trajectory of education. Though there is a list of primary specializations, the number of specializations may be infinite. Any subject may be studied at two levels: basic and specialized (advanced); i.e. specialization has become equated with the level of studies. To make a specialized school (class), it is sufficient to decide which of the main subjects will be studied at the basic level and which at the advanced level. (pp. 157–158)

And he goes on:

> This approach still maintains the idea that mathematical knowledge and skills should be arranged along a straight line and this line may be chopped up as necessary. Anything lying outside the straight line will be billed as "elective courses" and used in the manner of an optional condiment. (p. 158)

As far as can be judged, universal specialization has not yet arrived: a system in which students must, in effect, choose a profession at the end of ninth grade cannot be inculcated very easily, let alone a system of elective courses that assumes that there are people in schools who are capable of teaching such courses. Bashmakov's remark seems justified—it is easy to pour out more tea for some, less for others, stronger for some, more watery for others, but it is far more difficult to do so with mathematics.

What is important, however, is that the idea of specialization, which had once carved a path for itself with difficulty for a very small group of students especially interested in mathematics, now, on the contrary, is offered to all students, including those who are completely unprepared for such a choice. What is really at stake here is determining early on those students who will not receive a serious course in mathematics. In this way, individual wishes—first and foremost, the wish to study a little less mathematics—are indeed satisfied to a certain degree (at least, they may be considered satisfied), but how enduring such wishes are, and what should be done if they change, is not entirely clear (these very questions, as we have said above, were also being discussed decades ago).

11 Gifted Students

Identifying and supporting the gifted has come to be considered an objective of paramount importance. President Putin personally meets with gifted students at the specially formed Sirius Center; indeed, the government site kremlin.ru explains that:

> The Sirius Educational Center was formed on the basis of olympic infrastructure on the initiative of the head of state in December 2014. The goal of the center's work is the early identification, development, and further professional support of gifted children from all regions of Russia, who have manifested outstanding abilities in the spheres of the arts, sports, natural scientific disciplines, and also those who have achieved success with technological creativity.

And it continues:

> Each month, 600 children of ages 10–17, who are accompanied by over 100 teachers and coaches, receive free education at the center. The educational program, which lasts 24 days, includes both specialization-oriented classes and developmental recreation, master classes, creative talks with professionals recognized in their fields, a complex of health-promoting procedures, and during the school year—general education classes. (news from July 21, 2017)

Naturally, far from all of these children have manifested their giftedness specifically in the field of mathematics, but there are mathematically gifted children among them also. No expense is spared for these programs. The president has long talks with Sirius students, which are ubiquitously reported on in the press. Not only that, but it is reported that the very name "Sirius" was chosen by the president himself, in honor of the brightest star.

The times when specialized mathematics schools were almost officially referred to as a cancerous growth on the healthy body of Soviet education (Karp 2011) have gone. The government of Moscow, for example, now yearly gives out grants to the best schools, a large number of which specialize in mathematics (including specialized schools affiliated with major colleges). Nor is this surprising, since the metrics on which the selection is based include the results of Olympiads, the EGE, and the OGE (see the official site of the mayor of Moscow: www.mos.ru).[5]

The position of schools specializing in mathematics, however, is not all that sunny: all of the problems mentioned above remain; to some extent, the concept of the specialized mathematics school has been blurred; many specialized schools exist, and the differences between them are not always clear to everyone; the ties to the mathematics community, although they remain relatively strong in some places, nonetheless are clearly weaker than during the most flourishing years; the orientation of schoolchildren has changed, and a future as a mathematician is often no longer seen as anything enviable and desirable; and finally and most importantly, the weakening of mathematics instruction in the classes that precede specialized education is taking its toll, as we will see below (Karp 2011).

---

[5] Nor should we forget that the winners of Olympiads enjoy special privileges when enrolling in college: a successful Olympiad performance is taken to be the equivalent of top results on the EGE.

We should also note the appearance of a large number of various new forms of extracurricular work in mathematics, including mathematical competitions, and consequently the growth of the corresponding literature (Marushina 2016): this process began already during the preceding period, in the 1990s, but it continues successfully to this day. We might recall such books as Bratus' et al. (2003), Yashchenko (2005), Blinkov et al. (2007), Chulkov (2009), and Tikhonov and Sharich (2012). It is also important to point out the noticeable growth in the number of publications for children of elementary-school and middle-school age who are interested in mathematics (for example, Katz 2013, Kozlova 2008).

Special organizations and institutions for facilitating the development of mathematics education have appeared—the first of these that must be mentioned is the Moscow Center for Continuous Mathematics Education, MTsNMO (Sossinsky 2010). Initially organized as a kind of front to provide legal cover for the Independent University of Moscow, a scientific and academic mathematical center, MTsNMO has become a leading and influential institution working with hundreds and thousands of schoolchildren, supported and recognized by the Moscow and Federal governments, possessing its own publishing house, and exerting a noticeable influence on mathematics education across the country.

Another important development has been that, along with traditional competitions and problems, more attention in working with schoolchildren has started being devoted to problems that require prolonged thinking—research problems. Sgibnev (2013), for example, is a book devoted to such problems. As an example of a research problem, let us cite a question discussed in this book: how many ways are there of representing a fraction as a sum of fractions whose numerators are 1 (this question may be said to go back to ancient Egypt). This problem—with restrictions and in particular situations—was worked on by a sixth grader over quite a long period of time and with a certain degree of success. Such research activity is certainly no less useful than traditional "sports-like" Olympiad mathematics.

Successes in this "sports-like" mathematics, however, have started declining, despite the care and attention it receives on all sides. From 2010 until 2017, at the top International Mathematical Olympiad, Russia proved incapable of placing among the three prizewinners; indeed, in 2017, the country's national team took only the 11th place; and twice before that, the eighth. Of course, in each competition, a team's results reflect not only its efforts but also the efforts of the other participants; specialized preparation, based in no small measure on Russian methodology, is now widespread in other countries as well. Nonetheless, the yearslong lack of success also tells of existing problems. In 2018, Russia was able to return to one of the top three spots, being beaten only by the national team of the United States. This was preceded by consultations at the highest educational level—with the minister of education and the presidential assistant for education (Kak Rossiya vernulas'... 2018)—which reveals what kind of importance was assigned to the success and failure of the national team.

In the words of MTsNMO head, Ivan Yashchenko (Kak Rossiya vernulas'... 2018), a fundamental role was played by the precise and formalized selection process, whose rules were published in advance and required good performances on five Olympiads, including training Olympiads, the Russian National Olympiad, and the Romanian Master of Mathematics:

> This is also very important, because everyone understood that the system would be absolutely transparent. Both the children and the regional teachers knew the formula based on which the team would be selected. This was not a situation in which one had to "become friendly" with the coach in some way. There was no subjective element.

These changes were accompanied by changes in the way in which the preparation of the Olympiad was organized: "We sat down and agreed that our goal was to raise the Russian national team to a higher level, brushing aside the ambitions of particular regions, the ambitions of 'personal students,'" Yashchenko writes. "After all, if the best teacher starts preparing everyone, it may well turn out that his pupils will not make it onto the national team, while a stronger schoolchild from a different region, on the contrary, will make it, and precisely thanks to the higher-quality preparation, which this child would not have been able to obtain at home."

Kirill Sukhov, a teacher at the Presidential Physics and Mathematics Lyceum No. 239, and a representative of the St. Petersburg (Leningrad) school, was chosen as head coach for the team, and this choice paid off. Whether the existing problems, which may be surmised from Yashchenko's explanations, have been lastingly put to rest, and whether these were the only problems to blame, the future will show.[6]

It has been suggested that the causes of the difficulties lie deeper and cannot be reduced to various administrative and organizational oversights. Sergey Rukshin, who has spent his whole life working on Olympiads and preparing students for them, and who is the head of the office affiliated with school No. 239 from which the top coach for the national team had to be recruited, explains in a large interview (Rukshin 2018) that schools are not meeting the goal of preparing students in a way that would make it possible subsequently to identify them for mathematical work. Noting that the math circles of the Soviet period included few children from working-class families, he says that this was not due to a lack of ability on the part of such children:

> The families of such children simply didn't have a cultural tradition aimed at identifying a child's abilities in different fields. This was precisely the task of general education schools. We may not be able to teach everyone equally well. We don't need a cashier in a supermarket who knows how to take the integral. We have no need whatsoever for a worker in a supermarket, who arranges sunflower oil on a shelf, who remembers biology well.... But in order to identify who is capable of what, we need, at least at the initial stages, to teach everyone using the same curricula, to give them equal opportunities, and to select an elite from among people who had more or less equal starting positions....

Thus, even to him, a person who has spent his whole life working on Olympiads, the selection of this elite appears to be no simple matter: "Please show me the little gadget that, if you point it at a person, will show his giftedness or far-off successes," he exclaims. Further, he comes to the conclusion that conditions everywhere must therefore be identical, which means also identical textbooks and curricula; he

---

[6] In 2019, Russia took sixth place in the International Olympiad in Mathematics.

remarks that problems with education lead to problems with national security, and so on, employing a rhetoric that is quite consonant with official rhetoric. What is important, however, is that the actual official policy, for all its official concern with the gifted and with the elite, seems to him dangerous precisely for the education of the gifted—no Olympiads and no math circles will be able to help, if children have had no initial acculturation to mathematics and if no motivation to study mathematics has been instilled in them. It will not be possible either properly to designate certain children as gifted or even to select some deliberately narrow group among them. Ineluctably, very many capable individuals will be overlooked.

12 Mathematics Teacher Preparation

The Soviet system of teacher preparation included, first and foremost, education at a pedagogical college (4 or 5 years). By contrast with the usual practice in many other countries, in the USSR, the decision to prepare to become a teacher was made upon entering college; thus, all students in the mathematics department of a pedagogical institute prepared to become teachers of mathematics, and, say, courses in calculus were offered not to everyone wishing to study this subject in general but specifically to future teachers. The course of study included the study of mathematical disciplines—mathematical analysis, geometry, algebra, number theory, probability theory, and others; disciplines of the ideological type—the history of the Communist Party of the Soviet Union, philosophy, political economy, scientific atheism, and scientific communism; disciplines of the pedagogical-methodological type—general didactics and general psychology, age-specific psychology and pedagogy, methodology of mathematics instruction (of the *general* methodology, which covered general principles of instruction, and of a *specific* methodology, which covered the teaching of the basic topics of the school course; the course of methodology also included a course in school-level problem solving); and finally, additional disciplines—foreign languages, physical education, and the like. In addition, students conducted student teaching in schools (general information about mathematics teacher preparation and certain bibliographic references may be found in Stefanova 2010).

It should be noted that the courses in mathematics were quite broad, for example, the course in geometry using the textbook by Bakel'man (1967) included the axiomatic construction of Euclidean and non-Euclidean geometries, affine and projective transformations, differential geometry of curves and surfaces, and elementary topology. The course in geometry was taught over several semesters, and hundreds of hours were allocated to it, which included both lecture hours and practical problem-solving sessions. The course in mathematical analysis was very different from traditional calculus courses in the United States, if only because all theorems were strictly proved. In the courses on methodology, possible ways of structuring lessons were analyzed in detail, and possible mistakes that students could make and possible difficulties that they could encounter in each topic were examined—which could be done because the textbooks were stable, that is, they did not change for decades. The ideological courses, and frequently also the psychological-pedagogical ones, usually involved attending and studying lectures and analyzing various works (for example, by Lenin) in practical study sessions.

In general, the system was quite rigid, in the sense that it offered students very little free choice: the elective courses were very few in number—literally two or three over the whole time of study. It is natural to wonder to what degree the covered material was assimilated: this was tested by means of several tests in each semester-long course and usually by means of an oral examination in which students had to reproduce the proofs of randomly chosen theorems. It is no less difficult to say to what extent the study, say, of the topology of closed surfaces facilitated a deeper understanding of the standard school course, which graduates subsequently had to teach.

From the beginning of the 1990s, noticeable changes began to take place in the pedagogical colleges. First, they began to be renamed into universities—prior to this, a university (or as people now say, a classical university) was intended in its basic design to prepare research scientists (which does not mean, of course, that university graduates did not end up becoming school teachers). The universities were few in number; for example, Leningrad (St. Petersburg) had only one. Consequently, the renaming added prestige to the former institutes. Second, ideological courses disappeared—they were replaced (with certain organizational changes) by general courses in the humanities and social-economic courses, for example, the history of Russia (note that for quite a long time, these courses continued being taught by the same people who had worked in ideological or humanities departments previously). Third, in a number of pedagogical colleges, the numbers of students admitted to mathematics departments declined noticeably—other disciplines became more popular. Gradually, other changes also took place, which will be discussed below—briefly and without examining the entire process of change.

An important organizational step occurred when Russia joined the so-called Bologna Process, whose official goal consists in the harmonization of higher education across Europe and the formation of a common European educational space. Russia signed the requisite documents in 2003. Consequently, it became necessary to reconstruct teaching plans and to organize the teaching process in a new way, dividing the previously uniform preparation of specialist-teachers into the preparation of holders of bachelor's degrees and master's degrees.

These reforms were by no means always welcomed. Rukshin (n.d.), who has already been cited above, expressed himself in quite decisive terms, characterizing the Bologna Process as one of the more odious developments:

> The so-called "Bologna Process" and the transition to a two-tier system of "bachelors and masters" destroyed carefully designed and balanced teaching plans for the preparation of specialists and absolutely needlessly increased the duration of the courses of study at the universities.

There were discussions about the deep disparity between the system of bachelor's and master's degrees in the European countries and the United States, on the one hand, and in Russia, on the other. "In Russia, a first-year bachelor's student immediately begins acquiring knowledge in a narrow specialized field. In the physics faculty, he studies physics and higher mathematics, while in the philological faculty, he studies general linguistics and literary theory" (Vakhitov 2013). According to Vakhitov (2013), in this way, the idea of making bachelor's degree students far more free in choosing their own educational trajectory still fails to be realized, and what we have instead is simply a truncated version of the same specialized preparation that had existed before.

Already at the very beginning of the restructuring of higher education, Professor Testov (2005) of Vologda expressed doubts about its viability and worthwhileness, writing about the importance of a fundamental mathematics education and reasonable ways of combining it with professional education, in which Russia was alleged to have experience that the West was only beginning to acquire, and clearly perceiving a certain futility in the preparation of bachelor's degree holders, who, even after obtaining a fundamental education, would be completely unable to apply it in practice.

Ten years later, Dalinger (2015) from Omsk, offering a kind of overview of various (negative) views of the reforms, quoted from the proceedings of a conference on mathematics education that had taken place in 2007: "it can be confirmed that what the Bologna Process has brought to Russia so far has been mainly destruction; the illusions and unfounded hopes have been dispelled" (p. 397). However, he quotes V.P. Odinets:

> This is the fault not of the process itself, but of those persons who have overseen and are overseeing its implementation in Russia without thinking about the consequences or without understanding them. (p. 397)

The experience of the implementation of the new system (which, despite the objections, was implemented) is described in an article by Stefanova (2010). Bachelor's degree holders at the Herzen State Pedagogical University in St. Petersburg are prepared over 4 years, and upon completing which, those who have received a bachelor's degree acquire the right to work in basic schools (9-year school). In order to work in high schools (grades 10–11), they must obtain a master's degree. The course is divided into three stages: two 2-year stages for obtaining a bachelor's degree and one 2-year stage for obtaining a master's. Furthermore,

> The first stage may be characterized as the stage of general preparation. At this stage, students study subjects that represent all fields within a given specialization and select one field for subsequent study.
>
> The second stage is devoted to preparing students in the field which they have selected, and also to providing them with professional teacher preparation (in the case of future teachers of mathematics, this includes preparation for the basic school, i.e. grades 5–9).
>
> At the third stage, preparation in a specific field continues at a higher level, now with a certain degree of professional specialization. (Stefanova 2010, pp. 296–297)

Each of these stages includes the teaching of disciplines in the same, previously mentioned spheres—mathematical, pedagogical-methodological, and others. Federal state standards for higher professional education (FGOS 2019) require future teachers with bachelor's degrees to develop various competencies of a general cultural nature (for example, the holder of a bachelor's degree "possesses high-level reasoning ability, is capable of generalizing, analyzing, receiving information, setting an aim and choosing a means to achieve it"), of a general professional nature (for example, "is able to take responsibility for the results of his professional activity"), and of a professional pedagogical nature (for example, "is able to use the opportunities of the educational environment, including the information environment, to provide for a high-quality teaching-learning process").

Specialized knowledge is also not forgotten, of course, although the already cited Dalinger writes as follows (Dalinger 2017):

> One of the main criticisms of contemporary educational standards is the obvious disparity between the number of hours allocated to the study of a discipline, in this case, mathematics, and the amount of material necessary for the education of a future teacher of mathematics.

And to prove his point, he cites the following figures:

> In 1963, at the mathematics faculty of the Gorky Omsk State Pedagogical Institute, the teaching plan for the preparation of a specialist teacher of mathematics (with a course of study lasting four years) allocated 1000 h for the study of mathematical analysis and 192 for the study of additional chapters in mathematical analysis, while in 2016 the teaching plan for a bachelor's degree in the sphere of "Pedagogical Education," with the specialization "Mathematics Education" (with a course of study lasting five years) 540 h are allocated to the study of mathematical analysis (this is the workload,[7] of which 234 h are hours conducted at the college), and 108 h are allocated for additional chapters in mathematical analysis (this is the workload, of which 26 h are hours conducted at the college).

And he concludes:

> As experience shows, the sharp reduction in the number of hours for a bachelor's degree in the mathematical disciplines leads to students not developing either the celebrated subject knowledge, abilities, and skills, nor the competencies proclaimed by contemporary standards.

Without looking into how these figures were obtained, and without discussing how much they owe to the general standards (which merely require that the total number of lecture hours must not exceed 27 per week, without counting physical education classes) and how much to the specific college in question—which, to be sure, makes decisions under the pressure of existing circumstances—we will say that the widespread and not ungrounded plea for greater independence for the students can be naturally understood precisely as a plea for a reduction in the number of lecture hours and hours conducted at the college.

In keeping with the standards, programs of instruction and teaching plans are developed by each college separately (and subsequently approved by the ministry), for which reason a general analysis of the content of courses in mathematics and mathematics education in all pedagogical universities is hardly feasible. We should merely mention the attempts by Alexey Semenov, rector of Moscow Pedagogical University, to reduce specialized preparation in mathematics—virtually minimizing it, as critics claimed, to what is directly connected with

---

[7] As far as we can understand, the author means that the first number ("workload") is the total number of hours allocated for the study of the course, including independent work.

schools—while increasing the time for pedagogical practical training and psychological-pedagogical courses. These attempts were discussed in the press (for example, Privalov 2016), and Semenov himself (Semenov 2016) expressed his views as follows:

> Preparation in college must touch on the content of the school program; our graduates must get A-pluses in what they will teach in schools. Traditionally, pedagogical colleges are seen as inferior versions of ordinary universities—professors take ordinary courses, simplify them, and teach them to students, believing that the students will become smarter as a result and will begin to understand the school program. That's not so simple—I know that a large number of courses in mathematics have no relation to the school program, and as a result, graduates can solve school problems only at a grade C-level—because their professors have not taught them.

However, no discussion ensued. Semenov was fired from the post of rector, and it is not entirely clear what the real reason for this was. Questions about how teaching at a pedagogical college should be structured, in light of the new situation, undoubtedly remain.

In conclusion, we must also say a word about the system of professional development for mathematics teachers. In the USSR, it was usually implemented at special centers, for example, in Leningrad (St. Petersburg), such a center was first called the Institute for the Improvement of Teachers, then the University of Pedagogical Mastery, and now it is called the Academy of Postgraduate Pedagogical Education, which reflects not only a love for pretty names but also certain changes in the institution's activity. In the past, teachers regularly took courses here (usually, once every 5 years) in which difficult topics from school mathematics were analyzed, problems were solved, and so on, In addition, there were specialized courses, devoted, for example, to teaching certain curricula, or using certain textbooks, or teaching certain age groups (Karp 2004).

As far as can be judged, and without claiming to know everything taking place in all regions of the country, the system continues to be preserved to a certain degree, and among the aspects preserved are the courses offered every 5 years. At the same time, the overall number of hours of classes accessible to teachers has diminished somewhat, and such classes are also becoming less centralized and can be organized in different places and on different foundations (while remaining in most cases free for teachers).

13 Discussion and Conclusion

This chapter does not claim and cannot claim to give a complete description of what has happened in mathematics education in Russia over the last 30 years. We have no information about many regions of the country, about the real experience of change in each of them, about how the change has been perceived by all participants in the process, and much else, and moreover it appears that such information is not only not in this author's possession but to a large extent not in anyone else's possession either—it has not been collected.

And yet from the information we do possess, it is clear that the process has been a very contradictory one and that hoping to conclude its description with a flat verdict—that a catastrophe has occurred or, on the contrary, that conditions have been created for the harmonious development of every personality—is altogether impossible. It should not be forgotten that the development of mathematics instruction in the whole world has also gone through a certain crisis (Karp 2017), and probably one of the inevitably surprising aspects of what has happened in Russia is that the development of computer technologies, which has exerted an enormous influence on the style and problems of mathematics instruction in the West, has clearly had less of an impact on Russian education.

This does not mean that in Russia no thought has been given to using computers and that computers have not been used in practice. We can, without hesitating, name numerous books and articles about using computers and computer programs in mathematics classes (Dubrovsky and Bulychev 2017; Ovsyyannikova 2017), and even in what was discussed above, there are examples in which computers were used in one way or another—for instance, the research problems (Sgibnev 2013) that we have mentioned were, of course, solved with their help. Nonetheless, one can state affirmatively that the role and place of computer technologies in the teaching of mathematics have differed from what it is in many other countries.[8] Although the Russian school course in mathematics is changing, it continues to retain very many old features—this seems wonderful when it concerns such features as the orientation toward proofs and substantiation, but it raises doubts when it comes to certain parts of the content of the course. Ryzhik (2013) writes that requirements for college entrance examinations have led to:

> the appearance of a "mathematics for prospective college students," or more precisely a "pseudo-mathematics," which has been responsible for the appearance of specific topics in the school course. Its adoption was facilitated through the use of methodological literature of fleeting significance. Examples: absolute value problems; problems on the position of the roots of a quadratic trinomial; problems with parameters—in frightening quantity and frightening in quality; the height of absurdity—logarithmic equations (inequalities) with an unknown base, and in the form of a trigonometric function to boot. The deformation of the school course in mathematics, which came about inevitably from having an eye on college, has led to the disappearance of other useful sections of the course (for example, straightedge and compass construction problems have practically disappeared). (p. 7)

It is not difficult to notice that all this had already happened long before 1991, or that some of the listed types of problems (even listed above as the absurdities) might not be useless in certain circumstances, or that straightedge and compass construction problems also cannot be considered a novelty and they have disappeared from the programs of most countries in the world, even if for different reasons. Nonetheless, it is impossible not to agree that the Russian course—at least, the

---

[8] As an anecdote, we might recall that Minister of Education Dmitry Livanov once demanded that each textbook in the country have an electronic supplement (Ministerstvo 2014), which in practice led to the development of rather simplistic problem sets with a selection of answers in the style of the 1960s.

course oriented toward the stronger students—is overloaded with technical details and moreover that the new times have only brought several new types of problems, which have, however, entered the schools in a far more aggressive fashion than their predecessors, since their source, the EGE, plays a far greater role than separate college exams.

The discarding of old problems is seen by many, and possibly not without reason, as a betrayal of the classical traditions of thinking about difficult problems—and indeed, artificial and cumbersome logarithmic equations, with trigonometry thrown in as well, are at a minimum no less useful than ten one-step exercises of the same type, in which students must, say, calculate the base 2 logarithm of 4 or 8, such as may be easily found on school tests in certain Western countries. As for problems that are substantive without being cumbersome—problems that seem natural from today's point of view—not very many of those have appeared in recent decades in the textbooks.

One innovation in mathematics education over recent decades has been an increased attention in the upper grades to problems with real-world content; as has been shown above, such problems are now represented both on the EGE and the OGE and hence are also discussed in classes and in textbooks. These problems, however, are usually very simple (while the mathematics problems that actually arise in real life are usually very difficult—so difficult that it is usually impossible to give them in full versions in school). The result has been a kind of division: it is expected that weak students will solve these simple practical problems, while those who are stronger will be given classical Russian "theoretical" problems ("theoretical" in the sense that it is usually impossible or very difficult to invent a practical application for them, and they are given to students for their intellectual development). But here we come to the social problematics of the development of mathematics education.

As has already been repeatedly noted, it is difficult to talk about this topic, if only because of its politicized character. The changes that have taken place in the country—which have been called "democratization," even though they hardly merit such a name—nonetheless have undoubtedly given the country's citizens certain previously absent rights. Consequently, we can conceive of the present situation as one in which we are faced with the question: does democratization lead to the improvement of mathematics education or not? Consequently, criticism of the condition of mathematics education becomes part of the political discussion. It is no accident that in discussions of mathematics instruction, readers can often observe an unrestrained anti-Americanism and, more broadly, an anti-Western stance, often with the paradoxical conclusion that Americans are forcing American-style education onto Russia in order to enslave Russia more securely (although in that case, it would probably be more logical for them to restructure their own education along Russian lines at the same time).

In reality, everything is far more complicated. To begin with, the development of Soviet school mathematics itself does not fit into the anti-Western canon. It is not difficult to show that Russian prerevolutionary mathematics education, on whose traditions Soviet education was based after 1931, the year of the crackdown against revolutionary pedagogy (Karp 2010), developed under the strong influence of French and German mathematics education. Subsequently, Western education—for various reasons, which cannot be discussed in detail here—began to get restructured in one direction, with a decrease in attention to what was substantive-mathematical, proof-oriented, and generally "scientific" in schools, which became increasingly mass-scale; while Soviet education—in part under the pressure of competition with the West, in part due to the privileged position of the non-ideologized sciences of physics and mathematics, and in part by becoming hostage to its own ideological dogma of equality—extended the existing traditions practically to the level of universal education, which in turn led to methodological improvements.

Historians of education can point to other situations in which, as it were, the outskirts of the civilized world acquired knowledge from more advanced regions, only to surprise their former teacher-countries with this very knowledge decades later, when this knowledge had for various reasons waned in these latter countries (for example, we might recall how Charlemagne invited Alcuin from distant York; it is hardly possible to deduce from this English superiority over "Eastern"—meaning Italian—education.)

The artificial conditions created by the Soviet Union brought considerable successes to mathematics education. The same conditions prevented mathematics education from developing in a natural way, for example, by restricting the writing of new textbooks. When the Soviet Union collapsed and the situation began to change, mathematics education inevitably found itself worse off, if only because it ceased to be virtually the only kind of education that was truly unrestricted by the state. Many problems (for example, the same relatively archaic character of the course, or mass-scale grade inflation, or the leaks of copies of exams) existed in the USSR, but they by no means always became a subject for public discussion. Over recent decades, much has become more open, which cannot be equated with a worsening of education.

On the other hand, in the new situation, perhaps the main virtue of the Soviet system has been abandoned in a practically open manner: the orientation (even if it was often only a demagogic one) toward achieving equality of opportunities for all students, that is, toward providing all students with a relatively deep, proof-oriented course. In the new times, despite the discussions about democratization and democracy (which also took place during the Soviet period, however, even if with somewhat different overtones), the formation of an elite began to be openly talked about, with the further implication (not necessarily mentioned out loud) that membership in the elite was a hereditary quality, even if its doors were not closed to certain other potential members—"the especially gifted." The stratification observable at the very beginning of the period examined above became increasingly more noticeable, thereby also causing considerable damage to the selection of these very "gifted" children, as has already been discussed. In addition, the number of those sufficiently prepared to become successful teachers turns out to be insufficient (and the individuals who become teachers are usually by no means the most exceptional and talented from a mathematical point of view, as is quite understandable). This insufficient preparation of "ordinary" children is subsequently difficult to make up for in college classrooms.

There is no need to repeat that from an economic perspective, Russia has gone through a difficult period—naturally, when teachers, including mathematics teachers, went for months without receiving their salaries, as happened during certain years of the Yeltsin period, this could not but impact the state of education, including the prestige of the teaching profession, exerting a long-term influence on the development of teaching. But one should not attribute all problems in this field to these economic difficulties—in the "fat" Putin years, when high oil prices raised the standard of living, many problems in the teaching of mathematics remained.

At the same time, recent decades have also seen many successes and achievements. Education is based on traditions, and the traditions are still alive—there are still millions of people who cannot imagine textbooks almost without proofs and many thousands if not millions of people who assume that their children will attend math circles. Under abnormal Soviet conditions, people became teachers who probably would not have done so under other conditions, choosing other fields for the expression of their talents and knowledge. The coming of remarkable teachers into the schools helped to create special standards of teaching—there are still many people who had wonderful teachers or who became teachers under their influence.

As we have seen, in the Russian press (including the professional press), the ideal is usually conservative: let's go back to what we have lost, that is, to the Soviet model (as though it were possible to go back to it now, even if the Soviet political system were reconstructed). Reformers are seen as those who reproduce in Russia a system from other countries, that is, reject that which was created previously. Whether it will be possible to carry out genuine reforms while preserving the unique achievements and traditions of Russian mathematics education—and while enabling its free transformation and development, that is, its existence under the conditions of a genuinely democratic society—only the future will tell.

References

Abramov, Alexander M. 2009. Tri mifa kotorye my poteryali [Three Myths That We Have Lost]. *Nezavisimaya gazeta* (September 1).

———. 2010. Toward a history of mathematics education reform in Soviet schools (1960s–1980s). In *Russian mathematics education. History and world significance*, ed. Alexander Karp and Bruce Vogeli, 87–140. Hackensack, NJ: World Scientific.

Atanasyan, Levon S., Valentin F. Butuzov, Sergey B. Kadomtsev, Sergey A. Shestakov, and Irina I. Yudina. 1996. *Geometriya. Dopolnitel'nye glavy k shkol'nomu uchebniku 8 klassa* [Geometry. Supplementary chapters to the school textbook for eighth grade]. Moscow: Prosveschenie.

Bakel'man, Ilya Ya. 1967. *Vysshaya geometriya* [Advanced geometry]. Moscow: Prosveschenie.

Bashmakov, Mark I. 2010. Challenges and issues in post-Soviet mathematics education. In *Russian mathematics education. History and world significance*, ed. Alexander Karp and Bruce Vogeli, 141–186. Hackensack, NJ: World Scientific.

———. 2004. *Matematika. Uchebnoe posobie dlya 10–11 klassov gumanitarnogo profilya* [Mathematics. Textbook for grades 10–11 with a humanities profile]. Moscow: Prosveschenie.

Blinkov, A.D., E.S. Gorskaya, and V.M. Gurovits. 2007. *Moskovskie matematicheskie regaty* [Moscow *Mathematics regattas*]. Moscow: MTsNMO.

Bogomolova, E.P. 2014. Diagnoz: Matematicheski malogramotnyi [Diagnosis: Mathematically illiterate]. *Matematika v shkole* 4: 3–9.

Bratus', Tatiana A., Natalya A. Zharkovskaya, Alexander I. Plotkin, Tatiana E. Savelova, and Elena A. Riss, 2003. *Kenguru-2003. Zadachi, resheniya, itogi* [Kangoroo-2003. Problems, solutions, results]. St. Petersburg: IPO.

Bunimovich, Evgeny A. 2011. Combinatorics, probability and statistics in the Russian school curriculum. In *Russian mathematics education: Programs and practices*, ed. Alexander Karp and Bruce Vogeli, 231–264. Hackensack, NJ: World Scientific.

Butuzov, Valentin F., Yury M. Kolyagin, Gennady L. Lukankin, Eduard G. Poznyak, Yury V. Sidorov, Mariya V. Tkacheva, Nadezhda E. Fedorova, and Mikhail I. Shabunin. 1995. *Matematika. Uchebnoe posobie dlya uchaschikhsya 10 klassov obscheobrazovatel'nykh uchrezhdenii* [Mathematics. Textbook for tenth-grade students of general education institutions]. Moscow: Prosveschenie.

Chudovsky, Alexander N., Lidia A. Somova. 1995. *Sbornik zadanii dlya provedeniya pis'mennogo ekzamena po matematike v 9 klasse obscheobrazovatel'nykh ucherezhdenii* [Problem book for written exams in mathematics for grade 9 in general education schools]. Moscow: Mnemozina.

Chudovsky, Alexander N., Lidia A. Somova, and Vladimir I. Zhokhov 1986. *Kak gotovit'sya k pis'mennomu ekzamenu po matematike* [How to prepare for the written exam in mathematics]. Moscow: Prosveschenie.

Chulkov, Pavel V. Ed. 2009. *Vesennii turnir Archimeda* [Spring tournament of Archimedes]. Moscow: MTsNMO.

Committee on Education of St. Petersburg. 1994. *Sbornik normativnykh i metodicheskikh materialov* [Collection of Normative and Methodological Materials]. St. Petersburg.

Dadayan, A.A., E.P. Kuznetsova, R.I. Isaeva, A.V. Ivanel, A.A. Mazanik, and A.A. Stolyar. 1988. Oshibki na ekrane [Mistakes on the TV screen]. *Matematika v shkole* 4: 43–46.

Dalinger, Viktor A. 2015. Bolonsky protsess i rossiiskoe matematicheskoe obrazovanie v pedagogicheskikh VUZakh [The Bologna process and Russian mathematics education in pedagogical colleges]. *International. Journal of Experimental Education* 2: 396–400.

———. 2017. Podgotovka uchiteley matematiki v usloviyakh novykh gosudarstvennykh standartov po napravleniiu "Pedagogicheskoe obrazovanie", profil' "Matematicheskoe obrazovanie" [Mathematics teacher preparation under new state standards in the sphere of "pedagogical education," specialization: "Mathemetics education"]. *Sovremennye problem nauki i obrazovaniya*: 1.

Deminsky, V.A. 2010. EGE-2009 i uroven' matematicheskoy podgotovki studentov-pervokursnikov [The EGE-2009 and the level of mathematical preparation of first-year students]. *Matematika v shkole* 2: 62–65.

Dneprov, Eduard D.1998. *Sovremennaya shkol'naya reforma v Rossii* [Contemporary school reforms in Russia]. Moscow: Nauka.

———.1999. *Tri istochnika i tri sostavnye chasti nyneshnego shkol'nogo krisiza* [Three sources and three components of today's school crisis]. Moscow: Yabloko.

Dorofeev, Georgy V, Georgy K. Muravin, Elena A. Sedova. 1999. *Matematika. Sbornik zadanii dlya podgotovki i provedeniya pis'mennogo ekzamena za kurs sredney shkoly* [Mathematics. Problem book for preparing and conducting a written exam for the high school course]. Moscow: Drofa.

Dubov, E.L. 1994. Nuzhen ne standart, a urovnevaya differentsiatsiya [What we need is not a standard, but level differentiation]. *Matematika v shkole* 2: 12.

Dubrovsky, Vladimir N., and Vladimir A. Bulychev. 2017. MathKit and math practicum. In *Current issues in mathematics education*, ed. Alexander Karp, 13–27. Bedford, MA: COMAP.

Eidel'man, Tatiana N. 2007. God realizovannykh utopii: Shkoly, uchitelya i reformatory obrazovaniya v Rossii 1990 goda [The year of realized utopias: Schools, teachers, and education reformers in Russia in 1990]. *Novoe literaturnoe obozrenie* 1 (83): 350–367.

Firsov, Viktor V. 1998. K kontseptsii proekta standarta [Toward a conception of a blueprint for a standard]. *Matematika v shkole* 3: 2–9.

FGOS 2018. *(Federal'nyi gosudarstvennyi obrazovatel'nyi standard) osnovnogo obschego obrazovaniya. Izdanie 7* [Federal State educational standard for basic general education. Edition 7]. Moscow: Prosveschenie.

―――. 2019. *(Federal'nyi gosudarstvennyi obrazovatel'nyi standard) vyshego professional'nogo obrazovaniya po napravleniu obrazovanie i pedagogika* [Federal State educational standard for higher professional education in Education and Pedagogy]. http://fgosvo.ru/fgosvpo/7/6/1/5. Accessed 16 Jan 2019.

Galitsky, Mikhail L., Alexander M. Goldman, and Leonid I. Zvavich. 1992. *Sbornik zadach po algebre dlya 8–9 klassov. Uchebnoe posobie dlya uchaschikhsya shkol i klassov s uglublennym izucheniem matematiki* [Collection of problems in algebra for grades 8–9. Textbook for students of schools and classes with an advanced course in mathematics]. Moscow: Prosveschenie.

Gladky, Alexey V. 1994. Kakim ne dolzhen byt' standard [What a standard must not be like]. *Matematika v shkole* 2: 4–7.

Gnedenko, Boris V., and Dmitry B. Gnedenko. 1994. Standart obrazovaniya – Vzglyad v buduschee [A standard for education – A view of the future]. *Matematika v shkole* 3: 2–3.

Goldina, V.N., and S.A. Gil'derman. 2010. Sravnenie itogov testirovaniya pervokursnikov Moskovskogo avtomobil'no-dorozhnogo instituta s rezul'tatami EGE [Comparison between test results and EGE results of first-year students at the Moscow automobile and road construction state technical university]. *Matematika v shkole* 2: 38–42.

Institut obscheobrazovatel'noy shkoly. 1993. Standard srednego matematicheskogo obrazovaniya [Standard for secondary education in mathematics]. *Matematika v shkole* 4: 10–23.

Kaiumov, Oleg R. 2014. O problemakh, svyazannykh s mezhtsivilizatsionnymi zaimstvovaniyami v pedagogike [On problems associated with Intercivilizational borrowings in pedagogy]. *Vestnik Eletskogo gosudarstvennogo universiteta im. I.A.Bunina* 34: 7–12. Elets: EGU im. I.A.Bunina.

Kak Rossiya vernulas' v lidery Mezhdunarodnoy matematicheskoy olimpiady [How Russia Returned to Being a Leader in the International Mathematical Olympiad]. 2018. https://indicator.ru/article/2018/07/17/rossiya-mezhdunarodnaya-matematicheskaya-olimpiada-imo-2018/. Accessed 16 Jan 2019.

Karp, Alexander. 1993. *Sbornik zadach dlia 8–9 klassov s uglublennym isucheniem matematiki* [Mathematics problems for 8–9 grades of the schools with an advanced study of mathematics]. St. Petersburg: Obrazovanie.

―――. 2002. Math problems in blocks: How to write them and why. *Primus* 12 (4): 289–304.

―――. 2003. Mathematics examinations: Russian experiments. *Mathematics Teacher* 96 (5): 336–342.

―――. 2004. Conducting research and solving problems: The Russian experience of inservice training. In *The work of mathematics teacher educators: Exchanging ideas for effective practice*, ed. T. Watanabe and D. Thompson, 35–48. Association of Mathematics Teacher Educators (AMTE).

―――. 2007a. Exams in algebra in Russia: Toward a history of high-stakes testing. *International Journal for the History of Mathematics Education* 2 (1): 39–57.

―――. 2007b. The Cold War in the Soviet school: A case study of mathematics. *European Education* 38 (4): 23–43.

―――. 2010. Reforms and counter-reforms: Schools between 1917 and the 1950s. In *Russian mathematics education. History and world significance*, ed. Alexander Karp and Bruce Vogeli, 43–85. Hackensack, NJ: World Scientific.

―――. 2011. Schools with an advanced course in mathematics and schools with an advanced course in the humanities. In *Russian mathematics education: Programs and practices*, ed. Alexander Karp and Bruce Vogeli, 265–318. Hackensack, NJ: World Scientific.

―――. 2017. Reflecting on the current issues in mathematics education. In *Current issues in mathematics education*, ed. Alexander Karp, 1–12. Bedford, MA: COMAP.

Karp, Alexander and Vladimir Nekrasov. 2000. *Matematika v Peterburgskoy schkole* [Mathematics in St. Petersburg school]. Resource guide. St. Petersburg: Spetsial'naya literatura.

Karp, Alexander, and Bruce Vogeli, eds. 2011. *Russian mathematics education: Programs and practices*. Hackensack, NJ: World Scientific.

Karp, Alexander and Alexey Werner, 2000. *Matematika 10* [Mathematics for grade 10]: Textbook, Moscow: Prosveschenie.
———, 2001. *Matematika 11* [Mathematics for grade 11]: Textbook, Moscow: Prosveschenie.
Karp, Alexander, and Alexey Werner. 2011. On the teaching of geometry in Russia. In *Russian mathematics education: Programs and practices*, ed. Alexander Karp and Bruce Vogeli, 81–128. Hackensack, NJ: World Scientific.
Katz, Zhenya. 2013. *Pirog s matematikoy* [Pie with mathematics]. Moscow: MTsNMO.
Kozlova, Elena G. 2008. *Skazki i podskazki. Zadamiya dlya matematicheskogo kruzhka* [Tales and Hints. Tasks for the mathematical circle]. Moscow: MTsNMO.
Kuz'minov, Yaroslav. 2009. *Rektor VShE: Blagodarya EGE snizilas' korruptsiya v VUZakh* [Higher School of Economics Rector: Thanks to the EGE, Corruption in Colleges Has Declined]. https://www.yuga.ru/news/175140/. Accessed 16 Jan 2019.
———. 2012. *My khuzhe kitaitsev?* [Are We Worse Than the Chinese?]. *Argumenty i fakty* (July 4). https://www.hse.ru/news/1163611/55983879.html. Accessed 16 Jan 2019.
Kuznetsova, Galina M. 1998. *Matematika. 5–11 klassy. Sbornik normativnykh dokumentov* [Mathematics. Grades 5–11. Collection of normative documents]. Moscow: Drofa.
Kuznetsova, Liudmila V., Evgeny Bunimovich, Boris P. Pigarev, and Svetlana B. Suvorova, 2002. *Algebra. Sbornik zadanii dlya provedeniya pis'mennogo ekzamena po algebre za kurs osnovnoy shkoly. 9 klass* [Algebra. Problem book for conducting a written exam in algebra for the basic school course. Grade 9]. Moscow: Drofa.
Kuznetsova, Liudmila, Elena Sedova, Svetlana Suvorova, and Saule Troitskaya. 2011. On algebra education in Russian schools. In *Russian mathematics education: Programs and practices*, ed. Alexander Karp and Bruce Vogeli, 129–190. Hackensack, NJ: World Scientific.
Lednev, Vadim S., Nikolay D. Nikandrov, and Mariya N. Lazutova. 1998. *Uchebnye standarty shkol Rossii. Kniga 2. Matematika. Estestvenno-nauchnye distsipliny* [Leaning standard in Russian schools. Volume 2. Mathematics. Natural sciences]. Moscow: Sfera. Prometey.
Lukicheva, Elena Yu. and Irina V. Mushtavinskaya 2005. *Matematika v profil'noy shkole* [Mathematics in specialized schools]. St. Petersburg: Prosveschenie.
Malyshev, I.G. 2016. Matematicheskoe obrazovanie pod kolesnitsey FGOS [Mathematics education under the chariot of the FGOS]. *Matematika v shkole* 7: 3–6.
Marushina, Albina A. 2016, *Mathematical extracurricular activities in Russia. Unpublished doctoral dissertation.* Columbia University.
Materialy discussii ob itogakh i perspektivakh EGE [Discussion materials about the outcomes and prospects of the EGE]. 2004. *Voprosy obrazovaniya* 3: 273–287.
Ministerstvo obrazovaniya. 2014. Do 1 yanvaria sleduiuschego goda u kazhdogo uchebnika dolzhna poyavit'sia elektronnay versiya [Every Textbook Must Have an Electronic Version Before January 1 of Next Year]. https://xn%2D%2D80abucjiibhv9a.xn%2D%2Dp1ai/%D0%BD%D0%BE%D0%B2%D0%BE%D1%81%D1%82%D0%B8/4298. Accessed 19 Jan 2019.
MP (Ministerstvo Prosvescheniya) RSFSR 1985. *Sbornik zadanii dlya provedeniya pis'mennogo ekzamena po matematike v vos'mykh klassakh obscheobrazovatel'nykh shkol RSFSR* [Problem book for conducting a written exam in mathematics for grade 8 in general education schools of the RSFSR]. Moscow: Prosveschenie.
Nekrasov, Vladimir B., Dmitry D. Guschin, and Leonid A. Zhigulev. 2007. *Matematika. Uchebno-spravochnoe posobie (Gotovimsya k EGE)* [Mathematics. Learning-reference manual. (Let's prepare for the EGE)]. St. Petersburg: Prosveschenie.
Nesterenko, Yuri V. 2009. Nekotorye zamechaniya v sviazi s demonstratsionnoy versiey zadaniya dlya EGE v 2010 godu [Some remarks in connection with the sample copy of the EGE assignment for 2010]. *Matematika v shkole* 10: 67–76.
Novikov, Sergey P. 2018. Proizoshel raspad obyazatel'nogo znaniya [The decay of mandatory knowledge]. *Matematika v shkole* 1: 65–71.
Ot Glavnogo uchebno-metodicheskogo upravleniya obschego srednego obrazovaniya Gosobrazovaniya SSSR. 1991. O trebovaniyakh k matematicheskoy podgotovke uchaschikhsya srednikh shkol [On the requirements for the mathematical preparation of secondary school students]. *Matematika v shkole* 3: 3–4.

Ovsyyannikova, Irina. 2017. Using technology in mathematics education. In *Current issues in mathematics education*, ed. Alexander Karp, 39–43. Bedford, MA: COMAP.

Privalov, Alexander. 2016. O zavershenii odnoy reformy [About the conclusion of one reform]. *Expert* 43 (1005): 14.

Rukshin, Sergey E. 2018. *Interview*. https://www.youtube.com/watch?v=8LUX0drb1NM&t=30 78s. Accessed 16 Jan 2019.

———. n.d. *Neprodumannye reformy obrazovaniya i nauki kak ugroza natsional'noy bezopasnosti strany* [Ill-Conceived Reforms in Education and Science as a Threat to the National Security of the Country]. http://www.zavuch.ru/news/news_main/511/. Accessed 16 Jan 2019.

Ryzhik, Valery I. 2013. Krizis srednego matematicheskogo obrazovaniya glazami uchitelya [The crisis in secondary mathematics education through a teacher's eyes]. *Matematika v shkole* 10: 3–10.

———. 2014. Krizis srednego matematicheskogo obrazovaniya glazami Uchitelya [The crisis in secondary mathematics education through a teacher's eyes]. *Matematika v shkole* 1: 3–9.

Sadovnichy, Viktor A. Ed. 2003. *Obrazovanie, kotoroe my mozhem poteryat'* [Education we can lose]. Moscow: Moscow State University.

Sadovnichy: Okolo 60% pervokursnikov MGU, proshedshie po EGE, "provalili" pervuiu zhe kontrol'nuiu [Sadovnichy: About 60% of Moscow State University First-Year Students Who Passed the EGE "Failed" the Very First Test]. 2009. https://www.newsru.com/russia/28sep2009/60nemgu.html. Accessed 16 Jan 2019.

Sarantsev, Gennady I. 2003. Gumanitarizatsiya matematicheskogo obrazovaniya: Fantazii i real'nost' [The humanityzation of mathematics education: Fantasies and reality]. In *Formirovanie matematicheskikh ponyatii v kontekste gumanitarizatsii obrazovaniya*, ed. Gennady I. Saransev et al. Saransk: MGPI im. M.E.Evsev'eva.

Savvina, Olga. 2017. Priznaki krizisa otechestvennoy metodiki prepodavaniya matematiki [Signs of a crisis in the domestic methodology of mathematics instruction]. *Matematika v shkole* 2: 3–8.

Semenov, Alexey L. 2016. Semenov nepravil'no rukovodit MPGU, nado eto delat' inache [Semenov is steering Moscow State Pedagogical University the wrong way, things should be done differently]. Interview. *Meduza*. https://meduza.io/feature/2016/10/18/semenov-nepravilno-rukovodit-mpgu-nado-delat-eto-inacheon. Accessed 16 Jan 2019.

Sgibnev, Alexey I. 2013. *Issledovatel'skie zadachi dlya nachinaiuschikh* [Research problems for beginners]. Moscow: MTsNMO.

Shatalov, Viktor F. 1979. *Kuda i kak ischezli troiki* [Where and how did "threes" disappear?]. Moscow: Pedagogika.

———. 1980. *Pedagogicheskaya proza* [Pedagogical prose]. Moscow: Pedagogika.

———. 1987. *Tochka opory* [Support point]. Moscow: Pedagogika.

Shevkin, Alexander V. 2015a. Programma po matematike 2015 goda ili Torzhestvo neprofessionalizma [The mathematics curriculum of 2015, or the triumph of unprofessionalism]. *Matematika v shkole* 8: 3–11.

———. 2015b. Programma po matematike 2015 goda ili Torzhestvo neprofessionalizma [The mathematics curriculum of 2015, or the triumph of unprofessionalism]. *Matematika v shkole* 9: 3–17.

Smolin, Oleg, Sergey Kazarnovsky, Anatoly Kasprzhak, Svetlana Nikulina. 2009. Krizis otmeniaet EGE? [Does crisis cancel EGE?] *Echo Moskvy*, https://echo.msk.ru/programs/assembly/578535-echo/. Accessed 16 Jan 2019.

Sokolov, B.V. 2014. Analiz rezul'tatov EGE-2014 po matematike v Tomskoy oblasti [Analysis of EGE-2014 results in mathematics in Tomsk region]. In *Analiz rezul'tatov gosudarstvennoy itogovoy attestatsii vypusknikov 2014 goda obscheobrazovatel'nykh uchrezhdenii Tomskoy oblasti v forme edinogo gosudarstvennogo ekzamena. Informatsionno-analiticheskii otchet i metodicheskie rekomendatsii*, ed. P.I. Gorlov, 50–61 Tomsk: TOIPKRO.

Sossinsky, Alexey B. 2010. Mathematicians and mathematics educators: A traditions of involvement. In *Russian mathematics education. History and world significance*, ed. Alexander Karp and Bruce Vogeli, 187–222. Hackensack, NJ: World Scientific.

Standarty matematicheskogo obrazovaniya. Opyt provedeniya raznourovnevykh vypusknykh ekzamenov po matematike v Sankt-Peterburge [Mathematics education standards. Conducting multi-level graduation exams in mathematics in St. Petersburg]. 1993. St. Petersburg.

Stefanova, Natalya L. 2010. The preparation of mathematics teachers in Russia: Past and present. In *Russian mathematics education. History and world significance*, ed. Alexander Karp and Bruce Vogeli, 279–324. Hackensack, NJ: World Scientific.

Stefanova, Natalya, Natalya Podkhodova, and Viktoriya Snegurova. (Eds.). 2009. *Sovremennaya metodicheskaya sistema matematicheskogo obrazovaniya* [The contemporary methodological system of mathematics education]. St. Petersburg: RGPU.

Testov, Vladimir A. 2005. Matematika i Bolonskii protsess [Mathematics and the Bologna process]. *Vyshee obrazovanie v Rossii* 12: 40–42.

Tikhonov, Yuliy V. and Vladimir Z. Sharich 2012. *Komandno-lichnyi turnir shkol'nikov "Matematicheskoe mnogobor'e" 2008–2010* [Team and individual students tournament: Mathematical all-round]. Moscow: MTsNMO.

Timoschuk, M.E., and L.M. Nozdracheva. 1994. Nedostatki proekta standarta [The shortcomings of the blueprint for a standard]. *Matematika v shkole* 2: 11–12.

Vakhitov, Rustem. 2013. Bolonskii protsess v Rossii [The Bologna process in Russia]. *Otechestvennye zapiski* 4 (55).

Yashchenko, Ivan V. 2005. *Priglashenie na matematicheskii prazdnik* [Invitation to a mathematics festival]. Moscow: MTsNMO.

Yashchenko, Ivan V., Petr I. Zakharov 2014. *Matematika. Zadacha B8. Geometricheskii smysl proizvodnoy: rabochaya tetrad'* [Mathematics. Problem B8. The geometric meaning of the derivative: Workbook]. Moscow: MTsNMO.

Znak 2014. Patrioticheskoe vychitanie [Patriotic Subtraction]. https://www.znak.com/2014-04-08/pochemu_odin_iz_samyh_populyarnyh_uchebnikov_po_matematike_ne_proshel_gosudarstvennuyu_ekspertizu. Accessed 16 Jan 2019.

Zvavich, Leonid I., Dmitry I. Averyanov, Boris P. Pigarev, and Tatiana N. Trushanina. 1994. *Zadaniya dlya provedeniya pis'mennogo ekzamena po matematike v 9 klasse* [Problems for conducting a written exam in mathematics in grade 9]. Moscow: Prosveschenie.

Zvavich, Leonid I. and Leonid Ya. Shlyapochnik. 1994. *Zadachi pis'mennogo ekzamena po matematike za kurs sredney shkoly* [Problems of the written exam in mathematics for the secondary school course]. Moscow: Shkola-Press.

Chapter 6
Ukraine: School Mathematics Education in the Last 30 Years

Vasyl O. Shvets, Valentyna G. Bevz, Oleksandr V. Shkolnyi, and Olha I. Matiash

Abstract The aim of this chapter is to describe the history of Ukrainian mathematics education after the collapse of the Soviet Union. We understand this history as an integral part of the social history of Ukraine. We perform a comparative analysis of contemporary Ukrainian and Soviet approaches to mathematics teaching at school, and analyse the causes of similarities and differences that have emerged in the context of social changes in the country over the last 30 years. This chapter will concentrate on the main aspects of mathematics education: the organization of the learning process, educational tools, the assessment of students' achievements, the preparation and professional development of teachers, and so on.

Keywords Ukraine · Reform · Standards · Textbooks · External Independent Assessment · Teacher education · New Ukrainian School

1 Introduction

Ukrainian history starts in the ninth century in the time of the Kyivan Rus. After the Mongol invasion, these ancient lands became a part of the Lithuanian Dukedom, then the Polish-Lithuanian state, and eventually (since the seventeenth century) the Russian empire. In 1917, there were attempts to create an independent Ukrainian state, but eventually Ukraine became one of the 15 republics of the USSR. After the collapse of the Soviet Union in 1991, Ukraine became independent. Its recent political history has been turbulent. Since the mass protests against the pro-Russian government of Viktor Yanukovych in 2014, some parts of the country have been occupied by Russia.

V. O. Shvets · V. G. Bevz · O. V. Shkolnyi (✉)
Dragomanov National Pedagogical University, Kyiv, Ukraine

O. I. Matiash
Vinnytsia State Pedagogical University, Vinnytsia, Ukraine

© Springer Nature Switzerland AG 2020
A. Karp (ed.), *Eastern European Mathematics Education in the Decades of Change*, International Studies in the History of Mathematics and its Teaching, https://doi.org/10.1007/978-3-030-38744-0_6

Ukrainian education in general and mathematical education in particular have experienced different periods—yet in the seventeenth century, Ukraine was a gate through which western scholarship went further to the east. Later, in the Russian empire, Kharkiv, Odessa, and Kyiv (Kiev) became important centers of mathematical knowledge. In the Soviet period, Ukraine was unofficially considered to be second after Russia in the family of brother-nations. Many scientific centers and universities were located there, including centers of mathematics education. (However, this does not reject Nikita Khrushchev's well-known note that Ukrainians avoided being relocated from their land by Stalin only because there were too many of them and there was therefore not enough free space to send them somewhere else.)

This chapter attempts to discuss the development of Ukrainian mathematics education since 1991, of course, without describing all possible details but reporting on the aspects which appear the most important. It is useful to note at the very beginning that school education in Ukraine officially lasts 11 years (until grade 11) and is separated into elementary, grades 1–4; basic, grades 5–9; and senior years, grades 10–11.

The section on the goals and content of mathematics education in Ukraine is authored by Vasyl Shvets. The section "The organization of teaching mathematics at school and modern pedagogical tools" is authored by Valentyna Bevz. The section "The System of Assessment in Mathematics Education in Ukraine during the Period of Independence" is authored by Oleksandr Shkolnyi. Finally, the section "How and Why the Preparation of Mathematics Teachers Was Reorganized" was written by Olha Matiash.

2 General Information

As a part of the Union of Soviet Socialist Republics (USSR), Ukraine, like the other republics, participated in the implementation of mathematical education at school indirectly. The entire methodical system of school mathematics education (MSSME) (see Fig. 6.1) was developed by the Academy of Pedagogical Sciences of the USSR, relevant research institutions, the Federal Ministry of Education, and other institutions. In fact, the Ministry of Education of the Ukrainian Soviet Socialist Republic had only a limited role—it supervised how all decisions, programs, and instructions were implemented in practice.

After the collapse of the USSR in 1991, declaring itself as an independent state, Ukraine began to carry out its own educational policy. In particular, it began to work independently on the problem of school mathematics education for younger generations. The first step was to get rid of the communist ideology that permeated the entire educational policy of the Ukrainian SSR. In addition, it was necessary to create new state structures that would deal with all education-related problems.

These structures in Ukraine are the Ministry of Education of Ukraine (its activities and management units were reorganized); the newly formed Academy of Pedagogical Sciences of Ukraine (APS, established in November 1992), which is

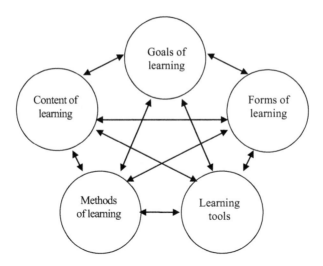

Fig. 6.1 Methodological system of school mathematics education (MSSME)

now called the National Academy of Pedagogical Sciences (NAPS) of Ukraine; the Research and Development Institutes (Pedagogy, Psychology, Vocational Education, and others), which are part of the NAPS of Ukraine; and other institutions.

Later, the NAPS of Ukraine became the main center for the restructuring of the educational system in Ukraine. Its leading role manifested in shaping the content, standards, and methods of preschool, school, higher vocational, and higher education, as well as the education of people with special needs. Specifically, it was a leader in the development of theoretical and methodological foundations for practical psychology, didactics, the theory of education, preparation and implementation of new national educational legislation, concepts and strategies for the modernization of education, etc. Additionally, as a result of developing the standards of education, it created the foundation for new textbooks. In cooperation with relevant ministries, parliamentary committees, public organizations, and scientists of the NAPS Ukraine and higher educational institutions, a cohort of practical teachers developed regulatory and legal documents of state importance, which were then adopted by the Supreme Council of Ukraine: the law of Ukraine "On Education" (1992), the law "On Vocational Technical Education" (1998), the law "On General Secondary Education" (1999), the law "On Out-of-School Education" (2000), the law "On Preschool Education" (2001), the law "On Higher Education" (2002), and some other documents that were previously absent in Soviet Ukraine.

Important for building a national education system in Ukraine were the following documents:

- The National Education Program ("Ukraine of the XXI Century"), which was approved by the First All-Ukrainian Congress of Educators in 1992 and approved by the Cabinet of Ministers of Ukraine in 1993 (see Nacional'na Programa 1993).

- The National Doctrine of the Development of Education in Ukraine, which was approved in 2001 by the second All-Ukrainian Congress of Educators and was approved by Leonid Kuchma's presidential decree (see Nacional'na Doctryna 2002).

The doctrine in particular suggested a transition to new content and a 12-year period of study for secondary education. In addition, senior schools were supposed to be specialized. However, political controversies in Ukraine led to the suspension of this transition in 2010 by the government of President Victor Yanukovych. After the regime of Victor Yanukovych was overthrown, the new government headed by President Petro Poroshenko resumed these transitions in 2017 with the new "Law of Ukraine on Education."

Even the brief information given above shows the tremendous work that the state of Ukraine carried out towards entering its education system into the European and world educational spaces, coordinating educational qualification levels and levels of education, and towards joining the Bologna process. Below we will focus on the MSSME reform in detail.

3 Goals and Content of School Mathematics Education in Ukraine

Obviously, school mathematics education reform should be studied within the framework of general reforms to the entire education system. Mathematicians, mathematics teachers, everyone who was connected to mathematics education, and the general public all came to the unanimous opinion that the whole MSSME should be reformed. That included the reformation of learning goals (defining what results should be achieved after studying the subject of Mathematics), the content of educational materials (defining what should be studied in order to achieve the planned results), teaching methods (defining what effective modern and traditional methods and activities should be used in order to achieve results), organizational forms of education, and mathematics learning tools.

The most thoroughly reformed components of the MSSME were learning goals, the content of educational materials, and mathematics learning tools. The other two components, teaching methods and methods and activities, were more conservative, had their own specific traits in relation to the others and therefore were reformed less completely. Instead, they were improved and supplemented by more efficient components.

Teaching objectives and the content of mathematics classes are described clearly in the Standards of Basic and Secondary Education (see Cabinet 2004). In these state regulatory documents, mathematics is represented as an independent branch of knowledge with a clearly allocated amount of study time (see the table below from Cabinet 2004) (Table 6.1).

Table 6.1 Distribution of academic hours

Educational branches	Total amount of academic hours								
	II level (5–9 grades)			III level (10–11 grades)			II + III levels (5–11 grades)		
	Week	Year	%	Week	Year	%	Week	Year	%
Invariant part									
Languages and literature	42	1470	27	14	490	18.4	56	1960	24.2
Social sciences	12	420	7.7	7	245	9.2	19	665	8.2
Aesthetic culture	8	280	5.1	1	35	1.3'	9	315	3.9
Mathematics	**20**	**700**	**13**	**6**	**210**	**7.9**	**26**	**910**	**11.25**
Natural sciences	26	910	16.7	11.5	402.5	15.1	37.5	1312.5	16.25
Technologies	8	280	5.1	4	140	5.2	12	420	5.2
Physical training and health	17.5	612.5	11.4	6	210	7.9	23.5	822.5	10.2
Total	133.5*	4672.5	86	49.5	1732.5	65	183	6405	79.2
Variable part									
Additional hours for educational industries, subject by choice, specialized training, electives, individual lessons, and consultations	21.5	752.5	14	26.5	927.5	35	48	1680	20.8
Maximum permissible workload of pupil	130	4550		66/72	2310/2520		196/202	6860/7070	
Total	155	5425	100	76	2660	100	231	8085	100

It should be noted that the developers of the basic and secondary mathematical education standard did not quickly come to an understanding of what a current educational standard is. Eventually, after much work and discussion, an educational standard in mathematics began to define not a curriculum, but only a basis on which a package of multilevel programs or curricula for different educational institutions could be developed. On this basis, new generations of textbooks, teaching aids, and other educational and methodological support could be created. The new curricula and textbooks provided an opportunity to expand and deepen the content of educational materials in order to achieve not only a standard (basic) level of mathematical preparation, but also an enhanced and in-depth one.

Thus, the educational standard for mathematics began to represent the minimum requirements for general mathematical education. It became the level any pupil was supposed to reach at whatever secondary education institution he or she studied. It was also assumed that if children had the intention to advance their intellectual and professional development further, they could do it from this base level.

The first state educational standard project in mathematics for basic and secondary education was begun in 1996 by a group of scientists led by the Academician of the Academy of Pedagogical Sciences of Ukraine, Doctor of Physics and Mathematics Mykola Shkil'.

When developing it, the group of authors assumed that the goals and tasks of studying mathematics in a secondary school should be:

- mastering by students of a system of mathematical knowledge, skills, and abilities necessary in everyday life and work activities and sufficient for the successful mastering of other branches of education and which would ensure the continuity of their education,
- the formation of the students' understanding of the issues and methods of mathematics, its role in the comprehension of reality, and the formation of the scientific worldview,
- the intellectual development of pupils (logical thinking, intuition, spatial representations, algorithmic and informational culture, memory, and attention),
- the growth of students related to their moral, career-oriented, economic, aesthetic, and patriotic education, including the formation of positive character traits (perseverance, intellectual endurance, purposefulness, tolerance for the environment, respect for the environment, etc.) (see Cabinet 2004).

Based on the group's formed goals and tasks, the educational standard emphasized certain content lines according to which the content of the school mathematics course was designed as follows:

Basic school

Numbers—natural, integer, rational, real numbers; ordinary fractions; decimal fractions; operations with numbers; percents, percent calculations, and proportions.

Expressions—degrees with natural and integer exponents.

Polynomial—operations with polynomials; fractions, operations with fractions; and identical transformations of expressions.

Equations and inequalities—equations and inequalities with one variable (linear, quadratic); equations with two variables; linear equations with two variables; systems of two linear equations with two variables; application of equations and their systems to solving problems.

Functions—functions; linear and inverse proportionality, quadratic functions; and numeric sequences.

Elements of combinatorics—the sets and combinatorial tasks.

The beginnings of probability theory and the elements of statistics—random events; the probability of a random event; ways to present data; frequency; and average value.

Geometric figures—the simplest geometric shapes on a plane; triangles, polygons, and circles; equality and similarity of geometric shapes; constructions by compass and straightedge; geometric transformations; coordinates and vectors; and geometric shapes in space.

Geometric values—the length of the segment and circle; measuring angles; area and volume; measuring and calculating linear and angular values, area and volume.

Senior school

Expressions— the generalization of the concept of degree; logarithms; and transformation of power, trigonometric, irrational, exponential and logarithmic expressions.

Equations and inequalities—trigonometric, irrational, exponential and logarithmic equations; indicative and logarithmic inequalities.

Functions—numeric functions; trigonometric, power, exponential, and logarithmic functions; the continuity of functions; derivatives and integrals; and application of a derivative and a definite integral.

Elements of combinatorics—connections without repetitions: permutations, placement, and combination.

The beginnings of probability theory and the elements of statistics—random events; the probability of a random event; conditional probabilities, independent random events; the idea of the law of large numbers; definition of probability; statistical tables of distribution and their visual representation; mode and median and average values.

Geometric figures—axioms of spatial geometry; the mutual arrangement of lines and planes in space; polyhedra and rotating bodies and their types and properties; constructions and transformations.

Coordinates and vectors.

Geometric values—distances; measures of angles between lines and planes; surface area and volumes of geometric bodies (see Cabinet 2004).

There were no fundamental changes in the content of mathematics courses, except for the new lines of content "Elements of Stochastics" (combinatorics, the beginnings of probability theory, and statistical elements), which was not a part of Soviet curriculum. Only the number of hours for studying individual lines and the order of the topics studied changed. The irrelevant, outdated sections of school mathematics were removed.

After lengthy discussions regarding the developed draft of an educational standard between scientists, teachers, and parents of pupils at various conferences, meetings, and congresses, in January of 2004 the new educational standard was approved by the Cabinet of Ministers of Ukraine and the Ministry of Education and Science. Additionally, the APS of Ukraine was instructed to develop new educational plans and syllabi for primary and senior school students, which was done quickly. New alternative textbooks and tutorials in mathematics were published and introduced on a competitive basis based on the established educational plans and syllabi.

Later, the reform of the education system continued. The existing normative documents that had already been adopted were supplemented by new ones: "The Concept of Mathematical Education at a 12-year school" (2001), "The Concept of

Specialized Education" (2013), and other related documents. Also, the paradigm of education was somewhat changed. Instead of the previous main goal of the educational branch "mathematics" which was seen in the formation of students' knowledge and skills, a different main goal was proclaimed. The new main goal was the formation of pupils' mathematical *competence* (the competence paradigm of education). Accordingly, there was a need to improve the educational standard in mathematics and to create new educational plans, curricula, textbooks, and teaching aids based on it.

The new improved State Standard of Basic and Complete Secondary Education was approved by the Cabinet of Ministers of Ukraine in November of 2011 (see Cabinet 2011). Even if the content taught in the "mathematics" branch had not changed much, the goals of the study of mathematics had been significantly changed. According to the new standard for mathematics, the goals and tasks of studying mathematics became the following competences:

- developing an understanding of the role and possibilities of mathematics in the comprehension and description of real processes and phenomena of reality and ensuring the understanding of mathematics as a universal language of the natural sciences and as an integral part of the general human culture,
- developing of the logical, critical, and creative thinking of students and their ability to clearly and convincingly formulate and express their opinions,
- ensuring students' mastery of mathematical language, their understanding of mathematical symbols, mathematical formulas, and models, which makes it possible to describe the general properties of objects, processes, and phenomena,
- forming students' ability to logically substantiate and prove mathematical statements, to apply mathematical methods to the process of solving educational and practical problems, to use mathematical knowledge and skills in studying related subjects,
- developing students' skills necessary for the use of textbooks, the independent study of mathematical texts, the search and use of additional information resources, and the evaluation of knowledge gained and the sources for obtaining it. Also, the ability to highlight main ideas, analyze, draw appropriate conclusions, and apply information obtained in personal life,
- the forming of the ability to assess the correctness and rationality of solving mathematical problems, to substantiate statements, to recognize logically incorrect reasoning, and to make decisions when information is incomplete, redundant, accurate, or probable (see Cabinet 2011).

Naturally, the next step of reformation was the creation of new educational plans, programs, criteria for evaluating student learning outcomes, alternative textbooks and teaching aids, and computer educational applications.

It should be noted that the process of reforming school mathematics education was a manifestation of the social, cultural, and economic life of Ukraine. And since this life was not proceeding smoothly (two revolutions, frequent changeover in government representatives, populist decisions made by the leadership, etc.), not all the segments of the population were happy with the results of the reforms. Some people praised the reforms, others criticized them, and there were people who condemned them (see, for example, Ksenz 2014, Pekar 2015, Kliuchnyk 2017).

The people who condemned the reforms were mainly those who studied and grew up under the Soviet regime. For them, the Soviet education system was perfect and better at providing for the deep knowledge and skills of students. Young Ukrainian citizens (especially who have had experience studying in Europe), we believe, typically praised the reforms. They supported changes to the educational system. In their opinion, the new education system should create a competent young person who is able to be successful under new economic conditions, competitively capable, ready to make the right decisions, calculate the risks of his or her activities, and be a patriot. The state understands that it is necessary to work for the future, so the reforms are continuing.

Until 2018, Ukrainian mathematics teaching in grades 5–9 (basic school) was conducted with a single curriculum, but using one of many textbooks written by different groups of authors. In grades 10–11, until 2018, senior school students had the opportunity to study mathematics in four programs: a standard level program, an academic level program, a specialized level program, and an advanced level program.

At the standard level, mathematics was studied as an integrated subject (without separating mathematical subjects). At the academic level, specialized level, and advanced level, geometry, algebra, and elementary calculus were studied separately (there were lessons of geometry and lessons of algebra and the beginning of analysis). There were some components clearly determined for each topic across all mathematics syllabi; among them were the content of the educational materials, the competencies that the pupils had to master, and the amount of hours to be devoted to studying. In addition, the explanatory note in each syllabus clearly spelled out the goals and objectives for studying each subject (mathematics, geometry, algebra, and algebra and elementary calculus). Each syllabus also contained the criteria for students' learning outcome assessment.

The process of mathematical education reform in Ukraine has not been finished yet. It is continued in connection with the 2017 adoption of a new law on education, declared by the Ministry of Education of Ukraine for reform under the name "New Ukrainian School" (see Cabinet 2017).

Since 2018, the study of mathematics in senior school (grades 10–11) has been conducted in two levels: the standard level (3 h per week allocated for studying mathematics) and the specialized level (9 h total per week allocated, 6 h for algebra and 3 h for geometry). In basic school (grades 5–9), pupils study two academic subjects. The first of the subjects is algebra (2 academic hours per week) and the second is geometry (also 2 academic hours per week). New changes are forthcoming.

4 The Organization of Teaching Mathematics at School and Modern Pedagogical Tools

4.1 The Organization of Teaching Mathematics at School

Below we will consider, in chronological order, the general characteristics of the changes to the educational organization of the younger generations in Ukraine after the formation of an independent state.

New types of schools—gymnasiums, lyceums, special institutions for gifted children, schools (and classes) with an advanced course in certain subjects, private schools, and specialized boarding schools and similar institutions—are rapidly developed in the country. The lyceums (grades 7–11 or grades 5–11) have been established at leading institutions of higher education. In these and many other specialized educational institutions, mathematics (algebra and geometry) is studied for 10–12 h, physics for 6–8 h, and English for 6 h a week. At the same time, gymnasiums and specialized schools with an advanced course in foreign languages and other subjects in humanities were established, and their courses for mathematics were substantially more modest.

In 1993, the Unified Junior Academy of Sciences of Ukraine (JAS) was established—an educational system that organizes and coordinates the research activities of pupils; provides conditions for their intellectual, spiritual, creative development, and professional self-determination; and contributes to building up the country's scientific potential. The research work of JAS is organized by scientific department and section according to the structure approved by JAS Presidium. There are three sections: "mathematics," "applied mathematics," and "mathematical modeling" under the scientific department of "mathematics."

Periodicals connected with mathematics education were created. The first attempts to print mathematical magazines for teachers were in Lviv and Poltava and then in Kyiv and Kharkiv. Since 1995, the magazine for students and teachers called *In the World of Mathematics* has been published in the capital, Kyiv. Since 1998, the magazine *Mathematics at School* and the newspaper *Mathematics* have been published in Kyiv. Since 2002, the magazine *Mathematics in Ukrainian Schools* and a book supplement have been published in Kharkiv.

These publications played a big role in popularizing the reform of school mathematics education. The syllabi for mathematics education for schools of different levels, including all kinds of planning, guidelines for organizing the study of mathematics at school, instructions for examinations, etc., were printed there. For nearly 10 years, on the pages of these magazines and newspapers, the authors of new mathematics textbooks presented the characteristics of their approaches to theoretical material, the methodological apparatus of the textbook, and described the system of tasks and methods for solving some of them. Such materials were demanded by teachers, and the circulation of these publications constantly increased. However, after 2010 the situation changed. The articles about textbooks ceased to be printed, because they were officially viewed as advertising. Teachers' access to the Internet expanded, and the opportunity to create associations of mathematics teachers at institutes of advanced training or in social networks had arisen. All of this led to a significant decrease in the circulation of periodicals for teachers.

Since 2001, the work on the creation of Conception for Specialized Education (an official and theoretical outline) and Conception of Mathematical Education in 12-year schools began in the Institute of Pedagogics of the Academy of Pedagogical Sciences of Ukraine. The first conception was approved by decision of the Board of the Ministry of Education and Science of Ukraine No. 10/2-2 of September 25, 2003. Later, this conception was modified and refined several times.

The Conception of Mathematics Education in 12-year schools was created with the aim of improving the mathematical preparation of secondary school graduates and suggested a new approach to the tasks, content, quality, and terms of school education. The priorities for the development of mathematical education, as well as the structure and content of school mathematics, are identified in the conception. The following structure of mathematical education was provided:

> Course A (general education). The course is offered to students of grades 1–9. It provides the basic mathematical training for students.
> Course B (applied) is offered to students of the 10th and 12th grades who have chosen for themselves areas of activity in which mathematics plays the role of an apparatus, a specific means for studying and analyzing the laws of the real world. The course is taught in classes with physical, technical, chemical, biological, ecological, agrobiological, and other specializations. This course is also taught at schools where no specializations are offered.
> Course C (general cultural) is intended for students in grades 10–12, for those who study mathematics only as a part of a general cultural education, for instance, students acquiring literary, sociohistorical, artistic, or aesthetic specialization.
> Course D (advanced) for students in grades 8–12 who plan to associate their future with mathematics (mathematical, physical and mathematical specialization, separate lyceums, colleges, specialized physical and mathematical schools, and classes in the advanced study of mathematics).

This structure of mathematical education with some amendments (for example, the names of courses were amended and schooling still lasts for 11 years rather than 12 years) is used now, but the conception itself was not approved and remained at the project level. A new conception is being developed.

Preparation for the creation of new teaching tools, including textbooks, has begun. In 2002, the Institute of Pedagogy of the Academy of Pedagogical Sciences of Ukraine established a collection of scientific works entitled "Problems of a Modern Textbook." The first issue of this collection was published in 2003, and the collection continues to highlight topical issues in the study of the theory of textbooks and the specifics of creating modern educational and scientific-methodical literature even now.

A new stage in the creation of methodological support for the educational process has begun. In addition to the traditional requirements for the compliance of textbook content with the curriculum, the differentiation of task material, and the development of the cognitive activity of schoolchildren, the new textbook regulations require the following aspects:

- The textbook should contain brief information on the history of science, culture, and technology in order to uncover the evolution of scientific ideas, discoveries,

the connections of science, industry, social practice, and the role of scientists, first of all Ukrainian scientists, in the discovery of scientific truths.
- The content of the textbook should provide a connection with real life, help to shape the personality of a schoolchild, and develop his/her abilities and talents.
- It is desirable to envision the use of information and communication tools in the educational process and the study of certain sections or subjects at school in general using computer technology to support the textbooks.

In order to implement the new approach, an action program for improving the quality of physics and mathematics education for 2009–2012 was approved. In particular, it proposed to:

- Bring the content of school physics and mathematics education into conformity with the modern development of the innovative needs of society.
- Provide a focus on applications when teaching the content of mathematics curriculum and natural sciences.
- Make sure that the content and sequence of study of the material in mathematics and the natural sciences (physics, chemistry, biology) are in compliance.
- Create elective courses in physics and mathematics both for pre-specialized and specialized education.
- Introduce in ninth grades the state final certification for mathematics (in integrated algebra and geometry).

In 2009, a new conception for specialized education in high schools was approved, but it was not fully implemented because the 12-year education plan had been so far abandoned. In 2011, the State Standard of Basic and Complete General Secondary Education was approved and a new conception for core education in high schools was developed. At this moment, the new conception has not received full practical implementation.

On April 23, 2012, the First All-Ukrainian Congress of Mathematics Teachers was held, where a new mathematics curriculum for grades 5–9 at general educational institutions was presented and discussed. At the congress, there were questions about the implementation of the state program for improving the quality of natural-mathematical education; questions regarding the need to reform the system of mathematics teacher education were also raised.

The working group responsible for curriculum development was assigned by the ministry to analyze all the shortcomings of the current curricula, to consider the content of the curricula of other countries, and to reduce the current syllabi in mathematics (including algebra and geometry) by 15%.

The overwhelming majority of speakers at the congress believed that it was necessary to have at least 5 h of mathematics weekly in each class for the implementation of the new approach and curriculum. Those proposals were not heard by the governing bodies either at the congress or in the future.

Since 2016, a new reform of the education system in Ukraine has begun. The ideology of the changes to the educational system in Ukraine is revealed in the Conception of New Ukrainian Schools. This is a long-term reform, which includes three phases (2016–2018, 2019–2022, and 2023–2029) that must be implemented in agreement with social changes. The first phase of this reform is being finalized. A new piece of legislation in Ukraine called "About Education" has been introduced; the state standard for primary education on a competency basis has been developed and approved; the commencement of work for an elementary school based on a new educational standard has been started; the curricula and syllabi for elementary and high schools has been revised.

In 2017, the curriculum "Mathematics, Grades 5–9" was approved for general educational institutions. For the first time, the debate regarding the curriculum was carried out through public discussion. The teachers and public could contribute to the improvement of the proposed project. No expected results were obtained, however. According to the results of the discussion, two versions of the curriculum were formed. The version which was close to the current program was approved. The main changes concerned the explanatory note and educational material for the eighth grade.

The explanatory note emphasizes that teaching mathematics in elementary school provides for the formation of a competency in mathematics, the essential description of which is presented in the section, "Expected Results of Educational and Cognitive Activity" of the curriculum. In addition, it emphasizes that mathematics education should make a definite contribution to the formation of core competencies. In this regard, the cross-cutting lines of key competencies such as "Environmental Safety and Sustainable Development," "Civil Responsibility," "Health and Safety," and "Enterprise and Financial Literacy" that are aimed at developing schoolchildren's abilities as well as the application of knowledge and skills in real-life situations were highlighted in the curriculum.

As follows from the above, the system of mathematical education in Ukraine over the past 30 years has become much stronger. The changes have affected not only the content, but also the style of its presentation and study. Let us consider in more detail how the pedagogical tools of teaching mathematics at school have changed during this time.

4.2 *Modern Pedagogical Tools for Teaching Mathematics at School*

At the beginning of the reform of school mathematics education, one of the main tasks was to create national mathematics textbooks. At first, the new textbooks differed a little from the Soviet ones. An exception to this were the experimental textbooks authored by Gregory P. Bevz, titled "Mathematics," and prepared for each of the 7–11 grades (Bevz 1994a; Bevz 1994b; Bevz 1995a; Bevz 1995b). Each book contained material on algebra (algebra and the elementary calculus) and on geometry (plane or space geometry) corresponding to the syllabus of the particular grade. These textbooks were black and white, printed on gray paper, had from 170 to 190 pages, and were used at schools for quite a long period of time.

Some interesting ideas were implemented in these textbooks. The study of algebra in Grade 7 began with the study of the topic "Equations" (in Soviet textbooks in different years, the study of algebra began with expressions or sets and functions).

The author of the textbook, Gregory P. Bevz, believed in the proposed approach because of the following reasons. The study of algebra can be built on different bases and begins with the transformation of expressions, the formation of the concept of a function, and the solution of equations. It is worth analyzing: how long ago people learned to solve various types of equations and why they did it, how the notation and methods for solving equations have changed and improved, and where the name "algebra" comes from as well as to compare these facts with the process of creating algebraic symbolism and the emergence of the concept of a function. An analysis of algebra in this way suggests that building a school course of algebra starting with algebraic expressions is too formal. In the system starting with algebraic expressions, students study degrees, monomials, polynomials, and actions on them for almost a year, but not understand why all of it is necessary.

The historical approach to solving this issue makes it possible to create a motivational base for the conscious study of algebra. By presenting the students a picture of historical development from the task to the equation and further to the expressions, the teacher can quite naturally and convincingly motivate the study of each topic. The equations are needed to solve tasks of a practical nature, and transforming expressions is necessary in order to solve complex equations. The knowledge of numerical and literal expressions and the types of equations and methods used for their solution are expanded in the students' course of study, and, consequently, the classes of tasks that students can solve become broader. Another idea utilized in the first Ukrainian algebra textbooks was the inclusion of the section "Elements of Applied Mathematics" into the textbook for grade 9 (Bevz 1996). This section included paragraphs on the following:

- Mathematical modeling
- Approximate calculations
- The first information about statistics
- Interest calculations.

The new experimental textbooks by Ukrainian authors were *Mathematics: Experimental Textbook for Grade 5* (Wozniak and Lytvynenko 1996a), *Mathematics: Experimental Textbook for Grade 6* (Wozniak and Lytvynenko 1996b), *Algebra, Grades 7–9* (Bevz 1996), *Geometry, Grades 7–9* (Bevz et al. 2001); *Algebra and the Beginning of Analysis, Grades 10–11* (Shkil et al. 1995), and *Geometry, Grades 10–11* (Bevz et al. 2004). Experimental textbooks for in-depth study of mathematics for all grades and schools were also printed—*Algebra, 9* (Kovalenko et al. 1998), *Geometry, 8–9* (Burda and Savchenko 1996), *Algebra and the Beginning of Analysis, 10* (Shkil et al. 1997), *Geometry, 10–11* (Bevz and Bevz 2000), and *Mathematics, Grades 10–11* (Burda et al. 2001).

Until 2000, many good textbooks were created in Ukraine for all grades and specializations. The teacher manuals, including time-planning schedules for each grade and for each textbook, were prepared, as well as recommendations for each lesson. All kinds of supplemental materials including workbooks and additional collections of tasks and the like were prepared. A real competition between authors and publishers had developed. Considering the mathematical magazines and newspapers, it is easy to conclude that never before had so many manuals for teachers of mathematics been printed in Ukraine. Although the young state experienced significant economic difficulties in the early years, although the teachers received no salaries for several months, and although Ukraine was in a deep and protracted financial crisis, the schools were still working; the students and teachers had the necessary textbooks, and teaching and didactic materials.

Unfortunately, a new kind of mathematics books appeared, which caused irreparable harm to mathematical education. We are talking about solution books (so called "Rozvyaznyky"), which provided the solutions to all the problems in the textbooks so that the lazy student could simply copy them without solving the tasks at home. These solution books (also called ready-made homework) have been published in enormous numbers for all grades (and not only for mathematics) behind the backs of the authors of the textbooks.

In 2004, a new stage in the creation of mathematics textbooks began—only the textbooks that had passed a special competition were allowed into schools. Not only textbook authors, but also publishers were taking part in the contest. Publishers were very interested in winning so that they could receive state orders for printing textbooks. Therefore, the textbooks became multicolored and beautifully designed; they become more attractive and could be compared to the best foreign ones.

According to the results of the first-generation contests, schools in Ukraine received 3–4 alternative textbooks for each of 5–9, written by the following authors:

Mathematics, 5, and *Mathematics, 6:*

- Gregory Bevz and Valentyna Bevz.
- Arkadyi Merzliak, Vitalyi Polonskyi, and Mykhailo Yakir.
- Halyna Yanchenko and Vasyl Kravchuk.

Algebra, 7; Algebra, 8; Algebra, 9:

- Gregory Bevz and Valentyna Bevz.
- Oleksandr Ister.
- Vasyl Kravchuk and Halyna Yanchenko.
- Arkadyi Merzliak, Vitalyi Polonskyi, and Mykhailo Yakir.

Geometry, 7; Geometry, 8; Geometry, 9:

- Halyna Apostolova.
- Gregory Bevz, Valentyna Bevz et al.
- Mykhailo Burda, Nina Tarasenkova et al.
- Oleksandr Ister.
- Arkadyi Merzliak, Vitalyi Polonskyi, and Mykhailo Yakir.
- Alla Yershova and Volodymyr Holoborodko.

These textbooks on mathematics are multicomponent and multidimensional teaching aids, with the help of which not only the educational content but also a model of the educational process is delivered; they also specify methods of implementation. The textbooks are intended not only for the formation of necessary mathematical knowledge and skill, but also for the development of the students' general culture. By studying with such textbooks, students will master their general cultural, mathematics, and key competencies.

For the first time, the interdisciplinary connection between mathematics and computer science was implemented explicitly in the textbook *Mathematics: Experimental Textbook for Grade 6* (Bevz and Bevz 2006). The students were asked to construct a pie chart using ICT in the "Chart" section. The textbook offered an algorithm for such a construction as well as the solution of a specific exercise.

The textbooks were written by a group of authors, under the leadership of Gregory P. Bevz, in observance of the most important didactic principles: scientific validity, accessibility, continuity, systematicity, practical orientation of training, and others.

The scientific validity of the textbook suggests that all the concepts and terms discussed in it are in compliance with modern mathematical science and scientific terminology. All of the rules and statements are scientifically correct so that in the future the students will not have to change or correct them. The practical tasks do not contradict science or life practice.

Accessibility is not understood abstractly (generally and for each age group), but in relation to specific topics and specific children. To make the presentation of educational material more accessible, charts, illustrations, tables, short notes, and additional clarifications are used. The content of the textbook is aimed at meeting the needs of boys and girls, rural and urban residents, strong pupils, and those who have low-level mathematical skills. The relatively difficult places in the textbook are set out in short sentences and are explained with specific examples. In some cases, computer assistance is used (for example, when constructing pie charts, introducing a degree with a real exponent, the limit of a function, etc.).

Systematicity is achieved by a well-designed presentation system for curriculum material, emphasizing the main pieces of information, as well as the constant combination of new and previously studied material combined with the life experience of schoolchildren. Textbooks are structured so as to provide opportunities for regular repetition of previously studied material, expanding and deepening it.

Continuity is ensured both with elementary schools and with each of the subsequent grades. A lot of questions are offered to students at first as a form of propaedeutic study and then at a higher level.

The practical orientation of textbooks means that they contain tasks with real and modern data that provide a holistic perception of the world and help students to realize the role of mathematics in the real world. The students are acquainted not only with abstract mathematical concepts and connections, but also with examples of their use in practice. Additionally, the way mathematics has developed and the names of its creators are discussed.

The educational and methodological apparatus of textbooks is aimed at stimulating all types of educational and cognitive activity in students. Modern and original techniques and tools for students are used in the textbooks.

Today, a textbook is expected to fulfill several functions—below we will list them to describe what is meant by the terminology in use, which will make the requirements of the textbooks and the expectations for them clearer.

The information function of the textbook is supposed to be realized by aligning the textbook's content with the mandatory content of the syllabus (the so-called invariant part of it) and by including additional topics from its optional part. The special tools (bold fonts, underlining, frames, dies, etc.) which allow the reader to select the most important pieces of text are supposed to be used to make the information better structured. The information to be memorized or needed for frequent use is to be given on the flyleaves.

The formative function. The theoretical material and the system of tasks and exercises in textbooks are to be designed to form the necessary competencies. The content and structure of the textbook should stimulate independent activity for schoolchildren and their creativity.

The building of personality function. The textbooks are supposed to develop a system of moral values and valuable personal qualities that focus on cultural aspects of human activity. The emphasis on cultural aspects suggests that geometrical forms are supposed to be illustrated by the works of artists and architects, the properties of the functions are to be revealed using ornaments, and the statistical data of the tasks should encourage students to reflect on preserving their own health and on saving human and natural resources.

The developmental function. The structure of textbooks, their content, and their methods of presenting educational material through the system of tasks, exercises, and illustrations—all of these are supposed to be aimed at developing a student's personality, in particular their intellectual abilities. The special tasks should be aimed at developing the students' attention and memory or logical and spatial thinking. Additionally, tasks with interesting plots and unexpected answers are to be included in textbooks to form an interest in the study of mathematics.

The managerial function is to be provided by creating opportunities for the teacher to diversify the forms and methods of teaching and to arrange for the permanent and thematic monitoring of the pupils' educational achievements. The textbooks should make possible close collaboration between the teacher and the students and fully serve both sides in the educational process. For the student, the textbook should be the source of content and the instrument for learning the material and forming the necessary competencies. The textbook should present the logic for learning, which in turn determines the teacher's ways and methods for organizing the educational process.

The structure of the modern textbook presupposes differentiation in education; for instance, the textbook should offer an extensive system of exercises and a variety of ways to present theories.

Typically, today each textbook begins with an introduction addressed to the students. In the textbooks, educational material is divided into sections and subsections. At the beginning of each section, the main content and brief motivational reasons for study are presented in a schematic. Illustrations (photographs, picture fragments, collages) and quotes from famous personalities help to create an interesting learning atmosphere.

In each section, there are additional subsections such as "Do You Want to Know Even More?" which can contain the rubrics: "Check Yourself," "Let's Do It Together," "Perform Mentally," etc. The main portion of the exercises and tasks for each section is typically presented in two sections: A and B. Each section ends with a list of tasks titled "Exercises for Repetition," whose purpose is to repeat the previously studied material or review reference knowledge for the next lesson.

In the 2011/2012 school year, some textbooks were published for students of graduating grade 11 classes which could be used for teaching at a few levels of study:

Mathematics (geometry, algebra and the beginning of analysis—the standard level):

- Olena Afanasieva, Yakiv Brodskyi et al.
- Gregory Bevz and Valentyna Bevz.

Algebra and the beginning of analysis (academic level and specialized training level):

- Gregory Bevz, Valentyna Bevz, and Nataliya Vladimirova.
- Arkadyi Merzliak, Vitalyi Polonskyi, Mykhailo Yakir et al.
- Yevhen Nelin and Olena Dolhova.

Geometry (academic level and level of specialized training):

- Halyna Apostolova.
- Gregory Bevz and Valentyna Bevz.
- Mykhailo Burda, Nina Tarasenkova et al.
- Alla Yershova and Volodymyr Holoborodko.

This approach has its positive and negative sides. On the one hand, in small schools, where there are few students in each grade, these integrated textbooks provided additional opportunities for organizing group and individual work with students whose interests are oriented towards mathematics. On the other hand, the textbooks increased in size. These textbooks contained a lot of material because after the publication of the tenth grade textbooks, it became clear that schools were not going to change to a 12-year education system as planned. Therefore, the educational material from the planned 12th grade moved to the 11th grade. This led to a significant increase in the size of textbooks for 11th graders and in the burden on students.

Overall, the conclusion can be made that a number of good-quality modern textbooks had been created in Ukraine by 2012. These textbooks, especially mathematics textbooks, could be used for at least another 10 years. But the process is still ongoing: a new competition has started and new mathematics textbooks have been created:

Mathematics, 5, and *Mathematics, 6:*

- Arkadyi Merzliak et al.

- Oleksandr Ister.
- Nina Tarasenkova et al.

Algebra, 7; Algebra, 8; Algebra, 9:

- Gregory Bevz and Valentyna Bevz.
- Oleksandr Ister.
- Arkadyi Merzliak et al.
- Nina Tarasenkova et al.
- Nataliya Prokopenko, Yuriy Zakhariychenko, and Nataliya Kinashchuk.

Geometry, 7; Geometry, 8; Geometry, 9:

- Gregory Bevz et al.
- Mykhailo Burda et al.
- Oleksandr Ister.
- Arkadyi Merzliak et al.
- Alla Yershova et al.

These new textbooks are more attractive in their design. However, their content has not changed significantly. In 2019, the contest for textbooks for grade 11 will come to an end, but the contest for elementary school textbooks is ongoing.

Let us dwell on mathematical problem-solving, which plays a leading role in teaching mathematics. Specifically, let us discuss the diversity of the system of tasks. Today, textbooks are supposed to contain various types of tasks and exercises (propaedeutic, providing practice, oral and written execution, logical, algorithmic and creative, simple and challenging). The tasks have roles and functions (educational, cognitive developmental, control, motivational). In addition to stimulating the cognitive interest and developing the intellectual skills and creative abilities of students, problems are given to ensure the implementation of cross-cutting lines of key competencies. These cross-cutting lines of key competencies are: "Environmental Safety and Sustainable Development," "Civil Responsibility," "Health and Safety," and "Enterprise and Financial Literacy." They are aimed at developing the ability to apply knowledge and skills in real-life situations. They try to call upon socially significant topics that help students to form ideas about society in general and develop the ability to apply their knowledge in various situations.

Today, one type of mathematical problem is becoming more popular—problems providing insufficient or redundant information—they can be associated with drawing but do not have to be. They can be word problems, which can be reduced to solving equations or systems of equations, the task of transforming expressions, determining individual elements of geometric figures, calculating perimeters or areas of figures, and the like. These tasks are sometimes called "open." To solve an open problem, the student should analyze it and determine which information is redundant or which is missing. After that, the creative process begins—the student independently creates a problem or problems in the context of a given one and solves them. The use of open tasks in the process of learning mathematics contributes to the development of logical and creative thinking, the formation of communicative experience, and the self-realization of students. To this end, the new textbooks on algebra and geometry support the inclusion of various types of open tasks.

In the reviews of textbooks, teachers indicate the need to include tasks that require integrating knowledge and applying acquired competencies in nonstandard conditions in school textbooks. It is advisable to offer such problems as repetition or at the end of the study of a topic or chapter.

Among the new types of tasks which appeared in the textbooks, the educational projects should be mentioned. For example, the following themes of the educational projects proposed for pupils in the grade 8 Geometrytextbook (Bevz and Valentyna Bevz 2016) are named here:

- Cutting and compiling quadrilaterals
- Similarity
- Right triangles in historical problems.

The textbook Algebra, 9 (Bevz and Bevz 2017), suggests the following topics for project activities:

- Interesting inequalities
- Functions around us
- Applications of mathematics.

More complex projects are offered to students in higher specialized classes.

We are not able to predict in detail the future in which current students will live. Therefore, school is supposed to help them to determine their own attitude towards the world, themselves, and their activities, in other words—to create their own system of values. Learning mathematics in school should make a certain contribution towards the development of young people and be oriented by the best national and international innovations and traditions.

Let us consider several examples where modern trends in mathematics education were introduced at schools in Ukraine.

For several successive years, at the initiative of the Institute for the Upgrading of the Content of Education, the remote winter and summer "STEM schools" have been organized twice a year. For 5 days, educators meet to discuss innovative educational technologies that presuppose broad and integrative views on the teaching of mathematics.

According to order no. 366 of the Ministry of Education and Science of Ukraine dated April 13, 2017, the State Scientific Institution, "Institute for the Modernization of Educational Content," offers to the entire country an innovative educational project on the topic "I am a researcher" for the years 2018–2021. The goal of this project is to create conditions for developing new methods using IT and STEM technologies. In mathematics, the manuals "I am a researcher, 5"; "I am a researcher, 6"; and corresponding teaching materials have been created.

Another example of a recent trend is the development of an online educational package for schools, "Mathematics and Critical Thinking Grades 5-9," on the Ukrainian platform GIOS (gioschool.com). This package includes an online course which can be used in different situations—for independent study and for a type of blended course in which online studies are supported by traditional classroom discussions. Using the platform makes the learning process more individualized and student centered. It allows students to gain knowledge at their own pace and at a time and place convenient for them to learn how to work with a new source of information—the Internet.

The modernization of the education system in accordance with the needs of the present and the near future is becoming an urgent problem all over the world. Ukraine is keeping up with this trend.

5 The System of Assessment in Mathematics Education in Ukraine During the Period of Independence

Let us consider the evolution of the Ukrainian assessment system for a student's educational achievements in mathematics during the period of independence. To begin with, we note that the concept of assessment will be considered in a wide sense, as a way of relating the student's level of knowledge, skills, and abilities (competencies) to a set of standards and requirements that are imposed on their level by a certain governing body.

In the USSR, the body establishing such requirements was, of course, the state, represented by the Ministry of Education. This ministry established the so-called assessment scale, which determined the various *levels of quality* of a student's mathematical preparation. Each of these quality levels was matched with a mark, expressed both qualitatively ("unsatisfactory," "satisfactory," "good," and "excellent") and quantitatively (by numbers from 2 to 5, respectively). For example, according to the standards of the USSR Ministry (see Wikipedia 2017), the mark "good" (4 points) could be given, when "The student knows all the material required by the program, understands it well and has firmly learned it, answers questions within the program without difficulty, knows how to apply the knowledge gained in practical tasks, uses appropriate language in oral answers and does not make mistakes, and has in written works only minor mistakes." We also want to note that a peculiarity of the Soviet rating scale was the presence of three positive marks ("excellent," "good," and "satisfactory") and only one negative mark ("unsatisfactory"). It was assumed that the negative mark, received by the student either during the lesson or on written work, should be replaced later by one of the positive ones by the student retaking or redoing the work. The negative mark, as an exception, could be received for the entire term and even be final, but in this case, it was supposed that later it would be changed to a positive mark. If a student received one or several "unsatisfactory" marks during the academic year, then by decision of the Pedagogical Council of the school, he or she had to take the same class again the following year. Also, in the USSR, negative marks could not be issued on the State Document on General Secondary Education (Certificate of Maturity).

In Ukraine, after gaining independence, the Soviet standards for the assessment of pupils' academic achievements were at first still applied. However, in 2000, in accordance with the Order of the Ministry of Education of Ukraine (see Nakaz MON 2000), the four-level and 12-point grading scale was introduced, in which there were no negative marks. It meant that any score of points could be final, was not subject to mandatory retaking (although the student still had the right to do so), and could also be put into the State Document on General Secondary Education. The requirements for the quality of mathematical training for the four levels of the new assessment scale ("high," "sufficient," "medium," and "initial") approximately corresponded to the same requirements for the marks of the Soviet scale ("excellent," "good," "satisfactory," and "unsatisfactory"). The following table compares the 12-point grading scale in modern Ukraine with the grading scale adopted during in the USSR (Table 6.2).

The public had a variety of reactions to the introduction of a new 12-point grading scale (discussion and criticism can be found, for example, in Opytuvannia 2005, Opytuvannia 2004, Blog 2006). Among the advantages of the new scale was that it was better for differentiating students' academic achievements, since previously one mark had to be used for significantly different quality levels of mathematical preparedness. It was also welcomed that pupils were no longer required to remain in the same class for a second year of study for having low annual marks. This, in fact, gave each student the right to be different from others and have personal strengths at studying different educational subjects.

Among the disadvantages, it can be noted that with this scale it was difficult for a teacher to explain the grades for students' oral responses, such as to justify why a student received, for example, 8 but not 9 points for solving a problem before the blackboard. Therefore, recently in Ukraine a lot of teachers have generally refused to perform oral evaluations for students, instead giving preference to marks for

Table 6.2 Comparing two grading systems

Mark according to the USSR scale	Levels of learning achievement in modern Ukraine	Points on the 12-point scale in modern Ukraine
5, "excellent"	High	12
		11
		10
4, "good"	Sufficient	9
		8
		7
3, "satisfactory"	Medium	6
		5
		4
2, "unsatisfactory"	Initial	3
		2
		1

students' written work in accordance with the assessment criteria (see Kryteriyi 2008). Work "near the blackboard" that is in front of the class continues to be an important element of teaching mathematics, but the evaluation function of it has now largely been removed. This has allowed students to be set free and significantly reduced the psychological pressure on them in connection to performing in front of the class.

To sum up, we note that by now the 12-point, four-level scale for assessing the educational achievements of secondary school students has firmly established itself in the Ukrainian educational space and in the mostly positive attitude of teachers, students, and their parents.

Let us now dwell on the peculiarities of conducting various forms of assessment for students' educational achievements in mathematics in Ukraine. Depending on the student's place in the educational process (see Shkolnyi 2015), we distinguish *current, thematic, and final assessment*, and depending on the final goal we also distinguish *diagnostic, formative, and controlling assessment*.

Current and thematic assessment is carried out in modern Ukraine at the teacher's discretion. Of course, there are general recommendations and norms for such types of assessments, but most of these norms are advisory in nature and provide the teacher with a certain amount of freedom to act within the limits of common sense. The methodology of the current and thematic assessment is also based on the teacher's professional knowledge that he or she received while taking courses in mathematics teaching methods and other teaching courses at a university and through the process of continuing education and self-education.

For example, depending on the audience or even a particular student, the teacher may not use marks at all for conducting current assessments or may use an individual rating scale (see, for example, Kolesova 2016, Stetsiuk 2015, Ministry 2017). Using computers for current, thematic, and final assessments has also spread considerably in modern Ukraine. In particular, offline and online testing systems using both stationary computers and mobile gadgets are becoming very popular (see, for example, SMIT 2018, Online Test Pad 2018, Plickers 2018).

For the organization and implementation of assessments for student achievements in mathematics (except Standardized Nationwide Final Assessment), no general requirements are usually formulated (see, for example, Ministry 2017). The same applies not only to organizational forms, but also to the means of assessments. The primary means used are problems and tasks. For example, teachers may use any mathematical tasks to carry out the current assessment, in accordance with its purpose, if they seem acceptable to them. If mathematics problems are exclusively used for *controlling* assessment, we will call them the *test items* (tasks). For the test tasks, the situation is somewhat different; rather strict formal requirements for them are given (see, for example, Baker 1985, Lord 1980, van der Linden 1997, and others). These requirements concern the features of both creating test tasks of a certain style (with alternatives, short answer, with full explanation, etc.) and their psychometric characteristics. If the problem performs not only a controlling function, but also, for example, a *teaching or developing one*, then compliance with these requirements is not always necessary. In this case, all tasks, regardless of the style they are presenting, can be considered as tasks with a full explanation, because the teacher can see not only the student's answer, but also the entire process used to solve it and can make necessary adjustments to the student's reasoning at any time. Making adjustments to students' work enables the teacher to achieve the ultimate goal of guiding the learning that is forming in the mind of the student concerning necessary mathematical knowledge, skills, and abilities (competencies).

Let us now consider the peculiarities of conducting the Nationwide Standardized Final Assessment in mathematics. Currently, there are two forms in Ukraine—the State Final Attestation (SFA) and the External Independent Assessment (EIA). The SFA in mathematics is intended solely for evaluating the final amount of knowledge that school students have in mathematics. Its purpose is to check the quality of a graduate's knowledge, skills, and abilities (competencies) acquired during the entire course of general mathematics education (as we will discuss later, it is administered at different stages of education). Students are selected to attend higher educational institutions according to their results on the EIA. In other words, the SFA in mathematics has a *controlling* function and demonstrates the result of schooling, but the EIA in mathematics is *diagnostic*; it shows how prepared the student is to study at a university.

In the USSR, the State Final Attestation in mathematics was carried out in the form of a final written exam. The problems on this examination were developed by the Ministry of Education. A similar situation was observed in Ukraine in the early years after the gaining of independence. The SFA in mathematics was carried out at the end of primary school (4th grade), basic school (9th grade), and senior school (11th grade). The Ministry of Education of Ukraine published special collections of problems for the examination (see, for example, Sobko and Romaniuk 1997, Lytvynenko et al. 1997), from which the problems for the written examination were chosen by the drawing of lots in a television studio. Unfortunately, the drawing was not conducted live, but was recorded the day before the examination, so some teachers and students knew the numbers of the examination assignments in advance.

Since 2000, in conjunction with the introduction of a 12-point grading scale, the structure of the SFA in mathematics was changed. As a result, the final examination was held in the form of a combined test, which contained test questions of various forms (multiple choice, short answer, and questions with full explanations required) and of various levels of complexity. Examination tests in mathematics were still published—now in the textbooks approved by the Ministry of Education of Ukraine—and now not individual problems but whole sets of tasks were chosen at random (see, for example, Sliepkan" 2004, Sliepkan" 2006).

Despite the changes to the form and manner of conducting the SFA in mathematics, society was dissatisfied with the high level of corruption surrounding the examination process. In particular, the sets of problems that would be on the examination continued to remain known to a certain people involved in the drawing. In addition, the students' written works were evaluated by examination boards consisting of teachers from the schools where the examination was held. These teachers were often under strong pressure from the administration to provide certain students with good grades.

There was a similar situation with the procedure for admission to universities of the country. Entrance examinations were conducted by examination boards, which were formed by the professors of that university. Of course, being significantly dependent on the administration of the university, the members of the board often could not resist the pressure exerted on them by the administration to provide some applicants with the marks they needed. These reasons, in particular, led to the need for significant changes to public policy relating to both the SFA and the admissions process to the country's universities.

Let us note that an attempt to radically reform both the SFA and the system of admissions to the universities of Ukraine was first undertaken in 1994, but it was unsuccessful. To reflect the chronology of events during this attempt, we should start with the adoption of the first law "On Education" (see Zakon 2017) in the autumn of 1991. By that law, it was established (article 42) that admission to all universities in Ukraine "is carried out on a competitive basis in accordance with ability," but the law did not indicate the method of organizing this competition. On November 3, 1993, the National Program "Education" (Ukraine of the twenty-first century) was adopted, which was mainly declarative by nature (see Nacional'na Programa 1993). As a result, the steps for the implementation of this program, prepared by the Ministry of Education and the Academy of Pedagogical Sciences, "To create a system of state tests … to control the quality of preparation at all levels of education" were also without any specifications.

Less than six months later, on March 15, 1994, the conditions for admission to Ukrainian universities (see Ministry 1994) were put into effect, in which paragraph 10 states the following.

> People who apply to university from April 1 to July 10 will be tested on general education subjects. For students who graduate from secondary schools in 1994, testing will be combined with their final examinations and conducted in their schools. For prior graduates of secondary school, college, technical school, and vocational educational institutions (regardless of their year of graduation), as well as for students of the pre-university training system, testing will be conducted in testing centers established by each university's admission committee or by several universities that operate a joint testing center. Testing Centers will be approved by the Ministry of Education of Ukraine until April 1, 1994. The mentioned committees will also be given the right, with the consent of students who did not pass the test, to use their grades from relevant educational disciplines found on their Education Certificate as their testing grade. In such cases, their equivalent test scores will be determined using a conversion scale, which is provided in the university's rules of admission. During the testing, secondary schools will follow the guide provided by the Instructions for Testing Procedure. Testing centers will be guided by the Instructions for Testing Centers and the Instructions for Testing Procedure for General Education Disciplines. These instructions describe the technical details of the testing procedure.

However, the next year the Conditions for Admission to Ukrainian Universities (see Ministry 1995) again became identical to 1993, in that there was no talk of any kind of testing. We can read these conditions in item 7: "The organization of admission to Ukrainian universities is carried out by a selection committee, which is approved by order of the rector (director) of the university." Why did the introduction of testing fail and the system return in 2002 to selection committees after only 8 years? The reasons seem to be the following:

Haste. There was less than six months between the decision to create a testing system and the order to introduce it. No one (universities, teachers, students, their parents) understood what these tests would be, how to prepare for them, and how to pass them. Thus, there was no trust in the testing system at any level of society. No explanatory work was conducted for the students and their parents; we could not find a single interview or press conference where this was mentioned. Tests were developed by mathematics educators who did not have special preparation and, therefore, could not make tests of high quality. Experiments on the introduction of testing on a significant number of samples were not carried out at all, so universities did not understand how they could trust this form of competitive selection.

Funding. To develop high-quality tests, it is necessary to train specialists, which requires significant funds. In addition, it is necessary to pay a large number of people who will administer the testing procedures both in schools and in testing centers at universities. But the years 1993–1994 were the peak of hyperinflation in Ukraine, when almost all educators experienced delays in the payment of their salaries for up to six months. During this period, there was also a massive outward flow of qualified personnel going abroad. In fact, the suggestion to introduce a testing system was only made out of enthusiasm.

Cheating and corruption. The testing in secondary schools was conducted by the same teachers who taught the graduating classes; therefore, they were directly interested in making sure that the students' annual final grades would coincide with their grades on the test. At the same time, control over the tests was given to school leaders who were also interested in the good results of their graduates. Furthermore, the tests administered at universities were carried out by admissions committees, the corruption of which has been known since Soviet times.

Lack of coordination and public control. No single structure was created that would deal with all aspects of the testing process. As a result, there was no one to coordinate the implementation of the new system and to organize public oversight. Administrators at the testing centers that were created according to the plan did not know how to interact with each other. So they did not communicate with each other. Therefore, different universities in the same year accepted certificates obtained from various tests. The results of these tests could hardly be compared, because they were created by different people, based on their personal positions and understanding regarding the essence of testing.

Inconsistency in educational policy. In the first years of Ukraine's independence, there was a continuous leapfrog in all areas of state life. The government changed very often; therefore, there was a constant changeover of ministers of education, each of whom had his own vision of a state education policy. The resolutions titled "Education" and "Conditions for Admission to Ukrainian Universities" in 1994 were adopted by Minister of Education Petro Talanchuk, who was a supporter of innovations in structure and an apologist for the test system. But in November of 1994 acting rector of the National University "Kyiv Polytechnic Institute (KPI)" Mykhailo Zgurovskyi, who had much more conservative views, became the new

Minister of Education. In addition, the KPI was always well known for its system of preuniversity training, which brought to the university, among other things, a lot of money. The system of statewide testing, in fact, went against the well-established system of preuniversity training at the best universities in the country, in particular, at the KPI, and the new Minister of Education did not like the disruption of this system.

These reasons contain objective and subjective components, which ultimately led to the failure in 1994 of attempts to introduce the SFA in mathematics as the national method of testing. So, let us return to the beginning of the twenty-first century. As already noted, in society, a discontent was formed against the existing system of conducting both the SFA and the university admission process. Based on this, in 2002, the Center for Test Technologies of the International Foundation "Renaissance" with the financial support of the Open Society Institute of George Soros initiated the development and implementation of the External Independent Assessment (EIA) in Ukraine, which included mathematics assessment.

The EIA system had to replace both the SFA and university entrance examinations. The chronology of the formation and development of this system is described in sufficient detail in other literature (see, for example, Shkolnyi 2015 or even Wikipedia 2018). From 2002 to 2007, a pilot experiment took place, during which time a separate structure was created under the Ministry of Education of Ukraine for carrying out the EIA. This structure was named the Ukrainian Center for Assessment of Education Quality (UCAEQ). For the development of their tests and their administration methods, domestic experts and international testing experts from the USA, Great Britain, Sweden, Latvia, Georgia, and other countries were involved.

As a result, starting in 2008, an independent testing system which performed a dual function was introduced in Ukraine. It both replaced the SFA and freed the university entrance examinations from local corruption. Since 2008, admission to Ukrainian universities is possible exclusively through the EIA. All internal university entrance examinations have been stopped. At the same time, Ukrainian language as well as either mathematics or history have become compulsory subjects of study for admission to all the universities of the country without exception. The third subject tested for admission is chosen by the university's admission commission from the list of subjects for which the EIA is available (foreign languages, physics, chemistry, biology, geography).

Until 2015, the SFA and EIA tests for mathematics in senior classes were conducted in unison. The final attestations were carried out at the same time, as before, in schools according to the sets of tasks chosen randomly from the test collections approved by the Ministry of Education. Independent testing was conducted separately by UCAEQ experts, the questions of which were not known in advance.

If students chose mathematics as their EIA compulsory subject, then they would not take the SFA test at school, but would instead use the two grades given in the UCAEQ certificate. The first grade is on a scale from 100 to 200 points, and it is used to enter universities. The second grade is on a 12-point scale, and students' scores are placed into their Certificates of General Secondary Education. Students who did not choose the EIA in mathematics as a compulsory subject would be obliged to take the SFA to get a final grade for the certificate. Since 2015, the separate SFA mathematics test has been discontinued. If students do not choose mathematics as a compulsory subject on the EIA, their final grade in mathematics on their certificate is created on the basis of an annual assessment of their last school grade.

The system of taking the State Attestation in mathematics at grades 4 and 9 has not changed; it is still carried out at schools and is checked by teachers of that institution. Due to the fact that according to the Constitution of Ukraine (see Konstytuciya 2016, article 53), "general secondary education is compulsory," the public attention given to the Intermediate Attestation (for grades 4 and 9) is much less than the final one (in grade 11). However, experts note the need to reform this component of the Ukrainian system of nationwide standardized assessment of pupils' educational achievements. In particular, the project "New Ukrainian School" (see NUSh 2017) provides for the introduction of the EIA system for assessing the quality of the mathematical preparation of graduates of primary and basic schools.

As for the mathematics test, its structure also evolved during the development of the EIA system. From 2003 to 2009, the mathematics test contained three selected parts. Part 1 included multiple-choice tasks (one correct answer from five proposed answers), which were worth one point each. Part 2 consisted of short answer problems (in decimal fractions or integer representation), which were evaluated at two points each. Finally, Part 3 was made up of questions requiring full explanations that were worth a total of four points each. Part 1 and Part 2 answers were checked by computer; Part 3 answers were checked by examiners who had undergone special training in the UCAEQ. The total number of items on the test increased from 25 (in 2003) to 33 (in 2009).

From 2010 to 2013, the mathematics test contained three types of tasks, such as multiple-choice questions, correspondence problems (finding logical pairs), and those with short answers using decimals or integers. The number of problems of each type varied slightly, and all answers were checked by computer. As one can see, questions requiring full explanations were completely excluded from the test. In the 2014 test, several short answer problems were replaced by structured problems, which consisted of two parts, the answer to each of which was evaluated at one point.

In 2015, a somewhat adventurous attempt was made to introduce two-level testing in mathematics without a preliminary introduction to the public. The need for its introduction naturally came from the fact that it was difficult to verify with one test the quality of a student's mathematical preparation for specialized physics and mathematics schools, for those who in the future planned to study mathematics professionally, and for students of secondary schools for whom mathematics acted as a toolkit. The new two-level test in mathematics consisted of two parts (the basic level and the advanced level) and the basic test was fully contained within the advanced one. The basic level test contained three kinds of questions: 20 multiple choice, 4 correspondence-based problems, and 6 short answer questions (two of which were structured). The advanced test level added to the above 4 more problems, 2 short answer, and 2 full explanation questions. The time limit for taking the basic test was 130 min, and 210 min was given for the advanced test.

It is quite natural that the innovation failed. The reasons for its failure were very similar to those described above in the attempt to introduce independent testing in 1994. We can highlight, in particular, recklessness (few people understood what a two-level test would look like), lack of adequate funding (due to the crisis caused by the fleeing of President Victor Yanukovych and the military conflict with Russia), and inconsistency in educational policy (frequent changes of management of the Ministry of Education and UCAEQ).

Therefore, during 2016–2018, the single-level test system returned, which now contains 20 multiple choice, 4 correspondences, 6 short answer (2 of which are structured), and 3 full explanation questions. The time limit for completing the test is 180 min. As one can see, problems requiring detailed answers returned to the test. This brought considerable relief to mathematics teachers, who criticized the testing system of 2010–2014 precisely for the lack of tasks that required students to write statements allowing them to show their ability to think logically and consistently. The detailed thematic structure of the current mathematics test can be found in UCOYaO (2018, p. 133).

The attitude of society to the current test of the EIA in mathematics is mostly positive (see, for example, Opytuvannia 2016; Onyshchenko 2018; TV Channel "24" 2018). However, given the transition to the National Program "New Ukrainian School" (see NUSh 2017), according to which senior school (10–12 grades) specializations are implemented, and in connection with the statements of the current Minister of Education Liliya Hrynevych regarding the introduction of mathematics as a compulsory subject of the EIA (see TSN 2018), in our opinion, the use of a one-level mathematics test is incorrect. It is more natural to transition to a two-tier system, but one that is more balanced and thought out than in 2015. Specific problems with introducing two-level testing in mathematics in Ukraine and the ways to solve them are considered in Shkolnyi (2017).

Thus, today in Ukraine, an assessment system for students' academic achievements in mathematics, which corresponds to societal requirements, has been almost completely formed and tested. In particular, the criteria for a current, thematic, and final assessment based on a four-level, 12-point scale have been developed, of which all quantitative marks are positive. The formation of the nationwide standardized assessment system is in its final stage. The SFA in mathematics for school leavers is held in the form of the EIA, which is conducted by a separate branch of the Ministry of Education and Science of Ukraine with the assistance of leading domestic and international experts. The results of the independent testing in mathematics are also used to determine university entrance in Ukraine. At the same time, all the data on the results of EIA, including the psychometric characteristics of the tests, are open and accessible to the public on the UCAEQ website. In the context of the implementation of the "New Ukrainian School" project, it seems natural to introduce the EIA in mathematics at the end of primary and basic school (in grades 4 and 9) and to upgrade the existing EIA test in mathematics for the graduating class.

6 How and Why the Preparation of Mathematics Teachers Was Reorganized

We will start by providing some background and basic information on mathematics teacher education and research in mathematics education at the time of the end of the Soviet Union and the beginning of Ukrainian Independence. As of the 1991–1992 academic year, there were 23 institutions for teachers in Ukraine that carried out mathematics teacher education. Teachers institutes and traditional universities provided a four-year course for those majoring in mathematics. A teachers institute or teachers university (or pedagogical university) is a multidisciplinary educational establishment. The central distinctive feature of the traditional university is its multidisciplinary teaching: ranging from the traditional departments of the humanities and engineering to medical schools and agrarian departments. Basically, the traditional university offers a wide range of majors as well as the possibility for students to gain wider knowledge and more fundamental training. Studying at such a university suggests that one is carrying out profound research or work.

In former times, top school graduates in Ukraine planning to become teachers of mathematics used to strive for admission to teachers institutes. Thus, among applicants competing for each place, there were lots of those who had left school with so-called gold medals for academic excellence. Gold medals were awarded to graduates who had excellent biannual and final grades in all the general educational subjects of the curriculum.

Based upon selection interviews and entrance examinations (for mathematics, there is an oral examination, and the examination in the Ukrainian language is a dictation assessment) at teachers institutes, more school graduates with gold medals were enrolled in mathematics and physics departments than in any other department. The required professional qualifications of mathematics teachers included having a high level of knowledge and skills in mathematics and methods of teaching. A special personality and professional strengths essential for working with younger generations were also required.

Theoretical foundations of preparation for mathematics education for prospective mathematics teachers were devised at the Laboratory for Teaching Mathematics and Physics at the Scientific Research Institute of Pedagogy of Ukraine which was headed by I.F. Teslenko. Defined was the scientific basis for: methods of teaching mathematics, shaping pupils' skills and knowledge, and developing teaching aids for teachers of mathematics. I.F. Teslenko was the first Doctor of Education (a second doctoral degree higher than the candidate of science, which corresponds to the PhD) in Ukraine specializing in methods of teaching mathematics (see Poliukhovych 2018).

Having defended a doctoral dissertation at the Academy of Pedagogical Sciences of the USSR in Moscow in 1987, Z.I. Sliepkan became the first woman not only in Ukraine but also in the Soviet Union to take a higher doctorate degree in education with a focus on methods of teaching mathematics. As the head of the program for mathematics and methods of teaching mathematics at Kyiv State Dragomanov

University, Z.I. Sliepkan actively involved the academic staff in the development of a system for the professional training of prospective mathematics teachers and later led the activities in this area in an independent Ukraine.

Every stage of the evolution of the reform of Ukraine's educational system and its priorities had an impact on the system for the professional training of mathematics teachers in the newly independent Ukraine.

1991–2001. As it has been described at the beginning of this chapter, at this time the Ministry of Education of Ukraine and the Academy of Pedagogical Sciences of Ukraine were set up, new educational laws were adopted, and new types of educational institutions began their operations. In November of 1995, a decree by the President of Ukraine "On Principal Directions of Ukraine's Higher Education Reformation" was issued.

The decree mentioned: a major crisis in higher education because of insufficient funds, the draining of teaching staff and educators, and a slump in the younger generation receiving fundamental education. A special emphasis was made on the drastically reduced requirements for educational attainment and background knowledge. Financial difficulties led to the shortage of textbooks, study guides, and other teaching aids. The statement further highlighted the necessity of system reforms in the field of higher education with the view of preserving the current system's potential, retaining the work of specialists, increasing state support for the priority areas of education and science, and adapting education at the home to fit the world's best achievements in science and technology as well as changes in the marketplace.

In 1997, the Cabinet of Ministers of Ukraine adopted, "The list of training programs and special subjects, in which preparation of specialists is carried out at higher educational establishments at respective education and qualification levels." The above document indicates that it is possible to qualify as a mathematics teacher upon getting an appropriate education in program 0101 "Teacher education," special subject—"Pedagogy and secondary education methods. Mathematics."

In the 1998–1999 academic year, fee-based education was introduced in Ukraine's higher education system, which meant that more than 50% of all students enrolled at universities began paying tuition. On December 23, 1998, "The Teacher Education Concept in Ukraine" was adopted according to which teacher education was supposed to prepare professionals who could facilitate the development of a child's personality, strive for personal and professional self-realization, and be willing to work creatively at learning institutions of various kinds. At the same time, teacher's institutes were transformed into teachers universities. The latter offered new special subjects and areas of specialization in a multilevel scheme for teacher education (Bachelor, Specialist, & Master) that was introduced.

2002–2012. In 2002, the educational standard for higher teacher training in special subjects "Pedagogy and Secondary Education Methods: Mathematics" was approved in Ukraine. The two parts of the educational standard are: the educational and qualification description of graduates and the education and training program. In the educational standard, the major activities of mathematics teachers are listed first. Skills essential for carrying out the above activities are then specified. For example, bachelor's degree holders in the special subject "Pedagogy and Secondary

Education Methods: Mathematics" got dual qualifications: as mathematicians and as teachers of mathematics at secondary schools. All teachers universities which trained prospective mathematics teachers used the curricula developed at the National Pedagogical Dragomanov University.

Research schools in the field of methods of teaching mathematics begin to work in Ukraine: in Kyiv was the school of Zinaida Slepkan, in Cherkasy was the school of Nina Tarasenkova, and in Donetsk was the school of Olena Skafa, among others.

The research school of Zinaida Ivanivna Slepkan is very theoretically and practically significant for developing methods of teaching in Ukraine. The research topics that interested the head of the school included mathematics lesson efficiency, ways of activating learning and cognitive activity of pupils studying mathematics at school, means for managing such activities, and the psychological and pedagogical basis of teaching mathematics. The research findings of Zinaida Slepkan were published in more than 200 academic papers and study guides. Among them were undergraduate textbooks and textbooks for schools as well as study guides for students, postgraduate students, and teachers.

The head of this research school also conducted investigations into the problems in the development of the teaching of mathematics, such as personality-centered education at secondary schools and higher educational establishments and developing creative thinking in students in the course of teaching mathematics. Zinaida Slepkan compiled her scholarly results from the field of the methods of teaching mathematics as well as from her longstanding teaching experience at higher educational establishments into her textbook *Methods of Teaching Mathematics* (2000, 2006). The textbook consists of two parts: general methodology and methods for teaching algebra and geometry. It includes various approaches to presenting and processing the educational content of the school mathematics curriculum, monitoring the learning achievements of students, and arranging out-of-school activities in mathematics. Professor Zinaida Slepkan was an academic advisor for five doctors habilitatus in the field of "theory and methodology of teaching mathematics" as well as more than 30 candidates of science, who joined in the scientific studies of their advisor and are now carrying out independent research work in the field of methods of teaching mathematics.

Another research leader, Nina Tarasenkova, worked out a scientific basis for using signs and symbols in teaching mathematics, theoretical propositions of the semiotic approach in teaching mathematics at school, fundamentals of the theory of competence tasks, and didactic principles for activating learning and cognitive activity of students under the lecture-tutorial system of teaching. Nina Tarasenkova is the author of more than 500 academic papers. She was an academic advisor for 3 doctors habilitatus and 13 candidates of science.

The research school of Olena Skafa deals with the issues of the theoretical and methodological bases of developing techniques for heuristic activities in teaching mathematics and modeling heuristic and didactic systems. Olena Skafa published 200 academic papers including six monographs (in 2009 one monograph was published in Bulgaria) and 43 training aids and study guides for mathematics teachers and students. Three doctoral and twelve candidate dissertations were defended under her supervision.

The improvement of the system of methodological training of future mathematics teachers in pedagogical universities from 2002 to 2012 was coordinated by Professor Zinaida Slepkan and Professor Vasyl Shvets, who supervised the study, "The System of Methodological Training of Future Mathematics Teachers in Accordance with the Goals and Objectives of the European Integration of Higher Education" (2007–2009) (see Shvets 2017).

The main problems to do with the education of future mathematics teachers during this period and the main directions of substantive and organizational changes in it are clearly visible in the scientific pedagogical research. From 2002 to 2012, six doctoral theses on mathematics teacher education were defended in Ukraine. These works investigated: the problem of the formation of mathematical competence in mathematics teachers educated using information technology (Rakov 2005), formation of the fundamentals of a professional culture of mathematics teachers (Mykhalin 2004), didactic and methodological foundations of professional training of future mathematics teachers (Motorina 2005), history of mathematics as a basis for teaching mathematical subjects in training future mathematics teachers (Bevz 2007), and developmental teaching in the system of methodological training of future mathematics teachers (Semenets 2011). In all these works, the attention is focused either on improving the process of learning the subjects from the teacher training curriculum or on the system of didactic and methodological support for the professional training of future mathematics teachers in Ukraine.

From 2002 to 2012, the subjects of doctoral (or candidate) studies pertaining to the professional education of future mathematics teachers were focused on the following themes:

- various approaches to the study of mathematical disciplines in the process of preparing future mathematics teachers (Nichyshyna 2008);
- use of multimedia and forming creativity and individuality in the independent work of future mathematics teachers (Konoshevskyi 2007);
- preparation of future mathematics teachers by utilizing different educational ideas (Karpliuk 2009);
- the readiness of future mathematics teachers for creative professional activity or to work with gifted students (Hnezdilova 2006);
- pedagogical conditions for monitoring the quality of mathematics teachers and their intellectual development (Niedialkova 2003);
- formation of professional competence in mathematics teachers and the development of cognitive activity (Voyevoda 2009);
- methodological training of mathematics teachers for particular contexts (Mykhailenko 2005);
- continuity of preprofessional and professional training within the "lyceum-pedagogical university" complex.

By 2012, the crisis in the political, economic, and social sector of Ukraine had left education devastated. Teachers and researchers were leaving to work abroad en masse, which negatively affected the formation of student populations at pedagogical universities and the quality of professional and pedagogical education of future mathematics teachers. A reduction in funding was observed for science and education that resulted in low wages and outdated material and technical support for educational institutions (outdated equipment, technical means of training, computer equipment, educational literature, etc.). There was a significant reduction of the number of students at all educational levels of training to become future mathematics teachers. This situation also resulted from the general state of the quality of higher education in Ukraine at the end of 2011.

As the results of the monitoring study (Finikova 2014, pp. 57-60) show,

> There is no systematic work being done to publish critical assessments of the educational activity and the quality of higher education in Ukraine. There is also no public monitoring and analysis of publications on this issue. A lack of culture of implementation of the needs and rights for educational process participants in Ukraine, in combination with outdated principles of educational administration, has led to the following negative phenomenon in quality assurance of higher education: a lack of motivation to improve the quality of education in students, teachers, employers, and governmental and non-governmental organizations.
>
> There is no association between a university's productivity and the amount of resources it receives from the state. The existing quality assurance procedures of universities are not transparent to internal participants or to the external observer. Information on the activities of universities, presented in various reports, is mostly redundant, irrelevant and often outdated. All players in the quality assurance system depend on the state authorities and are established and financed by the state; however, their activities do not generate sufficient confidence due to the high probability of biased decisions and actions on their part. The activities of university staff are associated with extremely high volumes of paper reporting. As a result, their time restraints do not allow for work spent on improving their quality assurance results.

This note refers to the conditions of mathematics teacher education in Ukraine in the period from 2002 to 2012.

2012–2018. During this period, a significant reduction of the number of students from all educational levels of mathematics teacher education continued (see Fig. 6.2).

The steady trend towards a reduction of the number of students studying mathematics teacher education is explained not only by the demographic and socioeconomic situation in Ukraine, but also by the growing popularity of professions related to IT technologies. Due to the latter, graduates of schools with a high level of mathematical education rarely choose the profession of mathematics teacher. For the first time, the Ministry of Education and Science (MES) of Ukraine openly admits that the state of physics and mathematics education in Ukraine and the preparation of future teachers of mathematics and physics is at a critical stage (from the speech of Oleh Sharov, Director of Higher Education Department, at a meeting of the Scientific and Methodological Council of the MES of Ukraine in 2016).

Since 2014, there has been a shift in the conception of quality control in education in Ukraine, granting more autonomy to higher education institutions and using recommendations of the European Higher Education Area (EHEA). The new version of the Ukrainian law "On Higher Education" (2014) contains general approaches to reforming modern higher pedagogical education, its goals, objectives, and structural requirements. The law On Higher Education focuses on the

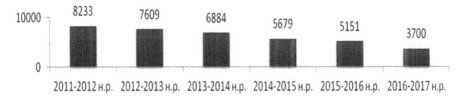

Fig. 6.2 Dynamics of the total number of students in mathematics teacher education (2011–2017)

preparing of specialists who can complete various academic programs: educational-professional, educational-scientific, and scientific in accordance with the educational level (primary, bachelor, master, etc.).

Since 2013, research on ways to improve the training of future mathematics teachers has also intensified. Work has been completed in the areas of: theoretical and methodological foundations for the formation of methodologically competent future mathematics teachers (Akulenko 2013), constructive approaches to the professional training of future mathematics teachers (Lenchuk 2013), and foundations to the formation of methodological competence in future mathematics teachers at teaching students geometry (Matiash 2014). On the premises of Vinnytsia State Pedagogical University, named after Mykhailo Kotsiubynsky, an International Scientific and Practical Conference titled "Problems and Prospects of Professional Training of Future Mathematics Teachers" (2012, 2015, 2018) was started. Scientists from different countries (Belarus, Bulgaria, Georgia, Russia, Poland, Moldova, Romania, and others) were invited to participate in the conference.

In 2015, Ukraine approved a New List of Areas of Knowledge and Specialties, according to which the training of mathematics teachers is carried out in Specialty 014: Secondary Education (Mathematics). In December 2016, the Conception of a New Ukrainian School was approved, according to which, in order to teach in a new way, teachers are encouraged to choose teaching materials on their own, improvise, and experiment. This freedom is provided by the new law "On Education" adopted in September of 2017. In 2018, the Conception of Pedagogical Education in Ukraine was approved. The main provisions of the Conception of Pedagogical Education are designed for the preparation and approval of new professional standards for teachers and educational standards for the training of future teachers in Specialty 014: Secondary Education. However, the trend in recent years towards a declining number of students studying mathematics teacher education has aggravated the problem of the shortage of mathematics teachers in Ukrainian schools. The Ministry of Education and Science of Ukraine and pedagogical universities are looking for ways to attract school graduates to the mathematics teacher profession.

As stated in the law of Ukraine "On Higher Education" (2014), the reform of the entire system of higher education should be based on a student-centered principle—maximal connection of future teachers' preparation with the nature of their future professional activity, skills, and ability to self-develop. The main emphasis of all the recent educational decrees and letters of the Ministry of Education and Science of

Ukraine are made based on the competence approach, pedagogy of partnership, and student-centered principle. As part of the implementation of the competence approach, a new system of measurement and evaluation of learning outcomes including mathematics teacher education should be established.

The current system of higher pedagogical education in Ukraine is represented by 22 pedagogical universities and three pedagogical institutes. The number of institutions engaged in educational and vocational training in Specialty 014: Secondary Education (Mathematics) for a bachelor's degree in Ukraine is only 39, including one private school. Education for a master's degree is provided by 31 state-owned higher education institutions and one private institution. Currently, the training of future mathematics teachers within the system of pedagogical education of Ukraine is given in two ways: a bachelor's degree program (4 years of education) or a master's degree program (1.5 years or 2 years). Together with the bachelor of education degree in secondary education (mathematics), graduates are awarded the professional qualification of teacher of mathematics with the right to teach mathematics in basic schools (grades 5–9). During the training, students have the opportunity to obtain the additional professional qualifications of teacher of physics or teacher of computer science. Having received the master of education degree, graduates receive the professional qualification of teacher of mathematics with the right to teach mathematics in profession-oriented schools and universities.

Currently, new licensing requirements for the implementation of educational activities in Ukraine, officially adopted in 2015, are being introduced, which encourage higher education institutions to create their own internal systems to ensure the quality of their educational services. In a setting of academic autonomy and a competitive educational environment, there is a need for flexibility and efficiency in responding to the different needs of seekers and users of higher education. There is a need to update and improve academic programs. Unfortunately, the processes to develop the education of mathematics teachers suffer from a lack of state documents directing them—educational and professional standards are still under development. Probably, there is no pedagogical university in Ukraine in which fierce debate concerning the contents of the educational programs for the preparation of future teachers in Specialty 014: Secondary Education (Mathematics) does not occur. The main recent differences in the education of future mathematics teachers in Ukrainian pedagogical universities (which are very significant!) result from the fact that universities are given a certain autonomy in the organization and conduct of the educational process.

Critical issues for discussion include: what should be the results of mathematics teacher education; what disciplines and how many hours of them should be included in the curriculum; what should be the length of teaching practice; and how can the educational program be improved, given that the level of mathematical preparation of students is low? The extreme positions in this discussion can be described as follows:

- The key point of mathematics teacher education is a fundamental education in higher mathematics; in particular, the central focus of this education should be mathematical analysis. Different pedagogical disciplines are secondary, including

teaching methods, so their number and hours devoted to them in the curriculum should be minimized.
- The main core of mathematics teacher education should be pedagogical, mathematical, and methodological education. A good mathematician is not always a good teacher. The mathematical education of future teachers should be focused on the issues of future professional activity and take into account the problems of the mathematical competence of students.

In the course of creating and approving educational programs, the constituent parts of which are study plans for prospective mathematics teachers, universities independently establish lists of overall and specific competencies as well as respective teaching outcomes, cycles, and the number of credits for studying various subjects. Let us consider a list of components of one of the educational programs for mathematics teacher education in Ukraine (bachelor's degree) (see Table 6.3):

In summary, it can be noted that the main changes in the goals, objectives, and content of mathematics teacher education in Ukraine over time were caused by changes that occurred in the Ukrainian system of teaching mathematics at schools. In particular, these include:

Before 1992	Recent period
• The process of teaching mathematics was built with a focus on an "average" student. • Unified mathematics program, the same mathematics textbook for each grade. • Basic forms of knowledge and skills assessment: spoken replies, individual work, tests, and exams. • Educational programs directed mathematics teachers on the formation of well-defined knowledge and skills for students.	• Level and profession-oriented differentiation of teaching mathematics at school is implemented. • Based on competitive selection, MES of Ukraine approves 4–5 multilevel textbooks, as well as several multilevel educational programs in mathematics for each grade. • In addition to conventional forms (test assessment of knowledge), state final exams (SFE) and external independent testing (EIT) systems for the assessment of students' achievements in mathematics are actively implemented. • Educational programs direct mathematics teachers on the development of students by means of mathematics and on the formation of mathematical competencies defined by the program.

Therefore, until 1992, teacher training universities pursued the following objectives: to produce future mathematics teachers able to efficiently teach mathematics in schools using unified programs and textbooks on mathematics and who were prepared strictly to follow the guidelines and implement learning technologies proven by study of the best educational experiences.

Today, the objective is to produce mathematics teachers who are able to work effectively among a variety of academic programs on mathematics and among differences of opinion on how to teach mathematics. They should also be able to effectively handle the freedom to choose textbooks for students from the alternatives in the context of competently using recent information technology according to methodology. A future teacher of mathematics should provide conditions for the

Table 6.3 Components of the educational program

Code c/d	Components of the educational program (academic disciplines, term papers [projects], practical training, qualifying paper)	Credits	Form of control
Compulsory subjects			
OK 1.	History of Ukraine	3	Credit test
OK 2.	The Ukrainian language for vocational purposes	3	Credit test
OK 3.	Philosophy	3	Credit test
OK 4.	Ukraine's cultural history	3	Credit test
OK 5.	Foreign language for vocational purposes	11	Credit test, examination
OK 6.	Health and basics of occupational safety	4	Credit test
OK 7.	Computer-oriented education technologies	4	Credit test
OK 8.	Linear algebra	5	Examination
OK 9.	Numerical methods	4	Credit test
OK 10.	Mathematical logic	3	Credit test
OK.11.	Number systems	4	Credit test
OK 12.	Differential geometry and analysis situs	3	Credit test
OK 13.	Probability theory and mathematical statistics, including a term paper	6	Examination
OK 14.	Multivariate analysis	4	Credit test
OK 15.	Differential equations	4	Examination
OK 16.	Nursing	4	Credit test
OK 17.	General psychology	2	Credit test
OK 18.	Developmental and educational psychology	2	Credit test
OK 19.	Social psychology	2	Credit test
OK 20.	Pedagogy	3	Examination
OK 21.	History of pedagogy	2	Credit test
OK 22.	Educational methodology	2	Credit test
OK 23.	Elementary mathematics	19	Credit test, examination, credit test
OK 24.	Mathematical analysis	17	Examination, credit test, credit test
OK 25.	Algebra and theory of numbers	6	Examination
OK 26.	Practical training session in solving problems of mathematics contest	9	Examination, credit test
OK 27.	Analytical geometry	5	Examination
OK 28.	Computer science	9	Credit test, credit test
OK 29.	Methods of teaching mathematics, including a term paper	16	Examination, credit test
OK 30.	Mathematics teaching techniques	3	Credit test
OK 31.	History of mathematics	3	Credit test
Total number of compulsory subjects		168	28 credit tests, 14 examinations, 2 term papers

(continued)

Table 6.3 (continued)

Code c/d	Components of the educational program (academic disciplines, term papers [projects], practical training, qualifying paper)	Credits	Form of control
Elective courses			
Elective block 1			
ВБ 1.1.	Economics	3	Credit test
ВБ 1.2.	Ethics and aesthetics		
ВБ 1.3	Intellectual property		
ВБ 1.4.	Political science		
ВБ 1.5.	Basics of educational measuring and monitoring of education quality		
Elective block 2			
ВБ 2.1.	Law	3	Credit test
ВБ 2.2.	Logic		
ВБ 2.3	Psychology of people with special needs		
ВБ 2.4.	Religious studies		
ВБ 2.5.	Sociology		
ВБ 2.6.	Inclusive education		
Elective block 3			
ВБ 3.1.	Ukrainian studies	3	Credit test
ВБ 3.2.	State-of-the-art IT solutions and media education		
ВБ 3.3	Ecology		
ВБ 3.4.	Rhetoric		
ВБ 3.5.	Information-oriented society		
Elective block 4			
ВБ 4.1.	Discrete mathematics	4	Credit test
ВБ 4.2.	Theory of graphs		
Elective block 5			
ВБ 5.1.	Functional analysis	4	Credit test
ВБ 5.2.	Integral equations		
Elective block 6			
ВБ 6.1.	Constructive geometry	4	Credit test
ВБ 6.2.	Non-Euclidean geometry		
Elective block 7			
ВБ 7.1.	Selected problems of computer science	3	Credit test
ВБ 7.2.	Computer mathematics systems		
Elective block 8			
ВБ 8.1.	Distant maintenance of school mathematics courses	4	Credit test
ВБ 8.2.	Experimental data processing		
Elective block 9			
ВБ 9.1.	Advanced educational experience	4	Credit test
ВБ 9.2.	Out-of-school activities in mathematics		

(continued)

Table 6.3 (continued)

Code c/d	Components of the educational program (academic disciplines, term papers [projects], practical training, qualifying paper)	Credits	Form of control
Elective block 10			
ВБ 10.1.	Mathematical modeling	4	Credit test
ВБ 10.2.	Application of algebra and theory of numbers		
Supplementary special subject—computer science			
ВБ 11	Object-oriented programming	11	Credit test, credit test, examination
ВБ 12	Basics of web programming	5	Examination
ВБ 13	Methods of teaching computer science	8	Examination
Total number of elective subjects		60	13 credit tests, 3 examinations
Academic and on-the-job practice		12	Credit test
Aggregate scope of the educational program		240	45 credit tests, 17 examinations

formation of general and mathematical competencies in students and for the encouraging of a cognitive interest and sustainable motivation for students to learn mathematics. Teachers should develop the personal qualities of students, in particular, their mental activity. These objectives form the basis for the design of modern academic programs for mathematics teacher education in Ukraine which, given the autonomy of higher education institutions, may differ significantly from one institution to another.

7 Conclusions

From the material presented above, one can draw the following general conclusions about the restructuring of school mathematics education in the independent state of Ukraine:

1. Over 25 years of reforms in Ukraine, communist ideology has disappeared from school mathematics education. The main goal of Soviet schools was to form a comprehensively developed, ideologically committed personality capable of actively participating in the renewal of communist society. Now, the main goal of Ukrainian schools is to develop a personality that possesses a creative capacity for study, shows initiative in self-development and self-education under contemporary conditions, and is capable of identifying itself as an important and responsible component of Ukrainian society, prepared to transform and defend the national values and interests of the Ukrainian people.
2. The knowledge paradigm of education (according to which each student needed to acquire a mandatory sum of knowledge, skills, and aptitudes), which predominated in Soviet schools, has been replaced by the competence paradigm, according

to which the end result of the teaching of school subjects must be the formation of competencies. Such competencies render students prepared and capable of applying their knowledge in real-life situations, bearing responsibility for their actions, and actively taking part in the life of society.
3. The content of school mathematics education has been harmonized with the content of school education in developed foreign countries. In particular, the applied component of education has been deepened; the study of probability-statistics, which had been poorly developed in the USSR, has been added to the mathematics curriculum.
4. At the high school level, uniform Soviet schools have given way to specialized schools. Specialization-oriented and level-based differentiation has become an important principle of pedagogy, including mathematics pedagogy.
5. The methodology for assessing students' academic achievements has changed in a fundamental manner. Each school grade in contemporary Ukraine has become an indicator of a student's achievement in the study of a school subject, rather than serving as a determination of the student's errors or failures. A system for the independent assessment of students' academic achievement, to be used both for final assessments and for competitive selection to the country's universities, has become a national concern and objective.
6. There has been a modernization of resources employed in students' education including: curricula, textbooks (including electronic textbooks), teaching manuals, pedagogical materials, and so on. By contrast with the USSR, in Ukraine, mathematics education makes use of alternative textbooks and teaching manuals and of corresponding curricula with various specializations and profiles. The methods and forms employed in mathematics education, including interactive ones, continue to be improved.
7. In tandem with the changes described above, the preparation of the new generation of school mathematics teachers has also been modernized. The function of the mathematics teacher is changing; from an authoritarian leader, organizer, and inspector (without completely giving up these functions), the mathematics teacher is gradually turning into a democratic moderator and facilitator.

The changes listed here support the conviction that the reform of school mathematics education in Ukraine is moving in a correct and relevant direction. The new reform program of reforms, which bears the resounding title "The New Ukrainian School," declares the intention to continue the reform process. We hope that this and subsequent reforms of education produce an educational system that is adequate to the demands of contemporary Ukrainian democratic society.

References

Akulenko, Iryna. 2013. *Teoretyko-metodychni zasady formuvannia metodychnoi kompetentnosti maibutnoho vchytelia matematyky profilnoi shkoly: avtoref. dys. doktora ped. nauk* [Theoretical and methodological principles in the formation of the methodological competence of future

mathematics teachers for specialized school: the author's abstract for a Doctor of Sciences thesis]. Kyiv: Dragomanov NPU Publishing.
Baker, Frank B. 1985. *The basic of item response theory*. Portsmouth: Heinemann Educational Books.
Bevz, Gregory. 1994a. *Matematyka: Probnyi pidruchnyk dlia 7 klasu* [Mathematics: Experimental textbook for grade 7]. Kyiv: Osvita.
———. 1994b. *Matematyka: Probnyi pidruchnyk dlia 8 klasu* [Mathematics: Experimental textbook for grade 8]. Kyiv: Osvita.
———. 1995a. *Matematyka: Probnyi pidruchnyk dlia 10 klasu* [Mathematics: Experimental textbook for grade 10]. Kyiv, Osvita.
———. 1995b. *Matematyka: Probnyi pidruchnyk dlia 11 klasu* [Mathematics: Experimental textbook for grade 11]. Kyiv: Osvita.
———. 1996. *Alhebra: Probnyi pidruchnyk dlia 7-9 klasiv* [Algebra: Experimental textbook for grade 7 - 9]. Kyiv: Osvita.
Bevz, Gregory, Valentyna Bevz. 2000. *Geometriya: Pidruchnyk dlya uchniv 10-11 klasiv z poglyblenym vyvchennyam matematyky* [Geometry for grades 10-11 for schools with an advance course of study in mathematics]. Kyiv: Vezna.
Bevz, Gregory, Bevz, Valentyna, Nataliya Vladimirova. 2001. *Geometriya: Pidruchnyk dlia 7 klasiv* [Geometry: Textbook for 7 – 9 grade]. Kyiv: Vezha.
———. 2004. *Geometriya: Pidruchnyk dlia 10 - 11 klasiv* [Geometry: Textbook for 10 – 11 grades]. Kyiv: Vezha.
Bevz, Gregory, Valentyna Bevz. 2006. *Matematyka: 6 kl* [Mathematics, grade 6] Kyiv: Geneza.
Bevz, Gregory, Valentyna Bevz, Nataliya Vladimirova. 2016. *Geometriya: Pidruchnyk dlya zagalnoosvitnih navchalnyh zakladiv. 8 kl* [Geometry, the textbook for secondary schools, grade 8]. Kyiv: Osvita.
Bevz, Gregory, Valentyna Bevz. 2017. *Algebra: Pidruchnyk dlya zagalnoosvitnih navchalnyx zakladiv. 9 kl* [Algebra: The textbook for secondary schools, grade 9]. Kyiv: Osvita.
Blog *akademiyi intelektu "Smartum". Siday, pyat'! Abo yaka systema ociniuvannia krashche?"* [The Blog of Academy of Intellectual Development "Smartum". "Sit down, five! Or which grading system is better?"]. 2006. https://smartum.com.ua/about_us/blog/parents/siday-p-yat-abo-yaka-sistema- otsinyuvannya-krashcha/. Accessed 22 Feb 2019.
Burda, Mykhailo, Leonid Savchenko. 1996. *Geometriya: Navchalnyi posibnyk dlia 8-9 klasiv* [Geometry. Grades 8-9: Manual for schools with an advanced course of study in mathematics]. Kyiv: Osvita.
Burda, Mykhailo, Olena Dubynchuk, Yurii Malovanyi. 2001. *Matematyka: Probnyi navchalnyi posibnyk dlya uchniv shkil, liceyiv ta gimnazij gumanitarnogo profilyu* [Mathematics, Grades 10-11. Experimental manual for students of schools, lyceums and gymnasiums with humanities specializations]. Kyiv: Osvita.
Cabinet of Ministers of Ukraine. 2004. *Derzhavnyi standart bazovoyi i povnoyi serednioyi osvity* [The state standard of basic and complete secondary education]. http://zakon2.rada.gov.ua/laws/show/24-2004-%D0%BF. Accessed 22 Feb 2019.
———. 2011. *Derzhavnyi standart bazovoyi i povnoyi serednioyi osvity* [The state standard of basic and complete secondary education]. http://zakon.rada.gov.ua/laws/show/1392-2011-%D0%BF. Accessed 22 Feb 2019.
———. 2017. *Concepciya Novoyi ukrainskoyi shkoly* [The Conception of the New Ukrainian School]. https://www.kmu.gov.ua/ storage/app/media/reforms/ukrainska-shkola-compressed.pdf. Accessed 22 Feb 2019.
Finikova, Tetiana. 2014. *Monitorynh intehratsii ukrainskoi systemy vyshchoi osvity v Yevropeiskyi prostir vyshchoi osvity ta naukovoho doslidzhennia: Monitorynh. Doslidzh. Analitychnyi zvit* [Monitoring of the integration of the Ukrainian system of higher education into the European space of higher education and scientific research activity: Monitoring. Researching. Analytical report]. Takson: Kyiv.
Hnezdilova, Kira. 2006. *Formuvannia hotovnosti maibutnoho vchytelia matematyky do zabezpechennia nastupnosti navchannia u zahalnoosvitnii shkoli i vyshchomu navchalnomu zakladi:*

dys.kand. ped. nauk [Formation of the readiness of future mathematics teachers to support the continuity of future teaching in secondary schools and higher educational institutions. Ph.D. Thesis]. Manuscript. Cherkasy.

Karpliuk, Svitlana. 2009. *Tekhnolohiia pidhotovky maibutnikh uchyteliv matematyky do orhanizatsii vzaiemonavchannia uchniv osnovnoi shkoly: dys. and. ped. nauk* [Technology for the preparation of future mathematics teachers to organize interactive teaching to students in secondary school. Ph.D. thesis]. Manuscript. Zhytomyr.

Kliuchnyk, Bogdana. 2017. *Reforma shkoly: yaki zminy chekayut na seredniu osvitu v Ukraini* [School reforming: what changes will be done in Ukrainian secondary education]. https://ua.112.ua/statji/reforma-shkoly-yaki-zminy-chekaiut-na-seredniu-osvitu-v-ukraini-362536.html. Accessed 22 Feb 2019.

Kolesova, Hanna. 2016. *Reytyngova systema yak instrument monitoryngu uspishnosti uchniv* [Rating systems as an instrument for monitoring students' performance]. https://www.pedrada.com.ua/article/297-qqq-16-m12-15-12-2016-reytingova-sistema-yak-nstrument-montoringu-uspshnost-uchnv. Accessed 22 Feb 2019.

Konstytuciya Ukrayiny [The Constitution of Ukraine]. 2016. http://zakon.rada.gov.ua/laws/show/254%D0%BA/96-%D0%B2%D1%80. Accessed 22 Feb 2019.

Konoshevskyi, Oleh. 2007. *Indyvidualizatsiia samostiinoi roboty maibutnikh uchyteliv matematyky zasobamy multymedia: dys. kand. ped. nauk* [The individualization of independent works of mathematics teachers by means of multimedia. Ph.D. thesis]. Manuscript. Vinnytsia.

Kovalenko, Volodymyr, Viacheslav Kryvosheev, Olha Staroseltseva. 1998. *Alhebra. 9 klas: Eksperymentalnyi navchalnyi posibnyk dlia shkol z pohlyblenym vyvchenniam matematyky i spetsializovanykh shkil fizyko-matematychnoho profiliu.* [Algebra. Grade 9: Experimental textbook for schools with an advanced course of study in mathematics and specialized schools of physics and mathematics]. Kyiv: Osvita.

Kryteriyi ociniuvannia pyc'movykh robit z matematyky [The Criteria of Mathematical Written Work Evaluation]. 2008. http://shkola.ostriv.in.ua/publication/code-27704C066F456/list-BD57D40B26. Accessed 22 Feb 2019.

Ksenz, Liudmila. 2014. *Kak spasti ukrainskuyu shkolu* [How can we save Ukrainian school]. http://www.dsnews.ua/world/kak-spasti-ukrainskuyu-shkolu-29072014052100. Accessed 22 Feb 2019.

Lenchuk, Ivan. 2013. *Teoretyko-metodychna systema navchannia evklidovoi heometrii maibutnikh uchyteliv na osnovi konstruktyvnoho pidkhodu: Avtoref. Dys. Doktora ped. Nauk* [The theoretical and methodical system of teaching Euclidean geometry to future teachers on the basis of a constructive approach: The author's abstract for a doctor of sciences thesis]. Kyiv: Dragomanov NPU Publishing.

Lord, Frederic M. 1980. *Application of item response theory to practical testing problems.* Lawrence Erlbaum Ass: Hillsdale N-J.

Lytvynenko, Grygotiy M., Leonid Ya Fedchenko, Vasyl' O. Shvets. 1997. *Zbirnyk zavdan' dlia ekzamenu z matematyky na atestat pro seredniu osvitu. Chastyna 2. Geometriya* [Collection of problems from the exam in mathematics for the certificate of secondary education. Part 2. Geometry]. L'viv: NTL Publishing.

Matiash, Olha. 2014. *Formuvannia metodychnoi kompetentnosti z navchannia heometrii maibutnikh uchyteliv matematyky: Avtoref. Dys. Doktora ped. Nauk* [Formation of the methodological competence in geometry teaching of the future teachers of mathematics: The author's abstract for a doctor of sciences thesis]. Kyiv: Dragomanov NPU Publishing.

Ministry of Education and Science of Ukraine. 1994. *Umovy vstupu do ukrayins'kyh VNZ v 1994 roci* [The Conditions for admission to Ukrainian universities in the year 1994]. http://zakon0.rada.gov.ua/laws/show/z0042-94. Accessed 22 Feb 2019.

———. 1995. *Umovy vstupu do ukrayins'kyh VNZ v 1995 roci* [The Conditions for admission to Ukrainian universities in the year 1995]. http://zakon.rada.gov.ua/laws/show/z0053-95. Accessed 22 Feb 2019.

———. 2017. *Nova Ukrayins'ka shkola. Recomendaciyi dlia vchytelia* [New Ukrainian School. Recommendations for teachers]. http://nus.org.ua/wp-content/uploads/2017/11/NUSH-poradnyk-dlya-vchytelya.pdf. Accessed 22 Feb 2019.

Motorina, Valentyna. 2005. *Dydaktychni i metodychni zasady profesiinoi pidhotovky maibutnikh uchyteliv matematyky u vyshchykh pedahohichnykh navchalnykh zakladakh: dys. d-ra ped. nauk* [Didactic and methodical principles of professional training for the future teachers of Mathematics in higher pedagogical educational institutions. Doctor of Sciences thesis]. Manuscript. Kharkiv.

Mykhailenko, Liubov. 2005. *Systema metodychnoi pidhotovky vchytelia matematyky u vyshchomu navchalnomu zakladi za zaochnoiu formoiu navchannia: dys. kand. ped. nauk* [The system of methodological preparation of the mathematics teacher in higher educational institutions for the part-time studies. Ph.D. thesis]. Manuscript. Vinnytsia.

Mykhalin, Hennadiy. 2004. *Formuvannia osnov profesiinoi kultury vchytelia matematyky u protsesi navchannia matematychnoho analizu: Avtoref. Dys. Doktora ped. Nauk* [The formation of the basis of the professional culture of mathematics teachers in the process of teaching mathematical analysis: The author's abstract for a doctor of sciences thesis]. Kyiv: Dragomanov NPU Publishing.

Nacional'na Doctryna rozvytku osvity [The National Doctrine of Education Development]. 2002. http://zakon.rada.gov.ua/laws/show/347/2002. Accessed 22 Feb 2019.

Nacional'na Programa "Osvita" (Ukrayina 21-go stolittia) [The National Program "Education" (Ukraine of 21-th Century)]. 1993. http://zakon.rada.gov.ua/laws/show/896-93-п. Accessed 22 Feb 2019.

Nakaz Ministestva osvity i nauky #428/48 "Pro vvedennia 12-bal'noyi shkaly ociniuvannia v systemi zahalnoyi serednioyi osvity" [The Order of The Ministry of Education and Science of Ukraine #428/48 "Regarding the introduction of a 12-grade scale for educational achievement evaluation for students in the general secondary education system"]. 2000. http://ua-info.biz/legal/baseap/ua-zmthct.htm. Accessed 22 Feb 2019.

Niedialkova, Kateryna. 2003. *Pedahohichni umovy intelektualnoho rozvytku maibutnikh uchyteliv matematyky u protsesi fakhovoi pidhotovky: dys. kand. ped. nauk* [Pedagogical conditions of the intellectual development of future teachers of Mathematics in the process of professional education. Ph.D. thesis]. Manuscript. Odesa.

Nichyshyna, Viktoriya. 2008. *Intehratyvnyi pidkhid do vyvchennia matematychnykh dystsyplin u protsesi pidhotovky maibutnikh vchyteliv matematyky: dys. kand. ped. nauk* [Integrative approaches to the studying of mathematical disciplines in the process of mathematics teacher education. Ph.D. thesis]. Manuscript. Kirovohrad.

NUSh. Oficiynyi sayt. [The New Ukrainian School. Official cite]. 2017. http://nus.org.ua. Accessed 22 Feb 2019.

Online Test Pad Company. 2018. *Testy z matematyky onlain* [Mathematical tests online]. https://onlinetestpad.com/ua/tests/math. Accessed 22 Feb 2019.

Onyshchenko, Oksana. 2018. *Analiz rezultativ ZNO 2018* [The analysis of EIA 2018 results]. https://dt.ua/EDUCATION/testi-zno-z-usih-predmetiv-buli-optimalnoyi-skladnosti-skladnoyu-bula-lishe-fizika-251645_.html. Accessed 22 Feb 2019.

Opytuvannia forumu "Maydan" "Shcho ne tak iz 12-bal'noyu systemoyu?" [A Poll of the "Maydan" forum, "What's wrong with the 12-grade system?"]. 2004. http://maidan.org.ua/arch/arch2004/1089540705.html. Accessed 22 Feb 2019.

Opytuvannia "Chy podobayetsia vam 12-bal'na systema ociniuvannia?" [The Poll "Do you like the 12-grade system of evaluation?"]. 2005. http://forum.ostriv.in.ua/theme/code-38D9886CAFE09. Accessed 22 Feb 2019.

Opytuvannia "Yak stavliatsia do ZNO yogo uchasnyky" [The Poll "How do the participants perceive the IEA?"]. 2016. https://www.oporaua.org/novyny/42752-opytuvannia-yak-stavliatsia-do-zovnishnoho-nezalezhnoho-otsiniuvannia-ioho-uchasnyky. Accessed 22 Feb 2019.

Plickers Company. 2018. *The system of mobile testing "Plickers"*. https://get.plickers.com. Accessed 22 Feb 2019.

Pekar, Valeriy. 2015. *Yakoyu bude reforma osvity v Ukraini* [What will be the reform of education in Ukraine]. https://hvylya.net/analytics/politics/valeriy-pekar-yakoyu-bude-reforma-osviti-v-ukrayini.html Accessed 22 Feb 2019.

Poliukhovych, Tetiana. 2018. *Do 110-richchia vid dnia narodzhennia Ivana Teslenka (1908-1994), ukrainskoho pedahoha, fakhivtsia z metodyky matematyky* [To the 110th anniversary of the birth of Ivan Teslenko (1908-1994), a Ukrainian teacher and professional in the methodology of mathematics]. Kyiv: Ped. muzei Ukrainy.

Rakov, Serhiy. 2005. *Formuvannia matematychnykh kompetentnostei uchytelia matematyky na osnovi doslidnytskoho pidkhodu v navchanni z vykorystanniam informatsiinykh tekhnolohii: dys. d-ra. ped. nauk* [Formation of the mathematical competence of mathematics teachers on the basis of research approaches to teaching with information technologies. Doctor of Sciences thesis]. Manuscript. Kyiv.

Semenets, Serhiy. 2011. *Teoriia i praktyka rozvyvalnoho navchannia u systemi metodychnoi pidhotovky maibutnikh uchyteliv matematyky: dys. d-ra. ped. nauk* [The theory and practice of developing education in the system of methodological training for future mathematics teachers. Doctor of Sciences thesis]. Manuscript. Zhytomyr.

Shkil, Mykola, Zinaida Slepkan, Olena Dubinchuk. 1995. *Alhebra i pochatky analizu: Pidruchnyk dlia 10 - 11 klasiv* [Algebra and the beginning of analysis, grades 10 – 11]. Kyiv: Vezha.

Shkil, Mykola, Tamara Kolesnyk, and Tamara Khmara. 1997. *Alhebra i pochatky analizu dlia 10 klas shkil z pohlybleny vyvchenniam matematyky* [Algebra and the beginning of analysis, for grade 10 schools with an advanced course of study in mathematics]. Kyiv: Osvita.

Shkolnyi, Oleksandr V. 2015. *Osnovy teoriyi ta metodyky otsiniuvannia navchal'nyh dosiahen' z matematyky uchniv starshoyi shkoly v Ukrayini* [The basis of the theory and methodology of educational achievement evaluation for mathematics in senior school students in Ukraine]. Monograph. Kyiv: Dragomanov NPU Publishing.

———. 2017. *Prospects for a two-level assessment of the quality of knowledge in mathematics in Ukraine. Evaluarea în sistemul educaţional: Deziderate actuale*, 194–197. Chişinău: Institutul de Ştiinţe ale Educaţiei Publishing.

Shvets, Vasyl. 2017. *Pid znakom integrala* [Under the symbol of the integral]. Kyiv: Dragomanov NPU Publishing.

Sliepkan', Zinaida I. 2004. *Zbirnyk zavdan' dlia DPA. Algebra. 9 klas* [Collection of tasks for SFA. Algebra. Grade 9]. Kharkiv: Himnaziya.

———. 2006. *Zbirnyk zavdan' dlia DPA z matematyky. 11 klas* [Collection of tasks for SFA in mathematics. Grade 11]. Kharkiv: Himnaziya.

SMIT Company. 2018. *Systema testuvannia "Testorium"* [The systems of testing "Testorium"]. http://www.znanius.com/5520.html. Accessed 22 Feb 2019.

Sobko, Mykhaylo S., Valentyna Ya Romaniuk. 1997. *Algebra. Zbirnyk zavdan' dlia pys'movoho ekzamenu v 9-mu klasi* [Algebra. Tasks for written examination in 9-th grade]. L'viv: NTL Publishing.

Stetsiuk, Liudmyla. 2015. *Matematychna skarbnycia vchytelia* [Mathematic treasure of the teacher]. http://stetsiukluida.blogspot.com/p/blog-page_13.html. Accessed 22 Feb 2019.

TV Channel "24". 2018. *Novyny ZNO* [EIA news]. https://24tv.ua/zno_tag1794/.

TSN. 2018. *Minisvuty zrobyt' ZNO z matematyky obovyazkovym dlia vsih vypusknykiv* [The Ministry of Education will make the EIA in mathematics mandatory for all graduates]. https://tsn.ua/ukrayina/minosviti-zrobit-zno-z-matematiki-obov-yazkovim-dlya-vsih-vipusknikiv-1204299.html?utm_source=push1&utm_medium=push1. Accessed 22 Feb 2019.

UCOYaO [UCAEQ]. 2018. *Thematic structures in mathematics tests*. http://testportal.gov.ua//wp-content/uploads/2018/08/ZVIT-ZNO_2018-Tom_2.pdf. Accessed 22 Feb 2019.

van der Linden, Willem J. 1997. *Handbook of modern item response theory*. New York: Springer-Verlag.

Voyevoda, Alina. 2009. *Formuvannia fakhovoi kompetentnosti maibutnikh uchyteliv matematyky zasobamy rozvytku piznavalnoi aktyvnosti: dys. ... kand. ped. nauk* [Formation of the professional competence of future mathematics teachers by means of the development of cognitive activity. Ph.D. Thesis]. Manuscript. Vinnytsia.

Wikipedia. 2017. *Sistema ocenivaniya znaiy v SSSR* [The System Of Knowledge Assessment in the USSR]. https://ru.wikipedia.org/wiki/Система_оценивания_знаний. Accessed 22 Feb 2019.

———. 2018. *Independent external assessment*. https://en.wikipedia.org/wiki/External_independent_evaluation. Accessed 22 Feb 2019.

Wozniak, Gregory, Gregory Lytvynenko. 1996a. *Matematyka: Probnyi pidruchnyk dlia 5 klasu* [Mathematics: Experimental textbook for grade 5]. Kyiv: Osvita.

———. 1996b. *Matematyka: Probnyi pidruchnyk dlia 6 klasu* [Mathematics: Experimental textbook for grade 6]. Kyiv: Osvita.

Zakon Ukrayiny "Pro Osvitu" [The Law of Ukraine "On Education"]. 2017. http://zakon0.rada.gov.ua/laws/show/1060-12 Accessed 22 Feb 2019.

Chapter 7
In Lieu of a Conclusion

Alexander Karp

Abstract This chapter attempts to compare and contrast what has taken place in mathematics education in the different countries discussed in this book while taking into account the political and social-economic changes of recent decades. The need for further studies is emphasized, since many topics are still difficult to discuss due to the absence of reliable data. At the same time, this chapter identifies certain similarities between what has taken place in different countries that have attempted to change approaches to mathematics education and that have sometimes attempted in certain respects (and sometimes in many respects) to return to the past.

Keywords Goals · Attitudes toward mathematics · Teacher education · Gifted education · Assessment

1 Introduction

In March 1946, Churchill made his famous pronouncement:

> From Stettin in the Baltic to Trieste in the Adriatic an "iron curtain" has descended across the continent. Behind that line lie all the capitals of the ancient states of Central and Eastern Europe. Warsaw, Berlin, Prague, Vienna, Budapest, Belgrade, Bucharest and Sofia; all these famous cities and the populations around them lie in what I must call the Soviet sphere, and all are subject, in one form or another, not only to Soviet influence but to a very high and in some cases increasing measure of control from Moscow.

In the following decades, the Iron Curtain continued to divide Europe (of the countries and cities named by Churchill, only Vienna and Austria ended up being neutral), and Churchill's metaphor even became a reality—in the form of an actual wall, separating West and East Berlin.

A. Karp (✉)
Teachers College, Columbia University, New York, NY, USA
e-mail: apk16@columbia.edu

© Springer Nature Switzerland AG 2020
A. Karp (ed.), *Eastern European Mathematics Education in the Decades of Change*, International Studies in the History of Mathematics and its Teaching, https://doi.org/10.1007/978-3-030-38744-0_7

Another wall arose—even if it was not made of stone—in the sphere of scientific and cultural activity, including education. What was happening in the West did not completely fail to penetrate to the East, but it usually came with a delay and often in a transformed state—sometimes deliberately distorted, but sometimes also inadvertently simplified and refined. Likewise, educators in London or New York learned about what was going on in Moscow or Sophia not immediately and not always accurately. In this case, the situation varied from place to place: a Muscovite, or even more so a resident of Novosibirsk, could only envy people in Warsaw or Belgrade, who had many more opportunities to find out about life in the West—but even those cities had their limitations. In our view, the communication channels and the manner in which ideas and approaches became transformed when they crossed national borders require further investigation.

What is practically indisputable, however, is that the mathematics education that was offered in the countries of the Soviet block was held in quite high esteem around the world. Not only during the age of the Sputnik, but also at other times, Eastern European mathematics education was always given its due—and not only by educators interested in mathematics competitions, in which the USSR, Romania, and Hungary were always strong, but also by those working on the education of "ordinary" children—despite the fact that the politics of these countries held no appeal to them whatsoever (Kilpatrick 2010; Roberts 2010).

And then, forty-odd years after Churchill's speech, the Berlin Wall collapsed, and the Iron Curtain, as it was proclaimed, also disappeared. Whether it disappeared in reality, and whether it disappeared everywhere, and whether it disappeared for good are questions that remain open to discussion. Clearly, for a certain time certain borders could not disappear, if only because of linguistic barriers: a Pole or a Russian in 1990 was more rarely prepared even for simple communication—let alone professional communication—with a colleague from Germany or the United States, than, say, someone from the Netherlands or Spain. But undoubtedly, the situation has changed considerably.

The goal of this book is to trace the precise manner in which the organization of mathematics education changed under the impact of both domestic political and economic developments and foreign influences. At this stage, it is probably difficult to make any global theoretical generalizations; in the suddenly altered style and rhythm of life, it was easy to lose track of the details and particulars of what transpired, and collecting and recording them remains a top-priority problem, in our view. Nonetheless, it must be acknowledged at once that by elucidating the meaning of the changes in mathematics education that have taken place in countries where tens and hundreds of millions of people live, we come to a better understanding not only of how mathematics is taught in these countries, but also of how the teaching of mathematics has changed under the impact of social-economic changes and of what the actual character of these changes was.

The problem to which this book is devoted can also be viewed more broadly. Schubring (2009, 2018) has studied the relationship between regional and general patterns in mathematics education, as well as the changes brought about by decolonization (Schubring 2017). Continuing in this vein, one may say that the regional development of mathematics education—in a country, or in a region that subsequently became part of a large country (such as Westphalia, studied by Schubring)—is subject to influences from larger formations and influences their mathematics education in its turn. This is especially true in empires, where government power is structured vertically (to use the phraseology of contemporary Russian propaganda, which enshrines the ideal of a "vertical of power"). The influence of the colonizing powers on the teaching of mathematics in the lands of the British, French, or Russian colonial empires is not questioned by anyone, just as no one questions that this influence was manifested differently in different regions, leading to different results—a topic that, in our view, has not been studied sufficiently. Also connected with this is the question to which the present volume is directly devoted: what happens to mathematics education when an empire collapses (the chapter by Karp et al. 2014 is devoted to similar questions). We can hardly expect to find a common answer to this question for all empires and for all of their parts. Nonetheless, the very possibility of looking at the processes that have taken place from a general point of view appears important; for example, the need to become part of a new larger community, which is felt by many mathematics educators in countries that have just emerged from the protection of a former big brother, is characteristic not only of recent times—something like it may be observed in different periods and on different continents.[1]

Below, the attempt will be made to identify certain similarities in the development of the countries of the former Soviet bloc.[2] To repeat, work on the interpretation and generalization of what has taken place is far from completion and must be continued.

2 On the Character of the Changes

Comparisons with the motion of a pendulum are not rare in educational literature and appear in the present book also. Indeed, education has many sides, and at different times, its different sides attract public and scientific attention, with interest subsequently returning to what has been temporarily forgotten. Analyzing the changes that have taken place in mathematics education in Eastern Europe over the past 30 years, we nonetheless observe something of a greater magnitude than usual—quite radical changes in educational policies, which are then repealed in an almost equally radical fashion.

---

[1] Fifteen years ago, the author of this chapter once observed a teacher in one of the former socialist countries begin a lesson by singing the praises of mathematics, without which spaceships would not fly. The speech was obviously one that he had been reciting for many years and bore a striking resemblance to an oft-quoted line from a famous Soviet film comedy "While our spaceships are furrowing the expanses of the universe..."—only for "Soviet spaceships," the astute pedagogue substituted "American spaceships."

[2] Naturally, the author of the present chapter bears sole responsibility for this attempt. Although he did rely on the other chapters in this book, this does not mean that the authors of those chapters necessarily agree with him in all respects.

Usually, this concerns matters that are common to the teaching of all subjects—the classical Soviet system provided a common curriculum, a single common textbook, and only one type of school. Of course, this system did not succeed in fully implementing itself everywhere. Indeed, from a certain time on, in the Soviet Union itself, somewhat different schools began to appear (all of them still public, but some of them, say, offering the in-depth study of certain subjects, i.e., different from other neighboring schools), and in other countries, other "liberties" were permitted as well. Nonetheless, probably all former socialist countries during the first period after the political changes of the late 1980s and early 1990s saw a wave of liberalization: schools were allowed to be private as well as public, and even in public ones, teaching plans could now vary widely, there could be many textbooks, and even the content of education in each subject was now regulated much less rigidly—the detailed curriculum that had to be followed was supplanted by far less rigid documents (for example, standards), whose requirements could be fulfilled in all kinds of different ways.

Although in itself the desire to open the door to the diversity that exists in life could only be welcomed, in reality everything turned out to be rather complicated. In some chapters in this book, the authors write about the reformers' idealism; one can also sometimes point to a lack of foresight in the steps they took. Naturally, conditions were different in different countries. East Germany merged with West Germany, and therefore, the direction of changes there proved more stable than in other places. Ukraine effectively went through two revolutions and a war during the period in question, which also inevitably impacted its education. Nonetheless, it is important to note the movement to repeal the innovations, which existed in different countries, and which proved successful to a certain degree. For example, the chapter on Hungary describes important decisions that limited the initially granted liberties—for example, the practical prohibition against publishing textbooks at private publishing houses. In Russia, the government entirely controls the granting of approvals for textbooks and their use in schools; consequently, even though textbooks may be published at publishing houses that are officially considered private, it cannot be said that the promised decentralization has taken place.

In the politics of certain post-Soviet countries, we can observe a return to former models or a movement toward some other type of conservatism and anti-liberalism. The return movement in education may be seen as being connected with these general processes, but undoubtedly criticism directed against reforms in education cannot be explained by general conservative tendencies alone. Reforms in education, including mathematics education, turned out to be very difficult to implement. This does not mean, of course, that reforms in the economy, say, have been easy, but education clearly has its own specific problems, and the processes of reform appear to have been difficult and painful everywhere. Naturally, economic difficulties, which in many countries have accompanied the transition to a new organization of life, inevitably impacted education—this self-evident observation has been confirmed in the history of the development of all of the former socialist countries. But the difficulties cannot be reduced exclusively to economic ones: there have been numerous purely educational problems as well. The chapter on Germany contains the noteworthy observation that, in the new federal states (lands) (that is, in the territories of the former East Germany), the best results on the national test have been found in those states in which the structure of education changed less than in others—naturally, this should not be seen as a proof that the changes were unnecessary altogether—the point is the difficulty of the process.

It must be acknowledged, however, that important changes have taken place even in those countries in which the return movement, which has just been mentioned, has been noticeably strong. Even if a struggle for centralized uniformity is effectively being waged in a country, this struggle is not being waged as it was a half-century ago: the educational space (to use an expression sometimes encountered in the professional literature) has become far more heterogeneous, which, however, sometimes entails a drop in the level of mathematics education under certain conditions.

3 New Official Goals and Changes in Rhetoric

As the authors of the chapter on Ukraine note, officially, the main goal of Soviet schools was "to form a comprehensively developed, ideologically committed personality capable of actively participating in the renewal of communist society." In mathematics classes, the role of direct ideological indoctrination was quite limited, no matter what Soviet pedagogy experts wrote (see, for example, Karp 2007), but the goal of preparing a harmoniously developed personality capable of carrying out the tasks that would be presented by the party and the leadership of the country undoubtedly did exist, and what it meant in practice was that schoolchildren had to successfully assimilate the content of a rather extensive course. Such was the case, with certain variations, in all of the examined countries.

The new times brought changes, and these are described in every chapter in the book. Now educators began to talk about the free individual capable of self-education, the discovery of this individual's potential, and the like. Consequently, instead of seeking to have students assimilate various kinds of knowledge and skills, educators began to talk about their competencies. It would be a separate problem, and one not without interest, to determine at what time it became popular to talk about competencies and by what particular pathways this term became widespread. It came to former socialist countries from the West, of course, but how it was assimilated and how it was applied is not always clear.

The chapter on Germany cites the following definition: competencies are "the cognitive abilities and skills available to or to be learnt by individuals in order to solve certain problems." This definition can be understood to mean (and was at least often understood to mean) that the goal should be not the assimilation of one or another section of the course in itself, but rather the ability to function successfully under certain life circumstances. The authors of the chapter on the Czech Republic state exactly that: "Success is not measured by the amount of mastered encyclopaedic knowledge but by the ability to solve problems and to respond adequately to unexpected situations."

The subtle point, however, consists in the fact that, as has already been noted, while educators know how to organize the assessment of how encyclopedic students' knowledge is, it is far more difficult to arrange the assessment of their ability to respond adequately to unexpected situations. It would be fair to say that the stated goals (which one cannot but share, of course) prove too general, for which reason educators in practice orient themselves less toward them than toward the high-stakes tests given to their students, which will be discussed below, and which can hardly always be considered ideal instruments for measuring the ability to orient oneself in nonstandard situation, the development of individual qualities, and so on.

The very appearance of such requirements as individualization, the acknowledgement of different cognitive styles, and so on, which are described in many of the chapters, seems important in and of itself. Such requirements to a certain extent have facilitated certain changes in teaching materials and the teaching process (or at least have accompanied them)—for example, the appearance of more varied problems in textbooks or different methods of presenting and illustrating the material.

The promotion of respect for human individuality in schools—even if it is not supported by the development of any procedures that might facilitate it—by itself exerts a certain beneficial influence on the educational process. Soviet schoolchildren were taught for decades that the separate individual ("unit") was nothing: "The unit is—what? The unit is—zero," as schoolchildren learned from a poem by Vladimir Mayakovsky. The change in rhetoric has been useful, and—let us repeat—it took place in all countries of the former socialist camp.

On the other hand, emphasis on the "human" role of mathematics, its connection with the surrounding world, has in some cases caused additional ideological demands to be placed on the teaching of mathematics, which had not been placed on mathematics education on such a scale before. Examples of this are found in the chapter on Russia, and certain other chapters also describe the problem of cultivating patriotism and the like (which might involve all kinds of content), which were perhaps taken less seriously before (even though these same goals were enunciated before as well).

4 The Content of Education: Applied and Pure Mathematics

The orientation toward preparing students for life and developing their ability to solve problems "from life" naturally leads to the popularity of real-world mathematics problems and all kinds of applications of mathematics. In this connection, it must be noted that the mathematics formerly taught in socialist countries was, as a rule, quite detached. Of course, we must not forget about the waves of polytechnization and connection to life that arose from time to time, nor about the place occupied by word problems, which additionally often carried an ideological burden that was considered useful by the authorities. Nonetheless, the school course was on the whole oriented toward the subsequent study of college-level mathematics, and the most subtle word problems aimed at cultivating a military-patriotic or atheistic outlook in reality turned out to be exercises in the application of a given schema for solving problems—introducing variables, formulating equations, solving equations, and so on. With the arrival of the new era, the situation changed.

Pure and applied mathematics are discussed most extensively in the chapter on Hungary; but in other countries as well, and often to an even greater degree, educators deemed it necessary to lighten the theoretical part and enlarge the practical part. Let us note once again that, as far as can be judged, at least based on the results of various tests, a school graduate who has been relatively successfully taught how to solve a trigonometric equation by substituting a variable and converting it into a quadratic equation, or even how to construct the graph of a cubic polynomial, nonetheless is by no means necessarily able to compute how much a vacation trip will cost a traveler, let alone how much salary will be paid after various taxes and deductions. Consequently, the idea of including such straightforwardly useful problems may have appeared attractive (the author of this chapter does not dispose of any data indicating how helpful the inclusion of such problems in the curriculum proved to be). These considerations, however, were not the end of the matter.

Let us again cite Schubring (2015), who noted that applied mathematics often and by no means only in the countries of Eastern Europe has been perceived as appropriate for the masses, while pure mathematics has been thought of as an elitist pursuit. Consequently, it seemed that by reducing the time spent on analyzing proofs and using this time for solving applied and practical problems, the system was reorganizing itself in the desired direction of democratization (nor should it be forgotten that in many countries, the time allocated for the study of mathematics was being reduced, as will be discussed below). We do not have sufficient data about all the studied countries to make any categorical assertions, but weighing the new practical problems against the necessary requirements for continuing mathematics education, we can say that at least in certain cases and in certain countries, the students have effectively been subdivided into those who are prepared for further, in-depth education and those who are not prepared, but who have some experience solving practical problems. In other words, democratization consists not in equal opportunities being offered to everyone, but in the fact that those who do not wish to obtain an in-depth education in mathematics are nonetheless cared for to a certain extent, and they are not considered to be merely rejects from the system who have not turned out as planned (which, unfortunately, was at least sometimes how it was under the Soviet system).

Here, however, it should be pointed out that equal opportunities in fact cannot always be offered: inevitably, in one way or another, any choice that is made rules out or very significantly complicates certain opportunities in the future—a graduate from the mathematics faculty of a university no longer has the opportunity to dance leading roles in classical ballets. Consequently, the question is *when* one or another choice can be made. As has been noted, for example, in the chapter on Russia, sometimes one hears the criticism that students are being offered to make this choice too early. Note that the quite aggressive manner in which practical mathematics is opposed to the mathematics of reasoning and preparation for further education, which has come to the former socialist countries, is unlikely particularly fruitful in and of itself. As Schubring (2015) writes, it is desirable to find "the necessary balance between two likewise valuable and legitimate goals" (p. 254).

Partly in connection with the saturation of the school course with practical problems, but more due to the fact that, as the authors of the chapter on Ukraine note, "the content of school mathematics education has been harmonized with the content of school education in developed foreign countries," the theory of probability and statistics has come to play a much greater role in the school course than they did previously. Previously, until the changes of the late 1980s and early 1990s, these fields were represented in the curricula of different countries in different ways—in Hungary, for example, much more extensively than in the USSR. Over the last three decades, we observe a noticeable growth in attention to these sections of mathematics. The theory of probability and statistics in the countries of the former Soviet Union have stopped being seen as specialized (and "college-oriented") sections and have come to be perceived (even if this is not altogether so in reality) as parts of general education. This process deserves separate study. It undoubtedly reflects the growth in the role of the corresponding methods (for example, statistical ones) in the life surrounding us, and, as has already been noted, the direct influence of the West and also, arguably, changes in the perception of the world—the rejection (even if not always explicit) of a deterministic viewpoint, which relegated the theory of probability to a role that was by definition modest.

Along with these changes, certain reductions in the "traditional" school mathematics studied by schoolchildren have also occurred: they are discussed in the chapters of this book to different, but usually not very great, extents. Nonetheless, such reductions have undoubtedly taken place, as the fact that the additions noted above have occurred also makes clear.

5 Attitudes Toward Mathematics and Its Study

Not all of the chapters discuss the changes in the position of mathematics in equal detail, but this phenomenon has inevitably affected everyone. The socialist world was technocratic, if only because the humanities were suspect—or rather, it would be even more correct to say that the key questions of the humanities were assumed to have been answered, while technical questions remained unanswered and could in no way be answered without mathematics. Mathematics was regarded as indispensable for an important objective—increasing combat capacity—and the government had a special attitude toward it. On the other hand, those who studied mathematics could to a large extent stay away from much that the government said or did, and this also could not help but appear attractive.

The changes that have taken place in society have inevitably affected the position of mathematics as a science and school subject. Even if the other subjects and sciences had simply been restituted the respect they deserved, mathematics would have lost part of the attention it had once enjoyed, but in fact much was said about how many useless engineers there were, and how lawyers, economists, psychologists, and simply capable entrepreneurs were in short supply—as indeed they were. Mathematics inevitably had to come down closer to the position that it occupies in the rest of the world. To these "natural" processes was also added the reformation of mathematics education that was taking place, which was by no means always met with enthusiasm.

The chapter on Hungary notes the continuous decline in the popularity of mathematics among schoolchildren after 1989. The chapters on Ukraine and Russia note the criticisms often aimed at the reforms and the position of mathematics. In the Czech Republic, as far as can be judged, the attitude toward the changes taking place has also not been unanimously positive. In Germany, similar critical views have become part of a general German criticism of mathematics education. The authors of the chapter on Poland believe that the decline in the prestige of the subject began already in the first half of the 1980s, when the ministry stopped requiring students to take mandatory graduation exams in mathematics and subsequently mandatory entrance exams as well. The times when success in mathematics was virtually unanimously considered a measure of human intelligence, and the teaching of mathematics an object of national pride, have passed or are passing.

As Marx (1970) noted long ago, "theory [or, as it is sometimes translated, an idea] becomes a material force as soon as it has gripped the masses"—the masses' love (or hatred) of an object, just as the masses' certainty (or uncertainty) in the quality of education in their own country, is a force that must be reckoned with. The chapter on Hungary notes (and in our view, this remark may also be applied to other countries) that a certain stratification of society has been taking place in connection with different attitudes toward mathematics: in some schools—and, we would add, social spheres connected with these schools—people still regard mathematics as they did previously; in other schools and social spheres, mathematics is seen in a new way. The restructuring of society in accordance with this factor deserves further study.

We would also note that the decline in the prestige of mathematics in a number of cases and countries has led to a decline in the number of lessons devoted to the study of this subject, which, naturally, has further lowered its prestige (to say nothing of other consequences).

6 The Teaching of the Mathematically Gifted

Several chapters discuss new possibilities in the teaching of those who are interested in mathematics (often called "the mathematically gifted," which is not quite precise). What a half-century ago was done, if not underground, then with definite circumspection, today is variously promoted, supported, and considered to be the pride of mathematics education. The old traditional forms of working with students interested in the subject continue to be successfully employed, but alongside of them new forms also emerge, including new mathematics competitions, which become popular. New books are published, which draw on both old and new experience. Useful Internet sites appear.

In the new world, the stratification of students is spoken of almost openly—some of them are now less pestered with demands to learn theorems and formulas, but on the other hand, others can be worked with more freely, preparing them to become, as is sometimes said, an intellectual elite. It must at once be said that the doors in an absolute majority of cases remain formally open, and nothing prohibits any given child from taking part in new or traditional contests or competitions. But the lack of a prohibition is not enough to attract those children who do not have in their immediate environment anyone who takes part or has taken part in them.

Inevitably, comparing what is taking place in the former socialist countries with what exists in the West, it is impossible not to note (without attempting a systematic comparison here) that no harmonization with school education in developed foreign countries, which was mentioned above, has taken place. The teaching of those who are interested and successful (who, to repeat, are not quite accurately called "the mathematically gifted") undoubtedly receives far less attention in the United States, for instance, than, say, in Hungary. The reasons for this include the fear of being accused of elitism, the simple shortage of interested and prepared teachers, and finally the lack of the corresponding traditions. In many countries of the former socialist camp, such traditions exist, and enriched (and not simply accelerated) mathematics education answers the desires of hundreds of thousands, if not millions, of parents.

One might ask how long these traditions will endure, given that they exist in a certain sense in isolation, unsupported by the system of education as a whole. This question is addressed in the chapter on Russia, and this problem is one that, in our view, does not pertain to Russia alone. At present, however, it is impossible not to rejoice at the indubitable efflorescence of all kinds of extracurricular activities in mathematics.

Let us also note here that the thinking of the leadership, at least in certain countries (Russia, Ukraine), goes beyond concern with (relatively) exceptionally performing children: the hope is that all mathematics education above the elementary school level will become specialized in the future, that is, will enable early decisions to be made about who will become what—who will become a chemist, who will become a machinist, who will become a mathematician. We would add that, so far, these plans have not been realized.

7 Assessment

Common to all the countries examined in this book was a desire to reform the system of assessing the outcomes of mathematics education. This was only natural: since educators were permitted to teach in different ways, there had to be some way of finding out what succeeded and where. The old system of school assessment in a very large number of cases was a sham, and this was due not only to all kinds of abuses, but also because the aim of finding out what the actual facts were was never formulated as a principle. To some extent, this was not the case with college entrance exams, but here too a great number of objections could be raised. Moreover, the reformers knew that in Western Europe, toward which many were oriented, exams that were to some degree centralized did exist (even if they were conducted in different ways).

The reforms began. Even before the changes of the late 1980s and the early 1990s, the situation was different in different countries, and the reforms were different. For example, the chapter on the Czech Republic relates that after several unsuccessful attempts at reform, in 2011, a system became established in which graduation exams consisted of one part that was composed at each school individually, and another part that was distributed in a centralized manner, and that instead of taking an exam in mathematics, students could also take an exam in a foreign language. In Poland, the same path was traversed, from graduation exams that at a certain time became nonmandatory to exams that were mandatory for high school graduates and conducted by a specialized central commission. In Hungary, graduation exams in mathematics were mandatory earlier as well—they are conducted at two levels. In Ukraine and Russia, exams in mathematics were, of course, envisioned as mandatory and completely centralized. The transformation of the system met with difficulties everywhere and was naturally accompanied by discussions that did not always perceive the changes in a positive light. Different topics came up in the debates, ranging from the general worldview to narrow professional issues, such as the form of the problems assigned to students. Let us repeat one more time that we have no room here to analyze and compare all viewpoints. We will focus in greater detail on one particular aspect.

The graduation exam, conceived as an objective means of measurement, inevitably becomes something far larger—a genuine teaching curriculum. It is justifiably customary to denounce studying for the exam, but it is difficult to imagine a situation in which students and teachers know about the importance of the exams' outcomes, but are still not particularly concerned about them. Can graduation exams force students to confront unexpected situations or must it ultimately inevitably be predictable, if only because society fears the unpredictable? To what degree is the exam updated annually, in the sense that new topics and new techniques appear on it? Such question, of course, can be asked about any exams (the author of this chapter knows of no studies that systematically and objectively investigate, for example, the second of them), but in countries that are only now making a transition to the new system, such questions appear especially urgent.

It may be supposed that a compromise system (such as the one described in the chapter on the Czech Republic), which combines different forms and parts, makes it possible to render school education more open to different influences than a system in which all requirements come from a single center, ensuring a degree of uniformity that did not even exist in the old days, when the entrance exams for which schoolchildren prepared did vary noticeably, after all, from one college to another.

On the other hand, it must be acknowledged that over the years that have passed, in one way or another, a new system of conducting exams has taken shape in the countries discussed in this book, which does actually function, in one way or another (even if it is criticized). Thanks to this system, we have acquired a better knowledge of the existing state of affairs and often have a better understanding of how poorly mathematics is learned by a large percentage of the students. Recognizing this, we should not automatically conclude that the outcomes have declined: previously, we simply did not possess the information that has now become accessible thanks to the new system (comparisons between the present and the past must rely on many sources of information).

To put it briefly, many chapters discuss not only graduation exams, but also other forms of assessment, ranging from ordinary grades given in class—which have also gone through changes—to centralized assessments at stages preceding graduation, as well as participation in international programs for the assessment of outcomes. Note that attitudes toward international studies (PISA or TIMMS) vary from country to country: in some places, they are considered very important; in others, public opinion virtually does not notice them.

8 Teacher Preparation

Teacher preparation naturally went through changes, in terms of both its organization and its substance. Suffice it to recall the Bologna Process, which affected all of the countries discussed in this book in one way or another. The preparation of teachers with bachelor's and master's degrees was something new for many of the countries and required a substantial restructuring of education. The relationship between and the sequencing of the general and the specialized in education varied from country to country and often differed from what they were in the West. A change in the system of higher education inevitably had to presuppose a change beginning practically at the school level, which could not possibly be accomplished in the time allocated for it anyway, even with the best intentions. But the best intentions themselves were by no means always present. Educational systems in general do not change easily or quickly, but, for example, for Russians, it was not always easy to understand why the existing system, with its familiar shortcomings and advantages, should be rejected for the sake of conforming to some general requirements, when hardly anyone would end up in circumstances in which such conformity might prove useful in any foreseeable future: the mobility of the labor force, which was one of the main objectives of the Bologna Process, is limited not so much by idiosyncratic systems for assessing achievements or courses that were established in the past in different countries, as by far more simple financial or political factors (the author knows of no attempts to estimate how many people, for example, in Russia benefitted practically or were able to find work as a result of Russia's participation in the Bologna Process). The chapter on Hungary notes that this country refused to participate in the Bologna Process. Naturally, on paper, this was not the case everywhere.

7 In Lieu of a Conclusion

But as for the declared objective of making mathematics teacher preparation similar in different countries, it is far from clear to what extent it was possible to approach this objective in reality.

From an organizational perspective, the acquisition of greater independence by colleges preparing teachers (not everywhere and sometimes only for a limited period) appears significant. This includes the appearance of the possibility of private preparation of future teachers of mathematics. In this connection, however, questions concerning the quality of such preparation immediately arose—as far as can be judged, it was by no means always ideal. However, to repeat, critical remarks that appear during a new period, including the harshest ones, do not by any means constitute a proof of decline, collapse, diminution of quality, and so on. The possibility of such criticism simply did not exist earlier. Divergent facts must still be collected and analyzed.

What seems more objective is the decline in the prestige of the profession of mathematics teacher, which is discussed in one way or another in different countries and which may be measured, for example, by comparing the numbers of students who enter the corresponding college departments and continue to study there. Once again, such a decline in prestige was to be expected to a certain extent—the structure of society prior to the changes was different, and the change has brought a kind of alignment with other countries.

Probably the most important changes have been substantive, pertaining to what future teachers are taught and for what purpose. The chapter on Ukraine describes a change in the position expected of the teacher, from an authoritarian leader, organizer, and inspector to a democratic moderator and facilitator. Such an objective will probably find favor among teacher educators from other countries as well (such a change is also described in many publications from other regions that are distant from the ones examined in this book). It is more difficult to describe how exactly this aim is expected to be met. One may suppose that a certain role in this respect will be played by textbooks and lectures that pose new problems and recommend new methods (some, albeit not many, examples appear in this book). As may be gathered from various chapters, questions concerning the relation between subject-specific content, subject-specific pedagogy, and general pedagogy in the preparation of teachers, which are widely discussed in the professional literature around the world, remain urgent in the former socialist countries as well, and the answers given to them vary.

9 West and East

A historian of American mathematics education can with relative ease identify the periods when mathematics was taught using British textbooks, when French textbooks became popular, and when domestic American textbooks finally arrived. In the twentieth and twenty-first centuries, for countries whose history was centuries old before they became parts of the Soviet empire, periodization cannot be so simple—the older brother had to be followed and imitated, but even so this process was more complicated. Also more complicated were processes that began after these countries acquired independence. (Note that political independence by no means necessarily implies a rejection of everything brought by foreign influences or even a desire for any such rejection—see, for example, Zuccheri and Zudini 2007.) As they went through reforms, these countries often wished not so much to follow their own paths, as to enter into a new community—a unified Europe, or even a unified West. The processes taking place are, of course, unfinished; nonetheless, it is natural to compare changes in the East and West.

Probably the most conspicuous difference between what is discussed in this book and what would have likely been the first thing to be pointed out by a researcher of Western mathematics education is that comparatively little attention has been paid here to technology and the changes connected with it. This does not mean, of course, that the east of Europe has not been affected by them. People on a bus in Moscow or Warsaw today hardly use cell phones less than they do in New York or Paris, and consequently schoolchildren also take pictures of the required pages, rather than copying them by hand. Much can also be said about far more sophisticated applications of modern technology in the teaching of mathematics in the former socialist countries.

The relatively modest place allotted in this book to technology in the schools stems in part from a conscious choice: we are more interested here in the social aspect of things, and from this point of view, even when speaking about technology, it is more important to us to note that, beginning at a certain moment, a mathematics teacher in, say, St. Petersburg was able to photocopy various assignments instead of copying them by hand, as had been done previously for decades; even though photocopying technology had long existed, it had been inaccessible, of course, since it was assumed that if schools were given such equipment, people would begin photocopying not just tests, but illegal literature as well.

In part, the book also devotes less attention to technology because the actual changes taking place are not sufficiently clear. Without a doubt, today a schoolchild from a small town in the middle of nowhere can in theory read various mathematical books online, which would have otherwise been inaccessible. It is not altogether clear, however, how many such schoolchildren actually do this. The potential openness of the world that is ushered in by the Internet by no means necessarily becomes an actual openness. Let us make this clear with a nonmathematical example: once, the Soviet Union blocked Western radio stations; today, of course, certain sites are also blocked on Russian territory, but there is no reason to think that other, similar sites, which are still permitted, enjoy mass popularity.

It is a fact that the use of graphing calculators among American teachers is very widespread—a fact that may be proved by citing assignments in textbooks, ordinary observations, and the requirements of various professional development courses as evidence. We do not possess sufficient information to make the same claim or the opposite claim about schools in the countries discussed in this book: the existence of a number of schools in which technology is actively used proves nothing—just as nothing is proved by all the reasons why the situation in the Russian city of Penza might be different from the situation in the American city of Cleveland—what is needed are detailed studies, which do not yet exist.

Other developments taking place in the West, as has already been said, may be easily compared with what we have observed in the countries discussed in this book. This pertains to the content of education, to new objectives, and to the conception of teachers as facilitators. Debates about the relation between mathematical content knowledge, pedagogical content knowledge, and pedagogical knowledge are going on in the West as well.

"Harmonization" with the West, as we have seen, has its limits. Each country has its history. Attempts to ignore which will fail in any case—absolute uniformity encounters opposition, sometimes openly, sometimes silently. And the need to destroy everything not made in the same mold is dubious anyway.

It is not only the practice of mathematics education that merits attention, but the science of mathematics education as well. This subject is treated in greatest detail in the chapter on Germany, where indeed it is easiest to trace the mutual influences between two scientific traditions, articulated in the same language. This chapter notes that the East German tradition is usually seen in a rather negative light, partly due to the difficulty of accessing dissertations from East Germany, partly due to criticisms leveled against the dependence of science on the state, and partly due to an unfamiliar research paradigm (one that includes a significant practical orientation). The analysis of relatively recent Russian dissertations (Karp and Leikin 2011) and the chapter on Ukraine in this book confirm this observation about the research paradigm: dissertations in these countries, both good and less successful ones, are written in a somewhat different style than in the West. This subject, of course, requires additional research, including studies of the corresponding materials in other Eastern European countries. Once again, since it is impossible to go into a detailed discussion of all the differences here, or to analyze how they came into being, let us merely say that if mathematics as a science today knows no borders, then in the science of mathematics, education borders, of course, do exist and can be observed not only in the different numbers of participants at various international conferences (which one can still attempt to attribute to economic reasons), but also in the different numbers of mentions and references to studies from Eastern Europe.

Probably the assertion encountered most frequently in this chapter is that additional studies are needed. And they are needed indeed. In fact, the main aim of this book is to help prepare the groundwork and program for future studies. Very many decisions, stages, changes, and details may become forgotten and lost for people—in the West and the East alike—who will desire to understand what exactly happened during these years.

Let us go back to what was said in the introduction to this book: it is strange that over the thirty years that have passed since the fall of the Berlin Wall, the mathematics education community has recognized neither what the mathematics education of "the ancient states of Central and Eastern Europe" (to use Churchill's expression) possessed, nor what has become of it. Meanwhile, we have been the witnesses of enormous transformations, which have directly affected the lives and education of tens of millions of people and indirectly all of mankind. We must sum up and assess what has happened.

References

Karp, Alexander. 2007. The Cold War in the Soviet school: A case study of mathematics. *European Education* 38 (4): 23–43.

Karp, Alexander, and Roza Leikin. 2011. On mathematics education research in Russia. In *Russian mathematics education: Programs and practices*, ed. Alexander Karp and Bruce Vogeli, 411–486. Hackensack, NJ: World Scientific.

Karp, Alexander, Charles Opolot-Okurut, and Gert Schubring. 2014. Chapter 19. Mathematics education in Africa. In *Handbook on the history of mathematics education*, ed. Alexander Karp and Gert Schubring, 391–404. New York: Springer.

Kilpatrick, Jeremy. 2010. Influences of Soviet research in mathematics education. In *Russian mathematics education. History and world significance*, ed. Alexander Karp and Bruce Vogeli, 359–368. Hackensack, NJ: World Scientific.

Marx, Karl. 1970. *Critique of Hegel's philosophy of right*. Oxford: Oxford University Press.

Roberts, David. 2010. Interview with Izaak Wirszup. *International Journal for the History of Mathematics Education* 5 (1): 53–74.

Schubring, Gert. 2009. How to relate regional history to general patterns of history? – The case of mathematics teaching in Westphalia. In *"Dig where you stand". Proceedings of the conference "On-going research in the history of mathematics education"*, ed. Kristín Bjarnadóttir, Fulvia Furinghetti, and Gert Schubring, 181–195. Reykjavik: University of Iceland – School of Education.

———. 2017. Mathematics teaching in the process of decolonization. In *"Dig where you stand". Proceedings of the conference "On-going research in the history of mathematics education"*, ed. Kristín Bjarnadóttir, Fulvia Furinghetti, Marta Menghini, Johan Prytz, and Gert Schubring, 349–368. Rome: Edizioni Nuova Cultura.

———. 2015. From the few to the many: On the emergence of *Mathematics for All*. *Recherches en didactique des mathématiques* 35 (2): 222–260.

———. 2018. Patterns for studying history of mathematics education: A case study of Germany. In *Researching the history of mathematics education*, ed. Fulvia Furinghetti and Alexander Karp, 241–259. New York: Springer.

Zuccheri, Luciana, and Verena Zudini. 2007. Identity and culture in didactic choices made by mathematics teachers of the Trieste Section of "Mathesis" from 1918 to 1923. *International Journal for the History of Mathematics Education* 2 (2): 39–65.

Author Index

A
Abramov, A.M., 175, 187, 203
Ács, K., 108
Ádám, P., 50, 105
Afanasieva, O.M., 246
Akulenko, I.A., 263
Alatorre, S., 25
Alcuin, 221
Alexits, G., 82
Andrews, P., 116
Anosov, D.V., 196
Apostolova, H.V., 243
Arnold, V.I., 196
Artigue, M., 26
Atanasyan, L.S., 187, 189

B
Bakel'man, I.Y., 214
Baker, F.B., 251
Balázsi, I., 102, 103, 118
Balkányi, P., 102, 103, 118
Balla, I., 104
Ballér, E., 117
Balogh, L., 106
Bánfi, I., 117
Bán, L., 105
Baranyai, J., 105
Barbarics, M., 101
Barrington, L., 80
Bashmakov, M.I., 174, 175, 191, 207, 210
Basl, J., 29
Báthori, Z., 85
Baumann, A., 63, 65
Baumert, J., 67

Bečvář, J., 4, 8, 11, 30
Bečvářová, M., 8
Bender, P., 46, 49, 58, 67
Beneš, P., 3, 31
Bevz, G.P., 241–244, 246, 247
Bevz, V.G., vi, 229–269
Birnbaum, P., 57
Blažek, R., 20
Blinkov, A.D., 212
Blum, W., 15, 67
Boček, L., 7
Bogomolova, E.P., 198
Bokhove, C., 24, 25
Bolyai, J., 78, 81, 82
Borneleit, P., 46, 57
Bradis, V.M., 82
Bragyisz, V.M., *see* Bradis, V.M.
Bratus, T.A., 212
Brislinger, E., 57
Broskyi, Y.S., 246
Bruder, R., vi, 45–69
Buchhaas-Birkholz, D., 48
Bulychev, V.A., 219
Bunimovich, E.A., 175, 191
Burda, M.I., 243, 246, 247
Butuzov, V.F., 190
Buzek J., 149
Bydžovský, B., 7

C
Cachová, J., 21
Carillo, J., 116
Cartier, P., v
Čech, E., 7, 9

Černochová, M., 25
Charlemagne (Charles the Great), 221
Chichigin, V.G., 82
Chudovsky, A.N., 193, 195
Chulkov, P.V., 212
Churchill, W., 275, 276, 289
Chvál, M., 20, 29
Clark-Wilson, A., 25
Clement, F., 116
Comenius, J.A., 4
Connelly (Connelly Stockton), J., 104, 106–108, 110
Cortina, J.L., 25
Cox, M.J., 25
Csahóczi, E., 106
Csapó, B., 100, 104, 117
Csatár, K., 106
Csépe, V, 99
Csicsigin, V.G., *see* Chichigin, V.G.
Csíkos, C., 117
Csorba, L., 105
Csüllög, K., 119

D
Dadayan, A.A., 176
Dalinger, V.A., 216, 217
Dárdai, Á., 95
Daxner, M., 52
De Corte, E., 116
Demeter, K., 83
Deminsky, V.A., 207
Depaepe, F., 116
Desai, N.R., 111
Deterding, S., 100
Dienes, Z., 82
Dixon, D., 100
Dneprov, E.D., 176, 177, 182, 184, 189
Döbert, H., 58
Dorofeev, G.V., 191, 194
Dragomanov, M.P., 258, 260
Dubov, E.L., 185
Dubrovsky, V.N., 219
Duda, R., 137
Dvořák, D., 20, 29

E
Eidel'man, T.N., 178
Einsiedler, W., 46
Engel, W., 64
Eötvös, J., 79
Eötvös, L., 79

Erdős, P., 108, 110
Eszterág, I., 89, 90

F
Fábián, M., 91
Fanghänel, G., 68
Fan, L., 24
Figueras, O., 25
Finikova, T., 262
Firsov, V.V., 185, 186
Forrai, T., 83
Frank, T., 79, 80, 87, 104
Fried, E., 118
Fried, K., vi, 75–123

G
Galitsky, M.L., 189
Gallai, T., 81
García-Campos, M., 25
Gauss, C.F., 78
Gazsó, F., 85
Geißler, G., 58
Gelfand, I.M., 188
Gergelová Šteigrová, L., 4
Giest, H., 52
Gil'derman, S.A., 207
Gladky, A.V., 185
Gnedenko, B.V., 185
Gnedenko, D.B., 185
Goldina, V.N., 207
Gordon Győri, J., vi, 75–123
Gosztonyi, K., 79, 84
Greefrath, G., 62, 65
Griesel, H., 57
Grimm, 201
Gruszczyk-Kolczyńska, E., 153
Guschin, D.D., 204
Gyarmathy, É., 111

H
Hajdu, S., 86
Hajnal, I., 82
Halmos, M., 84
Hankel, W.G., 4
Harnisch, W.C., 50
Hatch, G., 116
Hattie, J., 54
Hejný, M., 21, 24, 25, 29, 30, 35
Henning, H., 46, 49, 58, 67
Hersh, R., 78

Heymann, H.-W., 66
Hnezdilova, K.M., 261
Hódi, E., 82
Holoborodko, V.V., 243, 246
Horthy, N., 80
Horvay, K., 83
Hošpesová, A., vi, 1–39
Hrynevych, L.M., 257

I
Ister, O.S., 243, 247

J
Jahnke, H.-N., 46, 49, 51, 68
Jančařík, A., 25
Janík, T., 14, 17
Janoušková, S., 28, 29
Jelínek, M., 7
Jeřábek, J., 11
Jirotková, D., 29
John-Steiner, V., 78
Juhász, P., 108

K
Kabele, J., 8
Kádár, J., 77
Kaiumov, O.R., 199
Kalmár, L., 81
Kaposi, J., 93
Karp, A.P., v–vi, 84, 173–222, 275–289
Karpiński, M., vi, 131–168
Karpliuk, S., 261
Katz, Z., 212
Kertész, J., 105
Khaled, R., 100
Khrushchov, N.S., 230
Kilpatrick, J., 276
Kim, J., 105
Kinashchuk, N.L., 247
Kirsch, A., 49
Kiselev, A.P., 175, 176
Kis, G., 109
Klein, S., 86
Klieme, E., 67
Kliuchnyk, B., 237
Knecht, P., 17
Köhler, H., 54, 55
Kolesova, H., 251
Kolmogorov, A.N., 7, 175, 176, 188, 191, 196
Konoshevskyi, O.L., 261

Kosztolányi, J., 108
Kotásek, J., 9, 12, 13, 20
Kotsiubynsky, M.M., 263
Kovács, C., 106
Kovalenko, V.V., 242
Köves, G., vi, 75–123
Kozlova, E.G., 212
Krajčová, J., 10
Kramer, J., 63
Kravchuk, V.R., 243
Krummheuer, G., 46, 49, 51, 53, 68
Krygowska, A.Z., 134, 136, 138, 148
Krykorková, H., 17, 19
Ksenz, L., 237
Kudryavtsev, L.D., 196, 197
Kugel, M., 64
Kühnel, J., 50
Kuřina, F., 7, 21
Kurschak, J., 80
Kurucz, G., 117
Kuz'minov, Y.I., 202, 203
Kuznetsova, G.M., 186
Kuznetsova, L.V., 191, 195

L
Laczkovich, M., 117
Lajos, J., 91, 108
Lánczi, I., 118
Lang, J., 55
Lannert, J., 91, 104, 119
Larichev, P.A., 82
Laricsev, P.A., *see* Larichev, P.A.
Laufková, V., 27
Lazutova, M.N., 186
Lednev, V.S., 186
Leikin, R., 289
Leist, S., 53
Lénárd, F., 83
Lenchuk, I.G., 263
Lenin, V.I., 215
Lindgren, A., 201
Liskó, I., 88
Liu, A., 80
Lobachevsky, N.I., 78
Lokajíčková, V., 28
Lompscher, J., 52
Lord, F.M., 251
Louis the Great, King, 78
Lovász, L., 111
Lukicheva, E.Y., 209
Lytvynenko, G.M., 252

M

Malaty, G., 116
Malyshev, I.G., 202
Marciniak, Z., 156, 157
Mareš, J., 30
Marshall, G., 25
Maria Theresa, 78
Mariotti, M.A., 7
Maróthi, G., 78
Marushina, A.A., 212
Marx, K., 283
Matiash, O.I., vi, 229–269
Mavrou, K., 22
Mayakovsky, V., 280
Mécs, A., 109
Meletiou-Mavrotheris, M., 22
Melnikov, I.I., 196
Menter, I., 7
Merzliak, A.G., 243, 246, 247
Meyer, H., 48
Mikulčák, J., 4, 8, 11, 30
Milne, A.A., 201
Molnár, É., 119
Molnár, G., 100
Molnár, J., 21
Moraová, H., 24, 25
Morkes, F., 6
Morvai, É., 106
Motorina, V.G., 261
Müller, G.N., 17
Münich, D., 10
Muravin, G.K., 194
Mushtavinskaya, I.V., 209
Mykhailenko, L.F., 261
Mykhalin, H.O., 261

N

Nacke, L.E., 100
Najvar, P., 19
Neigenfind, F., 46
Nekrasov, V.B., 188, 204
Němec, J., 14
Németh, A., 81, 88
Nesterenko, Y.V., 207
Neubrand, M., 66
Neumann, J. (Johann von Neumann), 79
Nichyshyna, V.V., 261
Niedialkova, K.V., 261
Nikandrov, N.D., 186
Nikolsky, S.M., 196, 197
Niss, M., 65
Nótin, Á., 117
Novák, L., 86

Novikov, S.P., 198
Novotná, J., vi, 2–39
Nozdracheva, L.M., 185

O

Odinets, V.P., 216
Oláh, V., vi, 75–123
Olasz, T., 91
Onyshchenko, O., 257
Opolot-Okurut, C., 277
Op't Eynde, P., 116
Ostorics, L., 118
Ovsyyannikova, I., 219

P

Palečková, J., 28
Pálfalvi, J., vi, 75–123
Palincsár, I., 118
Pálmay, L., 88
Páskuné Kiss, J., 105, 117
Pekar, V., 237
Perrault, C., 201
Pestalozzi, J.H., 50
Péter, R., 81
Peterson, L.G., 201
Picker, B., 50
Píšová, M., 19
Podkhodova, N.S., 183
Pol, M., 36
Polonskyi, V.P., 243, 246
Pólya, G., 16
Porges, K., 54, 57
Pornói, R., 80
Poroshenko, P.O., 232
Pósa, L., 108, 118
Potužníková, E., 28
Potworowski, J., 143
Prescott, A., 54
Příhodová, S., 20
Privalov, A., 218
Prókai, E., 104
Prokopenko, N.S., 247
Pukánszky, B., 81, 88
Putin, V.V., 195, 196, 198, 211, 222

R

Rábainé Szabó, A., 118
Radnóti, K., 105, 106
Rakov, S.A., 261
Rambousek, V., 31

Raspe, E., 201
Rátz, L., 80, 81, 87, 100, 108
Rendl, M., 23
Rényi, A., 81, 108
Richter, W., 57
Riese, A., 50
Roberts, D., 276
Rockstuhl, H., 55
Rodari, G., 201
Rojano, T., 25
Romaniuk, V.Y., 252
Roth, J., 54
Roy, P., 111
Rukshin, S.E., 213, 215
Ruthven, K., 26
Ryzhik, V.I., 198, 219

S
Sadovnichy, V.A., 196, 207
Samková, L., 23
Sarantsev, G.I., 190
Savchenko, L.M., 242
Savvina, O.A., 198
Sayers, J., 116
Schmitt, O., 68
Schneider, S., 50, 51, 59
Schreiber, C., 53
Schubertová, S., 21
Schubring, G., 46, 51, 54, 276, 281
Šedivý, J., 7
Sedova, E.A., 194
Semadeni, Z., 140
Semenets, S.P., 261
Semenov, A.L., 217, 218
Senechal, M., v
Servais, W., 83
Sgibnev, A.I., 212, 219
Shadrikov, V.D., 175
Sharich, V.Z., 212
Sharov, O.I., 262
Sharygin, I.F., 196, 197
Shatalov, V.F., 176
Shevkin, A.V., 198
Shkil', M.I., 234
Shkolnyi, O.V., vi, 229–269
Shlyapochnik, L.Y., 194
Shulman, L., 31
Shvets, V.O., vi, 229–269
Sill, H.-D., 57, 59–61, 65, 67, 68
Sinka, E., 93
Skafa, O.I., 260
Slavík, J., 19
Sliepkan', Z.I., 252, 259, 260

Smirnov, V.I., 188
Smolin, O.N., 208
Snegurova, V.I., 183
Sobko, M.S., 252
Sokolov, B.V., 205
Solzhenitsyn, A.I., 196
Somfai, Z., 84, 117
Somova, L.A., 193, 195
Sorvali, T., 116
Sossinsky, A.B., 212
Stalin, I.V., 230
Starý, K., 27
Stefanova, N.L., 183, 214, 216
Steinbring, H., 17
Steinhöfel, W., 68
Stetsiuk, L., 251
Straková, J., 17, 19, 29
Strässer, R., 7, 33
Strzelecka, A., 144
Stuchlíková, I., 8
Sukhov, K., 213
Sun, W., 105
Suppa, E., 80
Surányi, J., 81, 118
Swetz, F., v
Swoboda, E., vi, 131–168
Szabó, C., 109
Szabó, P., 82
Szalay, B., 103, 118
Szalay, L., 105
Szász, R., 117
Szénay, M., 117
Szentgyorgyi, Z., 84
Szepesi, I., 103, 118
Széplaki, G., 106
Szeredi, É., 116
Szipőcsné K.-J., 118

T
Talanchuk, P.M., 254
Tarasenkova, N.A., 243, 246, 247, 260
Tatto, M.T., 7, 30, 33, 38
Teklovics, B., 111
Teller, E. (Edward Teller), 79
Teslenko, I.F., 258
Testov, V.A., 216
Tian, X., 105
Tichá, M., 8
Tikhonov, Y.V., 212
Tillmann, K.-J., 48, 59, 62
Timoschuk, M.E., 185
Titz, C., 62

Tkachenko, Y.V., 178
Tolnai, J., 82
Tomášek, V., 28, 29
Török, J., 116
Totyik, T., 102
Trencsényi, L., 91, 94
Tusk, D., 164

U
Ujváry, G., 80
Ulyakhina, L., 201
Urbánek, P., 29
Urban V.P., 78

V
Vadász, C., 103, 118
Vakhitov, R., 216
Van den Heuvel-Panhuizen, M., 13
Van der Linden, W.J., 251
Vaněk, V., 21
Vaníček, J., 26
Váňová, R., 19
Varga, A., 93
Varga, T., 81–84, 100, 121, 122
Vásárhelyi, É., 104
Vecseiné Munkácsy, K., 111
Verschaffel, L., 116
Vidákovich, T., 91
Virzsup, I., v
Vladimirova, N.G., 246
Vogeli, B.R., 187, 191
Vondrová (Stehlíková), N., 23
Voyevoda, A.L., 261
Vygotsky, L.S., 52

W
Waliszewski, W., 137
Walker, E., 111
Walterová, E., 5
Weber, K.-H., 57
Weinert, F.-E., 54, 65
Werner, A.L., 176, 190, 191
Wieschenberg Arvai, A., 104
Wigner, J. (Eugene Wigner), 79
Winkelmann, B., 67
Winter, H., 66
Wittmann, E.C., 24
Wozniak, G.M., 242
Wuczyńska, K., 137
Wuschke, H., 46, 57

Y
Yakir, M.S., 243, 246
Yanchenko, H.M., 243
Yanukovich, V.F., 229, 257
Yashchenko, I.V., 207, 212, 213
Yeltsin, B.N., 177, 182, 222
Yershova, A.P., 247

Z
Zabel, N., 46
Zakhariychenko, Y.O., 247
Zakharov, P.I., 207
Zambrowska, M., vi, 131–168
Zawadowski, W., 135, 137, 143, 145, 149
Zgurovskyi, M.Z., 254
Zuccheri, L., 288
Zudini, V., 288
Zvavich, L.I., 194, 195

Subject Index

A
Academic circles, 135
Academic level of education, 237, 246
Academy of Education, 184, 186
Academy of Postgraduate Pedagogical
  Education, 218
Accelerated mathematics education, 284
Accreditation Commission, 31, 32
Act of Education, 7, 9, 13, 14, 24, 28, 36
Adam Mickiewicz Poznań University, 155
ADAM Publishing, 148
Advanced level of education, 193
After-school care, 64
Algebra, 83, 90, 92, 93, 96, 110, 113–115,
  122, 139, 191, 193, 205, 208, 210, 214,
  237, 238, 240–243, 246–248, 260,
  266, 268
Algebraic symbols, 133, 192
Algebraic thinking, 155
Algorithmic thinking, 85
Analytic geometry, 120, 137, 192
"*And you will become Pythagoras*", 148, 149
Anti-Americanism, 220
  Applied mathematics, 119, 238, 242, 281
Archimedes PCs, 145
Arithmetic, 4, 50, 63, 82, 83, 90, 96, 97, 133,
  192, 200
Assessment, 2, 14, 19, 26, 27, 53, 58, 62, 68,
  86, 96, 98–101, 118, 154, 160, 201,
  202, 208, 230, 237, 249–252, 255–258,
  262, 265, 269, 280, 284, 286
Assessment criteria, 251
Assessment formative, 27
Assessment methods, 100, 142
Assessment scale, 249, 250
Assessment summative, 98
Association of mathematics teachers, 145
Association of Teachers of Mathematics
  (ATM), 145
Attitudes toward mathematics, 282, 283
Austro-Hungarian Compromise, 77, 79
Austro-Hungarian Monarchy, 3, 77, 79, 80

B
Bachelor of education, 264
Basic level of education, 133
Basic school, 6, 8, 10–12, 134, 139, 150, 154,
  179, 185, 186, 192, 200, 208, 216, 234,
  237, 252, 256, 258, 264
"The basic school", 176
Bear Maths, 109–110
Belgium, 116
Berlin Wall, 48, 276, 289
Blue Mathematics–KLEKS Publisher, 148
Bologna Agreement, 112, 113
Bologna process, 141, 215, 216, 232, 286
Bologna System, 113, 122
Bourbakists, 134
Budapest, 78, 80, 83, 84, 87, 105, 110,
  116, 275
Bund-Länder-Kommission for Educational
  Planning and Research Funding
  (BLK), 66, 67
Bydgoszcz Mathematical Bubble, 162

C
Calculus, 175, 188, 191, 193, 196, 214, 237, 241
Cascading trainings, 142

Central and regional examination commissions, 151
Central Council of Higher Education, 142
Central Examination Board, 157
Central Examination Committee, 150
Centralization, 27, 28, 37, 46, 58, 81, 82, 88, 89, 92, 94, 96, 122, 175, 194, 202, 218, 279, 285, 286
Centralized entrance exams, 27
Central Teacher Training Center, 142
Certificate of Maturity, 249
Charles University in Prague, 4, 149
Cherkasy, 260
Civic school, 11
Class "Zero" ("Zerówki"), 150
Classroom scenarios, 141
Cognitive styles, 21, 280
Combinatorics, 83, 90, 92, 93, 96, 108, 114, 120, 122, 138, 234, 235
Commission for the Assessment of School Textbooks, 154
Commission on current issues in mathematical education in the transition from school to University, 65
Committee of Experts on National Education, 136, 140
Committee of Mathematical Sciences of the Polish Academy of Sciences, 136–137, 140
Communicating in mathematics classes, 155
Communist party, 7, 77, 176, 196, 214
Competence approach, 264
Competence orientation, 48
Competences, 10, 16–19, 26, 27, 30, 63, 67, 90, 92, 98, 101, 102, 149, 152, 155, 161, 164, 199, 236, 260, 261, 263, 265, 268
Competitions, 8, 57, 80, 104–111, 118, 121, 122, 163, 164, 175, 186, 187, 191, 212, 221, 243, 246, 253, 276, 284
Complex Mathematics educational model, 121
Complex Mathematics Education Experiments (CMEE), 83, 86, 90, 113
Compulsory mathematics education, 134
Compulsory school attendance, 5–7, 10
Computer literacy, 84, 85
Computer programming, 85
Conception for the Development of Mathematics Education in the Russian Federation, 198
Conception of New Ukrainian School, 241
Constructivist notions of learning, 51, 54
Content of learning, 50, 53
Continuing teacher education, 155

Cooperative methods, 91
Core curriculum, 67, 86, 88, 136, 142, 149, 150, 152, 153, 156–159, 165–167
Creativity, 83, 91, 99, 122, 143, 199, 211, 245, 261
Critical areas of mathematics education, 23
Cross-cutting lines of key competencies, 241, 247
Curriculum (curricula), 4, 7, 9, 11, 12, 14, 19, 23, 26, 29, 30, 46, 53–55, 57–60, 65, 82, 89–91, 93, 94, 103, 114, 117, 120–122, 132, 135, 137, 138, 146, 150, 151, 153, 156, 174, 185–189, 194, 213, 218, 233, 236, 240, 241, 260, 269, 282
Czechoslovakia, 2–5, 7, 9, 28, 30, 46
Czech Republic, 2, 4, 9, 11, 12, 15, 16, 27–29, 31, 33–35, 37–39, 149, 279, 283, 285
Czech School Inspectorate (ČŠI), 20, 25, 27, 30

D
Dalton plan, 11
Database of Good Practices, 162
Decentralization, 19, 59, 89, 177, 278
Decolonization, 276
Deconstruction, 9, 10
Decree No. 1, 177
Department of Education, 263
Development of Diagnostic Measurements, 100
Diagnoses of the knowledge, 160
Didactica Mathematicae, 139
Didactic packages, 141
Didactics of mathematics, 7, 8, 49, 57, 65, 68, 132, 136, 138, 139, 142, 148, 149, 154, 155, 159
Didactic solutions, 135
Differential calculus, 137
Differentiation, 10, 20, 52, 59, 68, 84, 86, 91, 117, 121, 177, 192, 239, 245, 265, 269
Discovery learning, 53, 54
Discrete mathematics, 138, 175, 191, 200, 267
Discussions on educational issues, 34, 36
Diversity, 151, 247, 278
Doctor habil., 154, 155, 260, 261
Donetsk, 176, 260

E
Early childhood education, 63, 161
Early education levels, 135
East, 4, 46, 52, 56, 58, 59, 61, 62, 67, 68, 105, 132, 230, 275, 276, 278, 279, 287–289

Subject Index

Eastern Bloc, 118, 149
Educational achievements in mathematics, 249, 251
Education Act, 10, 30, 32, 85, 87, 178, 182
Education Act of 1992, 177
Educational area of Mathematics, 21
Educational programs, 201, 211, 264–268
Educational Research Institute (IBE), 159–162
Educational Social Service Nonprofit Kft., 91
Educational Staff Law, 37
Educational standards for mathematics, 65
Educational Theoretical Research Group, 100
Education and Upbringing (Oświata i Wychowanie), 135
Education system FRG, 63
Education system GDR, 63
Electronic Diagnostic Measurement System (eDia), 100
Elementary school, 20, 30, 50, 78, 138, 143, 152, 162, 167, 179, 212, 241, 247, 284
Elite, 6, 55, 77, 186, 213, 214, 221, 284
Empire, 3, 5, 9, 80, 229, 230, 277, 287
Empirical turn, 48
Enriched mathematics education, 284
Enthusiasts of education, 146, 159
Entrance examinations, 87, 91, 95, 98, 99, 137, 151, 157, 188, 194, 195, 203, 207, 219, 253, 255, 258
Entwurf
  Entwurf der Organisation der Gymnasien und Realschulen in Oesterreich, 79
Eötvös Loránd University, 78
Equations, 106, 133, 180, 184, 192, 193, 197, 204, 208, 219, 220, 234, 235, 242, 247, 266, 267, 281
ERASMUS, 116
Erdős Pál Camp, 110
Erdős Pál Matematikai Tehetséggondozó Program, *see* Erdős Pál Camp
Ericsson Hungary Ltd., 108
European Girls' Mathematical Olympiad (EGMO), 111
European Social Fund, Human Capital Operational Program, 160
European Union, 4, 90, 143, 158–160, 162, 164
Exit examinations, 27, 96, 151, 156, 157
Expert Commission on Mathematic Education in Hamburg, 63
External examination, 150, 158, 160
External Independent Assessment, 252, 255
Extracurricular programs, 104, 122
Extracurricular work in mathematics, 212

F

Fasori Street Lutheran Gymnasium, 79, 80, 87
Fasori Evangélikus Gimnázium, *see* Fasori Street Lutheran Gymnasium
Fazekas Mihály Fővárosi Gyakorló Általános Iskola és Gimnázium, *see* Fazekas Mihály Gymnasium
Fazekas Mihály Gymnasium, 84
Federal Council of Experts, 187
Federal State Educational Standard, 199, 202
Final assessment, 202–209, 251, 252, 257, 269
Finland, 116
First All-Ukrainian Congress of Mathematics Teachers, 240
First wave of reforms, 134
Folk High School Act, 79
Foreign language teaching, 143
Formal logic, 137
Formal mathematics, 142
Fractions (ordinary and decimal), 133, 177, 197, 212, 234, 256
Frame curricula, 89, 93–95, 101, 103, 119–121
Framework for Education Programmes (FEP), 14, 16–22, 29, 30, 34, 36
Functions, 5, 19, 23, 53, 60, 67, 69, 83, 87, 90, 93, 96, 102, 104, 110, 114, 115, 133, 136, 151, 180, 186, 198, 200, 204, 208, 210, 219, 234, 235, 242, 244, 245, 247, 248, 251, 252, 255, 269, 279, 286

G

Gazeta Wyborcza, 151, 152
Gdańskie Wydawnictwo Oświatowe (GWO), 147
Gdansk notebooks, 148, 149
General pedagogy, 142, 287
General school plan, 133
Generic model, 24, 25
Geometric thinking, 155
Geometry, 4, 6, 22, 23, 57, 60, 78, 83, 90, 92, 93, 96, 104, 108, 110, 114, 115, 120, 122, 133, 139, 187, 189, 191, 193, 205, 207–209, 214, 235, 237, 238, 240, 241, 243, 246–248, 260, 263, 266, 267
Germany, 3, 45–69, 155, 207, 276, 278, 279, 283, 289
Gifted children, 64, 81, 211, 238
Gifted from Pomerania, 163
Gifted/talented mathematics education, 84, 105–106, 111, 112

Goals, 2, 6, 13, 14, 17, 19, 20, 25, 30, 32, 39, 49, 51–54, 63, 64, 66, 82, 92, 100, 101, 139, 161–163, 174, 177, 183, 188–190, 196, 200, 211, 213, 215, 230, 236, 248, 251, 252, 261, 262, 265, 268, 276, 279, 280, 282
Goals of learning, 53, 54, 232
Goals of teaching, 142
Gorky Omsk State Pedagogical Institute, 217
Government, 6, 30, 32, 34, 35, 64, 76, 77, 80, 81, 85, 88–92, 95, 99, 112, 113, 132–135, 140, 141, 147, 149, 155, 157, 164, 165, 177, 178, 186, 198, 211, 212, 229, 232, 237, 254, 277, 278, 282
Graduation examinations, 66, 193, 194, 202
Graphisoft Ltd, 108
Graphs, 90, 92, 93, 96, 108, 120, 122, 133, 180, 192, 200, 208, 210, 267, 281
Great Britain, 45, 116, 143, 255
Gymnasium (Grammar school), 79, 80, 87, 150–152, 157, 160, 164, 205, 238

H

Habsburgs, 3, 77, 78
Herzen State Pedagogical University, 216
Heuristic approaches, 138
Higher education, 5, 31–33, 78, 136, 137, 142, 143, 145, 149, 152, 154, 192, 203, 215, 216, 231, 238, 259, 261–264, 268, 286
Higher Education Act, 31, 141
Higher vocational studies, 144
High school, 55, 133, 134, 137, 140, 150–152, 156–158, 167, 184, 185, 188, 199, 210, 216, 240, 241, 269, 285
High-stakes testing, 26
History of science and mathematics, 93
Human Capital Operational Program, 159, 163
Humanities-oriented classes, 190, 193, 194
Humanityzation of education, 177, 189
Hungarian Academy of Sciences, 78, 108, 115
Hungarian Educational Authority (Oktatási Hivatal), 98, 99, 102, 103
Hungarian Mathematical and Physical Society, 80
Hungarian Olympiad Team, 122
Hungarian Socialist Labor Party, 77
Hungarian Students' Competition in Mathematics, 80
Hungary, vi, 3, 4, 8, 46, 75–123, 276, 278, 281–286

I

Impact factor index, 154
Implementation, 10, 11, 14, 17, 19, 20, 23, 35, 54, 65, 113, 133–135, 140, 142, 148, 152–154, 159, 161, 162, 164, 184, 196, 199, 201, 216, 230, 231, 240, 247, 251, 253–255, 257, 262, 264
Individualization, 21, 280
Individualized teaching, 141
Inquiry-based model, 23
In-service mathematics teacher education, 33, 34, 38, 81, 99
Institute of the General Education School, 184
Instruction design, 53
Integrated teaching, 150, 153, 165
Internal differentiation/initial differentiation, 52, 68
International Association for the Evaluation of Educational Achievement (IEA), 103, 117
International comparative studies, 17, 67
International Mathematical Kangaroo Competition, 163
International Mathematical Olympiad, 8, 80, 107, 212
International Mathematical Olympiad Foundation (IMOF), 80
I Play, You Play, We Play (Grasz, gram, gramy), 164
Iron Curtain, 275, 276

J

Jagiellonian University in Kracow, 155
Journal of Laws, 137, 141, 149, 150, 152
Junior high schools, 147, 150–153, 157, 160, 161, 164, 165, 167

K

Kharkiv, 230, 238
Kindergarten, 35, 54, 87, 98, 144, 150, 154
Klebelsberg Center, 88
Kolmogorov reform, 7, 175, 191
Kolozsvár (Cluj Napoca), 78
Kommunist, 176
Košice (Slovakia), 3, 4, 154
Középiskolai Matematikai Lapok, see Mathematical Journal for Secondary Schools
Krakow's school of didactics of mathematics, 136
Kultusministerkonferenz (KMK), 46, 61, 65, 67
Kyiv, 230, 238, 254, 258, 260

Subject Index

L

Law "On Education", 87, 91, 102, 108, 116, 253
Law "On Higher Education", 231, 262, 263, 286
Learner-centered model, 8, 23
Learning tools, 232
Leaving certificate, 137
Leningrad (St. Petersburg), 174, 175, 183, 186, 189, 194, 203, 206, 213, 215, 216, 218, 288
LEROPOL, 143
Licensed examiners, 157
Lifelong learning, 33, 90
Little Abacus (Bydgoszcz), 164
Little League of Tasks, 164
Local government, 99, 144, 160, 162
Logarithmic tables, 138
Logic, 8, 18, 31, 60, 83, 90, 96, 114, 122, 245, 267
Lviv, 238
Lyceum No. 239, 213

M

Magyar Tudományos Akadémia Rényi Alfréd Institute of Mathematics, 108
Maria Grzegorzewska University in Warsaw (Akademia Pedagogiki Specjalnej w Warszawie), 155
Master of Education, 264
MaTech, 111
"Matematyka v shkoli" (Mathematics at school), 62, 118, 230, 238, 241, 260, 265
Mathematical and Physical Journal for Secondary Schools, 79
Mathematical Commission for the Transition from School to University, 65
Mathematical competitions, 212
Mathematical education, 2, 9, 153, 155, 163, 168, 230, 233, 235, 237, 239, 241, 243, 262, 265
Mathematical Journal for Secondary Schools, 79
Mathematical knowledge, 22, 24–26, 51, 52, 65, 90, 92, 103, 112, 116, 119, 120, 160, 198, 210, 230, 234, 236, 244
Mathematical logic, 92, 93, 120
Mathematical Marathons, 163
Mathematical Messages (Wiadomości Matematyczne), 138
Mathematical modeling, 22, 85, 210, 238, 242

Mathematical Olympiad, 8, 57, 105, 111
Mathematics, 2, 4–11, 13, 16–18, 21–31, 33–36, 38, 45–69, 76, 78–95, 98, 100–105, 107–123, 132–168, 183, 210, 211, 217–222, 235, 246–248, 251, 252, 254–258, 261, 276–289
Mathematics 2001, 147–149
Mathematics around us, 147, 149
Mathematics Connects Us Association, 109
Mathematics contests, 266
Mathematics Education Traditions of Europe (METE), 116
Mathematics for all, 134, 242
Mathematics teacher education, 122, 140, 154, 240, 258, 261–265, 268
Mathematikai és Physikai Társulat (Association of Mathematics and Physics), 79
Matriculation examination, 95–97, 151, 157
Matura-secondary school examination, 151, 156
Medve Experience Days, 109
Medve Matek, *see* Bear Maths
Medve Outdoor Mathematics Challenge, 109
Medve Summer Camps and Mathematics Weekends, 109
Mental representation, 23, 51
METE, *see* Mathematics Education Traditions of Europe (METC)
Methodical system, 230
Method of supportive abstracts, 176
Methodology, 31, 46, 49–51, 53, 59, 68, 82, 84, 100, 105, 114, 115, 136, 168, 175, 198, 203, 212, 214, 251, 260, 265, 266, 269
Methods of learning, 231
Millward Brown SA, 166
Minimum curriculum, 88, 141, 142, 146, 150
Ministerium des Cultus und Unterrichts, 79
Ministry of Education (Hungary), 11, 76–80, 82–85, 89, 95, 96, 98, 101–105, 107–113, 115–119, 123, 178
Ministry of Education of the RSFSR, 174
Ministry of Education of Ukraine, 230, 237, 250, 252, 253, 255, 259
Ministry of National Education, 141, 143, 144, 146, 149, 151, 159, 165, 167
Ministry of Religion and Education (Hungary), 81
Ministry of Science, 137, 149, 154
Minor Education Act, 5
Mixed mental exercises, 52
Modern textbook, 239, 245, 246

Montessori method, 11
Moscow, 132, 174, 178, 185, 194, 196, 202, 203, 207, 211, 212, 258, 275, 276, 288
Moscow Center for Continuous Mathematics Education, MTsNMO, 212
Moscow Pedagogical University, 217
MTA Rényi Alfréd Matematikai Kutatóintézet, *see* Magyar Tudományos Akadémia Rényi Alfréd Institute of Mathematics
Multiple choice tasks, 102, 256
Municipal school, 9, 11, 12

N
Nagyszombat (Trnava), 78
National Academy of Pedagogical Sciences of Ukraine, 231
National Accreditation Authority for Higher Education, 32
National Competency Measurement, 100–102, 119
National Core Curriculum (NCC), 86, 88–95, 101, 119–122
National Council of Teachers of Mathematics (NCTM), 65
National Development Plan, 91
National Institute of Education, 88
National Pedagogical Dragomanov University, 260
National Talent Program, 107
Natural sciences, 30, 61, 64, 78, 79, 105, 110, 117, 120, 153, 155, 157, 158, 178, 196, 197, 210, 236, 240
Nauczyciele i Matematyka–Teachers and Mathematics (*NiM*), 145
NCC, *see* National Core Curriculum (NCC)
Németh László Secondary School, 87
New Math, 7, 49, 51, 59, 83, 175
Nizhny Novgorod, 203
Non-public universities, 143
Non-state institutions, 141
Notebook for mathematics for grade 4, 147
Novosibirsk, 203, 276
Number theory, 92, 93, 96, 108, 110, 114, 115, 214

O
Odessa, 230
OECD countries, 28, 37, 103, 158
Oktatási Hivatal, *see* Hungarian Educational Authority
Oktatáskutató és Fejlesztő Intézet, 93, 120, 121

Olomouc (Czech Republic), 154–155
Olympiad mathematics, 212
Omsk, 216
Open ended tasks, 102
Oral examination, 96, 98, 193, 215, 258
Organization for Economic Cooperation and Development (OECD), 12, 19, 20, 28, 29, 37, 65, 91, 102, 103, 158
Országos Pedagógiai Intézet (OPI), *see* National Institute of Education
Ottoman Empire, 77
Outline of Didactics of Mathematics ("Zarys dydaktyki matematyki"), 136

P
PajDej, 164
Parametric evaluation of scientific institutions, 154
Partial solutions, 194
Partial stabilization, 9
Pázmány Péter University, 78
Pedagogical content knowledge, 31, 142, 289
Pedagogical University in Kracow, 155
Pedagogy, 11, 31, 55, 114, 136, 214, 231, 269, 287
Pendulum, 49–54, 121, 277
Percentage calculations, 133
"Percentomania", 175
Performance assessments, 57
Petition "Save the Kids and Older Children Too", 165
Phases of development after 1989, 9–20
PhD's degrees, 139, 259
Piagetian theories, 136
Podkarpakie voivodship, 164
12-Point grading scale, 250, 252
Polish-American Freedom Foundation, 156
Polish Academy of Arts and Sciences (Polska Akademia Umiejętności), 154
Polish Mathematical Society, 134–136, 139, 163
Polish Mathematics Teachers Association, 145
Polytechnische Oberschule, 46, 57
Popularity of mathematics, 117, 118, 283
Post-graduate studies, 155, 156
Post-Soviet countries, 278
Pre-school teachers, 141
Pre-service mathematics teacher education, 30–32
Primary education, 11, 81, 82, 87, 121, 133, 135, 140, 241
Probability, 60, 83, 90, 93, 96–98, 102, 114–116, 120, 122, 137, 138, 185, 190,

Subject Index 303

191, 197, 208, 214, 234, 235, 262, 266, 282
Problem solving, 14, 16–19, 22, 23, 51, 66, 68, 83, 91, 106, 107, 109, 110, 113–116, 119, 121, 122, 138, 164, 214
Product-oriented view of mathematics, 67
Professional secondary school, 10
Program for International Student Assessment (PISA), 48, 158
Project method, 91
Proof, 120, 122, 155, 183, 188, 190, 196, 197, 215, 219, 222, 279, 281, 287
Proportionality, 234
Proving, 21, 23, 52, 91, 120, 155
Psychology, 31, 33, 55, 136, 214, 231
Pure mathematics, 49, 119, 280–282

R
Radnóti Miklós Secondary School, Budapest, 105
Ratio Educationis, 78
Real world problems, 101
Reconstruction, 10, 13–17, 20
Reform, 3, 5, 7, 17, 19, 36, 54, 58, 59, 77, 82–86, 88, 92, 119, 122, 132, 134–140, 144, 147, 149–151, 157, 158, 164, 167, 175, 176, 183, 196, 215, 216, 222, 232, 235, 237, 238, 240, 241, 253, 256, 259, 263, 268, 269, 278, 283–285, 288
Reformist ideas, 5
Relations, 12, 18, 20, 22, 83, 90, 91, 102, 116, 132, 140, 143, 149, 153, 200, 202, 208, 218, 232, 244, 287, 289
Research, 7, 8, 17, 20, 22, 23, 25, 28, 31, 35, 38, 48, 49, 51, 54, 57, 58, 60, 66–69, 79, 82, 105, 106, 108, 116, 117, 119, 123, 136, 138, 139, 142, 148, 149, 154, 155, 159–162, 166, 168, 176, 187, 188, 195, 197, 212, 215, 230, 231, 238, 258, 260, 261, 263, 289
Research paradigm, 289
Research problems, 212, 219
Reunification, 46–48, 55, 56, 59
Revolution and War of Independence, 77, 79
Richter Gedeon Plc., 108
Russia (Russian Federation), 173, 199, 205
Rzeczpospolita, 166

S
Scholarship Foundation of Wroclaw Mathematicians, 163
"School-college" system, 188, 193, 194

School Curriculum Institute, 146
School Education Programme (SEP), 14, 17
School inspectorate, 53
School leaving exam (Maturita), 5–7, 9, 10, 27, 28, 30, 34, 36, 37
School of Mathematics Didactics, 139
Schools (classes) with an advanced course in mathematics, 8, 51, 60, 83–85, 87, 94, 95, 98, 104, 105, 109–111, 113, 117, 120–122, 220, 279
Science camps, 163
"Scientification", 197
Scientific discipline, 132, 136, 138, 142, 154, 211
Secondary education, 7, 16, 18, 33, 37, 55, 62, 81, 104, 121, 149, 175, 231–234, 240, 249, 250, 255, 256, 259–260, 263, 264
Second wave of reforms, 134
"Secrets of Mathematics", 164
Section for Didactics of Mathematics (Institute of Mathematics of Czechoslovak Academy of Sciences), 7, 8
Senior school, 232, 235, 237, 252, 257
Set of tasks, 24, 156
Sets, 8, 10, 12, 24, 30, 31, 36, 50, 51, 59, 64, 77, 79, 82, 83, 87, 89, 90, 92, 93, 96, 99, 109, 110, 114, 120, 122, 133, 135, 137, 139, 140, 144, 148, 150, 153, 162, 176, 184, 188, 191, 201, 204, 208, 234, 242, 244, 249, 251, 252, 255, 259
Seven tasks of general education schools, 66
Shadow education, 117
Sirius Educational Center, 211
Śląskie Voivodeship, 164
Slovak Republic, 2, 3
SMART, 143
Socialist countries, 278–281, 284, 287, 288
Society for Dissemination of Scientific Knowledge (Tudományos Ismeretterjesztő Társulat /TIT), 79
SOCRATES, 116
Soviet Army, 77
Soviet block, 276
Soviet Union, 133
Soviet Union (USSR), 3, 5, 7, 8, 82, 105, 173, 174, 200, 203, 214, 218, 221, 229, 230, 249, 250, 252, 258, 269, 276, 278, 282, 288
Spain, 116, 276
Specialization, 49, 105, 155, 209, 210, 216, 217, 239, 242, 257, 259, 269
Specialized level of education, 205, 209, 237
Specialized (profiled) high schools, 150

Special mathematical classes (*specmat.*), 84, 108
Special schools for mathematics and natural sciences, 64
Sputnik, 276
Sputnik shock, 51, 82
Standard level of education, 237, 246
Standards, 20, 31, 55, 62, 64–67, 77, 96, 145, 150, 151, 154, 158, 180, 181, 183–186, 189, 195, 199–202, 207, 215–217, 222, 231–237, 240, 241, 249, 250, 259, 263, 264, 278
State Final Assessment (GIA), 208
State Final Attestation, 252
State Higher Pedagogical School in Krakow, 136
State Standard of Basic and Complete Secondary Education, 236
Statistics, 60, 83, 90, 93, 96, 114, 116, 120, 122, 190, 191, 197, 205, 234, 235, 242, 266, 282
St. Petersburg University, 194
Strategy 2020, 20
Strategy of Digital Education 2020, 26
Stratification, 180, 221, 283, 284
"Subject-centeredness", 197
Subject matter didactics, 49, 51, 68
Subject matter knowledge, 142
SuliNova Kht., 91
Szczytno, 145
Szeged University, 100

T

Táltos, 105
Teacher education (preparation), 10, 17, 63, 66, 85, 103, 105, 134, 140, 141, 154–156, 161, 166, 188, 203, 214, 216, 231, 253, 262, 281, 286, 287
Teachers' assessments, 27, 99
Teacher's Studies Bulletin (Biuletyn Studiów Nauczycielskich), 145
Teachers College, Columbia University, 156
Teachers colleges, 144, 145
Teacher Training College(s) Faculty of Eötvös Loránd University, 31, 144, 145
Teacher Training College of Budapest, 116
Teacher Training Department of the Ministry of National Education (DKN MEN), 144
Technical school, 7, 9, 55, 150, 156, 167, 253
Technological devices, 22
Technology, 21, 25, 26, 31, 57, 132, 191, 192, 204, 219, 233, 239, 240, 248, 255, 259, 261, 262, 265, 266, 288

TEMPUS, 116, 143, 145
Ten-year-schooling, 134, 135
Textbook, 2, 4, 7, 8, 10, 14, 24, 25, 30, 54, 57–60, 78, 81–84, 86, 89–92, 94, 95, 108, 122, 132, 135–141, 147–149, 153, 154, 161, 165, 166, 174, 175, 182, 186–192, 201, 205, 213–215, 218, 220–222, 231, 233, 235–248, 252, 259, 260, 265, 269, 278, 280, 287, 288
Textbooks for Humanities-Oriented Classes, 189–191
Textbooks' language difficulty, 25
Textbooks' mathematical content, 82
Textbooks' textual and non-textual elements, 25
Theoretically oriented mathematics teaching, 260
Thinking method, 90, 92, 93, 96, 114
Tomsk, 205
Trade Union Solidarity "Solidarność", 132
Transmissive style of teaching, 141
Trends in International Mathematics and Science Study (TIMSS), 48, 158
Trigonometric formulas, 138
Trigonometry, 96, 193, 209, 210, 220
Types of lessons, 142

U

Ukraine, 232, 237, 246, 248–259, 278, 279, 282–285, 287, 289
Ukrainian Center for Assessment of Education Quality (UCAEQ), 255–257
Undifferentiated (unified) school, 81, 87, 98, 120, 207, 265, 288
Unified Junior Academy of Sciences of Ukraine, 238
Uniform State Exam (EGE), 202–208, 211, 220
Union of Czech mathematicians and Physicists, 22, 34–36
Union of Czechoslovak mathematicians and Physicists, 5, 9
University of Debrecen, 78
University of Pannonia, 110
University of Rzeszów, 155
University of Warsaw, 140, 143, 156
University of Wroclaw (Institute of Mathematics), 7, 108, 155

V

Varga Tamás Methodological Days, 81
Velvet revolution, 2, 10
Vinnytsia, 263
Vocational school, 5, 7, 9, 10, 55

W

Waldorf pedagogy, 11
West, 46, 56, 58, 61, 67, 68, 118, 132, 143, 145, 149, 216, 219, 221, 275, 276, 278, 279, 282, 284, 286–289
West London Institute of Higher Education, 143
White Paper, 10, 12–14, 19, 20
Word problems, 12, 23, 91, 134, 180, 247, 280
World War I, 3, 10, 77
World War II, 3, 5, 76, 77, 79, 81, 84, 85, 87, 98
WSiP (Wydawnictwo Szkolne i Pedagogiczne-School and Pedagogic Publishing House), 146

Y

Yekaterinburg, 203
York, 221, 276, 288

CPSIA information can be obtained
at www.ICGtesting.com
Printed in the USA
LVHW021016100521
686964LV00001B/43